MOVING IMAGES: FROM EDISON TO THE WEBCAM

MOVING IMAGES: FROM EDISON TO THE WEBCAM

Edited by John Fullerton and Astrid Söderbergh Widding

John Libbey
JL
LONDON • PARIS • ROME • SYDNEY
Aura

Cataloguing in Publication Data

Moving Images: From Edison to the Webcam
(Stockholm studies in cinema)

1. Cinematography

II. Fullerton, John II. Söderbergh Widding, Astrid

791.4’3

ISBN: 1 86462 054 4 (Hardback)

Published by

John Libbey & Company Pty Ltd, Level 10, 15–17 Young Street, Sydney, NSW 2000, Australia.
Telephone: +61 (0)2 9251 4099 Fax: +61 (0)2 9251 4428
e-mail: jlsydney@mpx.com.au

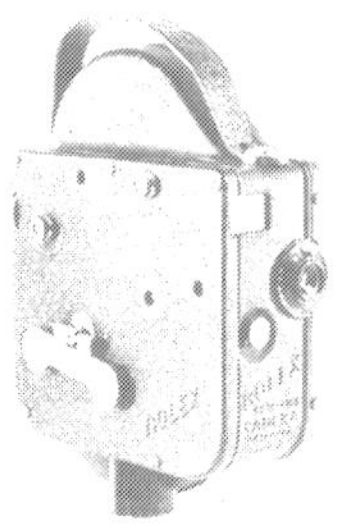

This volume constitutes volume 5, numbers 3/4 of *Aura. Film Studies Journal.*

Printed in Malaysia by Kum-Vivar Printing Sdn Bhd, 48000 Rawang, Selangor Darul Ehsan.

Contents

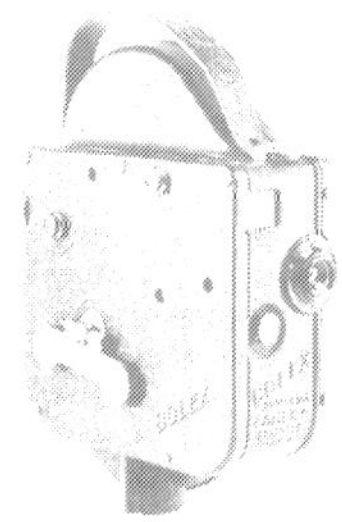

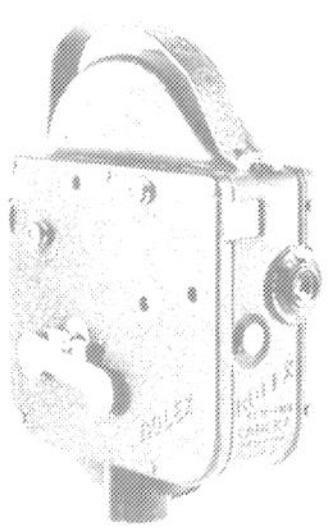

Foreword

Moving Images: From Edison to the Webcam* is the outcome of a conference held in the Department of Cinema Studies at Stockholm University from 6–9 December 1998. Organised in association with the Institute for Futures Studies, the conference showcased thirteen keynote addresses and almost sixty papers covering aspects as diverse as intermediality, indexicality, prosthesis, film and stage, screen practices and reception studies, documentary, film, history, memory, film and changes in the modes of subjectivity, and transformations in the public and private spheres.

With receptions at City Hall and at The Technical Museum, and with a demonstration of a virtual environment improvisation by researchers at The Royal Institute of Technology, Stockholm, the conference also proved to be an exciting social as well as memorable academic gathering. As in all selections, there are regrets about what has been passed over or not been available for some reason, but I believe the essays included in this volume reflect the range of issues and excitement that delegates from eleven countries generated in four intense days of debate in December 1998. I hope that this same spirit will be transmitted to readers who could not attend the conference.

Jan Olsson
Chair, Conference Steering Committee

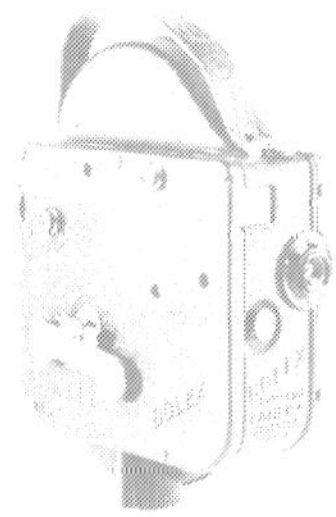

Acknowledgements

We would like to thank all those whose work before, during and after the *Technologies of Moving Images* conference contributed to its success. Thomas Fürth, formerly of the Institute for Futures Studies, Stockholm, Sten Frykholm at The Swedish Television Archive, Stockholm, Folke Sandgren at The Royal Library, Stockholm, and Marianne Lundqvist at The Technical Museum, Stockholm, who were supportive of the conference from the outset, and who variously provided financial and material support. Without their commitment, the conference would not have become a reality. The City of Stockholm for providing a sumptuous reception for conference delegates on the first evening of the conference. Kjell Karlsson, David Pettersson and Rigmor Söderberg of The Technical Museum and Pelle Lönndahl of the Swedish Film Institute, Stockholm, for their assistance in preparing an exhibition of cinematographic apparatus, nineteenth-century lanterns, lantern slides and stereoscopic slides, and Peter Pluntky for the presentation of his magnificent collection of Swedish magic lantern slides at a reception in The Technical Museum. John Bowers, Sten-Olof Hellström, and Kai-Mikael Jää-Aro at The Royal Institute of Technology, Stockholm, for a dazzling virtual environment improvisation, *Performing Infinite Cinema*, and Edward von Past for piano accompaniment to screenings. Elaine King, conference co-ordinator and editorial assistant in preparing *Moving Images: From Edison to the Webcam*, for her ever-resourceful support and formidable organisational abilities. Marianne Sigmond, Lena Sunnegårdh, amd Hannu Luntinen for administrative support, the conference technical team of Bart van der Gaag, Björn Thuresson and Bertil Friberg for their unstinting and magical handling of a battery of apparatus, and the team of student volunteers – Gylla Ersson, Eva Esseen, Mats Isfors, Per Lundberg, Robert Sington – who helped to make the conference run smoothly. Archives du Film du Centre National de la Cinématographie, Bois d'Arcy, Cinémathèque française, Paris, Department of Film and Video, The Museum of Modern Art, New York, National Film Center, The National Museum of Modern Art, Tokyo, National Film and Television Archive, London, Swedish Film Institute, Stockholm, and Swedish Television, Stockholm, for outstanding screenings during the conference. Finally, we are indebted to Jan Olsson whose vision was the motor force behind the conference.

John Fullerton and
Astrid Söderbergh Widding

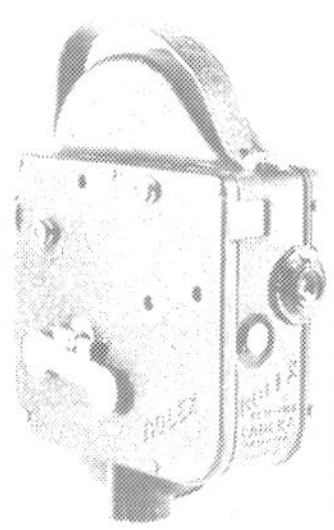

Notes on the Contributors

William Boddy teaches media studies and film studies at Baruch College and at the Graduate Center of the City University of New York. He is the author of *Fifties Television: The Industry and Its Critics* and has published articles in *Screen, Cinema Journal, Media, Culture and Society, The Historical Journal of Film, Radio and Television*, and *The International Journal of Cultural Studies*. He is currently working on a social history of electronic media in the home.

Warren Buckland is lecturer and Head of Screen Studies at Liverpool John Moores University, UK. He is editor of *The Film Spectator: From Sign to Mind* (1995), and author of *Film Studies* (1998) and *The Cognitive Semiotics of Film* (2000). He is currently writing a book on Spielberg's blockbusters.

Carlos Bustamante is Professor in the Theory of Audio-Visual Design and Realisation at the Institute for Timebased Media, Hochschule der Künste Berlin. He has contributed essays to *Griffithiana, Les vingt premières années du cinéma français*, and *Cinéma sans frontièrs 1896-1918*, and as a filmmaker and cinematographer, his recent short experimental videos include *Aus dem Archiv des Verflossenen Instituts* (1997), *The Upper Bay* (1999) and *Licht in der Malerei, eine Einführung* (1999).

Valeria Camporesi is Professor of Film and Television History in the Art History Department of the Universidad Autónoma of Madrid. She has worked extensively on the definition of national identity in mass media history with regard to different national cases. Her publications include a series of essays on BBC reactions to Americanization and *Para grandes y chicos. Un cine para los españoles, 1940-1990* (1993), which deals with national identity in Spanish film production. She is currently working on cinema-television interrelations during the early years of television.

Bent Fausing is Associate Professor and Head of Department in the Department of Aesthetics and Visual Communications, Institute for Nordic Philology, University of Copenhagen, and is currently Research Professor for the interdisciplinary research project, *The Visual Construction of Reality*. He has published widely on fascination, visual communication, visual aesthetics, images of dreams, silence and visuality, and is author of *Emotion in Motion. On Affect and Images* which has recently been published in Danish and is soon to appear in English.

Oliver Gaycken is a graduate student in the Department of English at the University of Chicago. He is currently engaged in doctoral research on early scientific cinema.

Alison Griffiths is an Associate Professor in the Department of Speech, Baruch College, City University of New York, where she teaches media studies and film. Her doctoral dissertation *Origins of Ethnographic Film* won the 1999 Society for Cinema Studies Dissertation Award, and will be published by Columbia University Press. She has published widely in *Wide Angle, Visual Anthropology Review*, and *Film History*, and has contributed to numerous anthologies. She is currently researching the use of new media in museums of natural history.

Chris Hales is Senior Lecturer, Interactive Media, in the Faculty of Art, Media and Design at the University of the West of England. As a doctoral candidate in the Film and Television School at the Royal College of Art, he is working on a project, *Making Interactive Mov-*

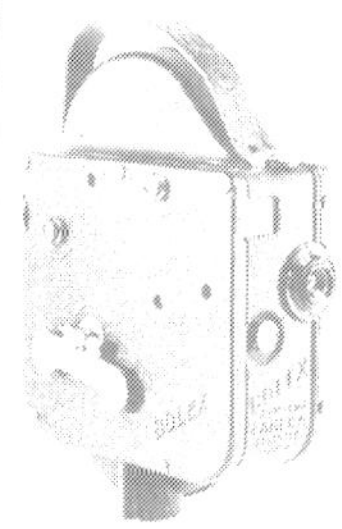

ies Work. Twelve, an interactive CD-ROM, was published in 1996.

Jan Holmberg is a doctoral candidate in the Department of Cinema Studies, Stockholm University where he is currently completing his dissertation on close-ups in early cinema.

Solveig Jülich is a doctoral candidate in the Department of Technology and Social Change, Linköping University, Sweden. She is currently completing her dissertation on radiology, X-ray imaging and visual culture in Sweden 1896-1928.

Frank Kessler teaches film and television studies at Utrecht University. He is co-editor of *KINtop. Jahrbuch zur Erforschung des frühen Films* and the author of numerous articles on film history and film theory which have appeared in *iris*, *Montage/AV*, *Versus* (Nijmegen), *Kinoschriften*, *La licorne*, *Revue belge du cinéma* and *Art Press*.

Jay Moman currently lives in Tucson, Arizona. After graduating from the University of California, Berkeley he worked with the San Francisco International Lesbian and Gay Film Festival and served on the Board of Directors for Frameline. He completed a Master's Degree in Film and Television Studies at the University of Amsterdam in 1998, and has recently served as adjunct instructor in Media Arts at the University of Arizona.

Sheila C. Murphy is a doctoral candidate in Visual Studies at the University of California, Irvine. Her current research projects include a history of lurking in analogue public spaces, the ontological project and public personae of the Teletubbies, and media intertextuality and violence.

Pelle Snickars is a doctoral candidate in the Department of Cinema Studies, Stockholm University. He is currently completing his dissertation on spatial discourses in early non-fiction film, and has published in *Häften för kritiska studier* and *Filmhäftet*.

Paul C. Spehr is the former Assistant Chief of the Motion Picture, Broadcasting and Recorded Sound Division at the Library of Congress, Washington, D.C. He is working on a biographical history of William Kennedy Laurie Dickson, and has written several articles about Dickson and the Biograph and Edison companies.

Björn Thuresson is a doctoral candidate in the Department of Cinema Studies, Stockholm University and is currently working in the Centre for User Oriented IT-design (CID) at the Royal Institute of Technology, Stockholm, where he is researching projects on social interaction in communities on the web, workplace/workspace, the use of sound in 3-D environments, and interfaces for digital TV. He has been involved in the production of several educational CD-ROMs on film production and film analysis.

Åke Walldius is a doctoral candidate in the Department of Cinema Studies, Stockholm University and in the Centre for User Oriented IT-design (CID) at the Royal Institute of Technology, Stockholm. His doctoral project is entitled *The Documentary meets Digital Media* and examines factual moving digital images from the perspective of genre interplay. In 1998 he published a report for CID, *Visual diaries in networked communities: design patterns and the concept of genre as tools for media analysis*.

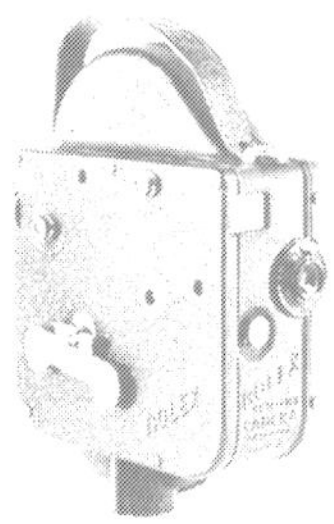

Introduction

John Fullerton and Astrid Söderbergh Widding

On 17 October 1888, Thomas Alva Edison filed a caveat in which he announced that he was 'experimenting upon an instrument which does for the Eye what the phonograph does for the Ear, which is the recording and reproduction of things in motion, and in such a form as to be Cheap, practical and convenient'. Just as work on the development of the instrument to which Edison referred, a precursor of the Kinetoscope, instances an apparatus that was framed in terms of the known technologies of the phonograph and the microscope, the essays in this collection variously address the contexts of innovation and reception that have framed the development of moving images in the last one hundred years. Three concerns are of particular interest: the relation of moving image technologies to the contexts of innovation, reception and popular imagination; the role of the observer whose vision and cognitive processes define some of the limits of enquiry and insight, and the role of new media which, transforming the traditional dichotomy between public and private sphere, engage the domestic sphere as cultural interface.

In the first part of the collection, a number of essays are brought together which consider specific innovations in cinematographic technology or discuss the reception of technology in terms of the popular imagination. In the opening essay, Paul C. Spehr charts the innovation and dissemination of 35mm film as the standard gauge for filmmaking. Arguing that in an industry renowned for competing and often conflicting technologies and patents, the introduction of 35mm film provides an unusually stable example of how the industry innovated a common format. The following two essays present a broad ranging discussion of technology. Oliver Gaycken investigates the development of a pre-cinematic apparatus, the ophthalmoscope. Drawing upon Hermann Helmholtz's discussion of the development of the optical instrument, Gaycken argues that the ophthalmoscope was developed neither as a device for seeing the self seeing, nor as a device for seeing another's gaze. As an instrument, however, that stages the impossibility of the occurrence of such an event, the ophthalmoscope may be understood to presage issues that also characterise the historiography of early cinema. The concern with the metaphorical implications of technological development characterises Frank Kessler's discussion of the nineteenth-century stage tradition of the *féerie* and its incorporation in the work of Georges Méliès. In a detailed examination of the place of fairies and the *féerie* in the late nineteenth-century imaginary, Kessler explores the links between the genre and the representation of modernity and scientific progress in which context, Kessler proposes, the *féerie* functioned as a dialectical and emblematic Other of technology. Sharing a concern for the popular understanding of technologies of vision, Solveig Jülich addresses the issue of dark adaption in X-ray imaging and the cinema at the turn of the nineteenth century. Noting that scotopic vision in both radiology and cinema calls into question the subjectivity of the observer, Jülich examines the tensions that arose in the developing institutions of radiology and cinema, and goes on to argue that both institutions sought to ensure public confidence and authority in institutions which were popularly associated with spectacle and entertainment rather than scientific objectivity. Jülich concludes that while the film

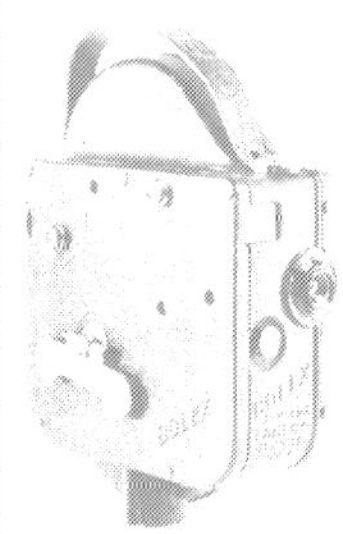

censor 'stands as watchdog guarding the values of mechanised science, the observer of scientific images celebrates the art of judgment'. In the final essay in the first part of the collection, Carlos Bustamante surveys the development of the Bolex 16mm camera from the Bol Cinegraph in the 1920s to its use in the heyday of American avant-garde cinema in the 1950s and 1960s. Through close analysis of Maya Deren's *Meshes of the Afternoon*, Bustamante identifies how the camera's technical features were employed in this classic example of American independent cinema.

As Jay Moman notes in his contribution, no anthology of cultural theory texts in the 1990s 'is complete without a contribution to the increasingly pervasive discourses on the body'. This part of the anthology, devoted to a consideration of the observer, and developing a concern announced in the earlier part of the book for the epistemology of vision, covers a broad range of topics from early cinema to digital images. In 'Sore Society', Bent Fausing discusses the representation of the body in digital images and, given its capacity for change and exchange within an increasingly digital culture, its truth value. With the performance artist Orlan as a central example, Fausing analyses images of the fragmented or destroyed body as instances of the fissures in contemporary society where the experience of discontinuity is perceived to have overtaken the experience of identity. In Fausing's view, the very boundary between exterior and interior which, traditionally, has been used to define the body, is challenged so making the body an image that is continually 'created, dissolved, and created anew'. If Fausing examines a dichotomy prevalent in contemporary culture, Jan Holmberg, in 'Closing In', discusses the apparent paradox of distance and proximity that is crucial to an understanding of modernity, a dialectical relation which, Holmberg proposes, is also central to Heidegger's concept of de-distancing. The paradox of modernity, according to Holmberg, lies in the fact that the technologies of approaching used both to magnify the world and the body, to bring them closer, also serve to maintain their invisibility by becoming too large, too detailed. Epistemological issues relating to vision are also central to Alison Griffiths' essay, 'We Partake, as it Were, of His Life', where Griffiths discusses early ethnographic film in the context of analysing Alfred Cort Haddon's films of Torres Strait Islanders and Australian Aborigines on Mer Island from an 1898 expedition, as well as Carl Lumholtz's films shot in central Borneo during a four-year expedition between 1913 and 1917. With this analysis as a point of departure, Griffiths concludes that technologies of vision have 'by no means achieved a secure home within anthropology', and that questions of ethnographic evidence and the anthropological object remain to be solved. Griffiths also provides a brief discussion of the possibilities new digital technologies offer the field of anthropology. The body in architectural space forms the focus of Pelle Snickars' essay on architecture as a technology of the moving image. Comparing Eisenstein's understanding of architecture with Le Corbusier's 'architecture of motion', Snickars considers the work of contemporary architects such as Bernard Tschumi and Jean Nouvel, and characterises the organisation of space in contemporary deconstructivist architecture which, Snickars argues, is close to cinematic experience. In this regard, a common genealogy for an architectonics of seeing is proposed in the intersection of modernity and early cinema. While twentieth-century technologies have given greater access to the unseen, Jay Moman deals with the historical discourses of the body reconsidered within a post-modern framework that addresses politics, economics and culture as central concepts. Proposing the term, *physioanalysis* to cover the different approaches to the body as encoded identity, Moman traces an emerging discourse on the concept of virus which is based on the cinematic surveillance of the internal body. Contrary to Foucault's thesis that the surface of the body has been displaced in favour of the unknown territories of the unseen interior body, Moman argues that the surface of the body 're-emerges as a text, a screen inscribed with the information of its own subdermal terrain'.

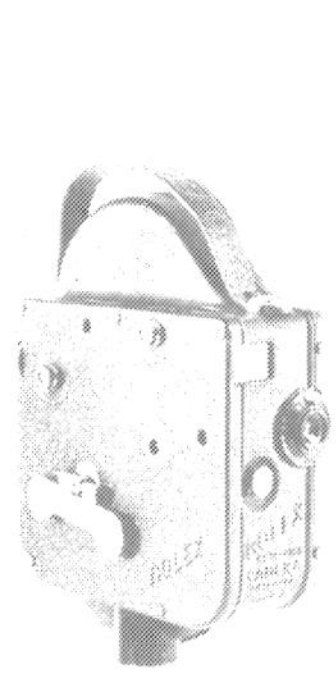

In the third and final part of the collection, the introduction of new media in the domestic sphere forms the principal focus of discussion. Charting the uncertain future that attends the introduction of high-definition television in the US domestic market, William Boddy characterises how the political and regulatory institutions in the US differ from public policy debates in the UK, and examines some of the myths of national identity which are currently being constructed in the US and the UK. Examining the take-up of new media in the domestic sphere, Boddy also identifies the ambivalence such technologies currently engender. The representation of television in Spanish films from the 1960s provides the focus for Valeria Camporesi's discussion of the introduction of television in Spain. With close reference to their narrative concerns, Camporesi argues that Spanish films contributed to the popular stereotyping of socially received views regarding television in the 1960s. The emergence of digital narrative in the contemporary context provides the background for Warren Buckland's examination of the impact of interactive video games on cinema. With Luc Besson's *The Fifth Element* as example, Buckland argues that the psychologically-motivated, cause and effect narrative logic typical of classical film is being displaced in favour of video game rules. Proposing that interactive digital media should be conceived in terms of ritual and ceremony, Buckland argues that classical filmic pleasure is firmly rooted in the society of the spectacle, a configuration which interactive media challenge. The impact of interactivity on narrativity is further developed by Björn Thuresson in his discussion of spatiality, temporality, causality, dramaturgy and personification in computer games. A concern with designing digital environments also leads Thuresson to propose a number of innovations as to how interactive environments may be developed in the near future. Interested in the new mode of cultural engagement which webcams offer, Sheila C. Murphy characterises a technology and mode of interactive surveillance which has not only become a dominant cultural logic but, in some cases, a positive means for sub-cultural expression. Attentive to the emancipatory possibilities of the webcam, Murphy argues that webcams not only represent a shift from a Foucauldian view of culture to a post-modern culture in which panopticism becomes a part of everyday life, but instance a technology which, contrary to the voyeuristic dynamics of a psychoanalytical account of film, celebrates looking and lurking as socially acceptable activities. Other ways in which digital technologies are revitalising genres associated with film are central to Åke Walldius' investigation of whether a cinema of meditation will evolve in parallel with the new cinema of digital attractions. Taking the film diary as his example, Walldius proposes that generic concerns may respond to medium specificity. *A Diary for Timothy*, a video diary, and a CD-ROM diary from The MIT MediaLab provide examples with which Walldius charts the shifts in perspective that attend technological and formal innovation. In the final essay of the collection, Chris Hales provides an overview of the development of interactive film, and discusses how his work using desktop Apple Macintosh technology experiments with structure and interface to open up a middle ground which is neither film nor game. Distributed as a CD-ROM or set up as a touch-screen installation, Hales' interactive films provide an example of how new media may not only institute new patterns of ownership and distribution, but also transform the domestic sphere as cultural interface.

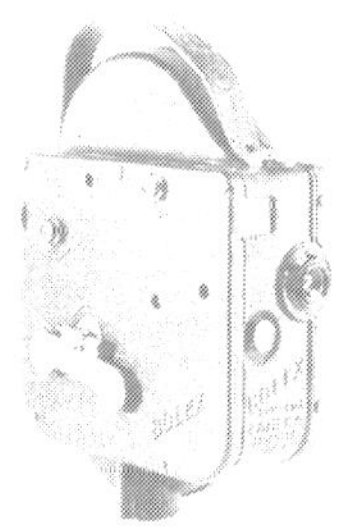

1

The Apparatus

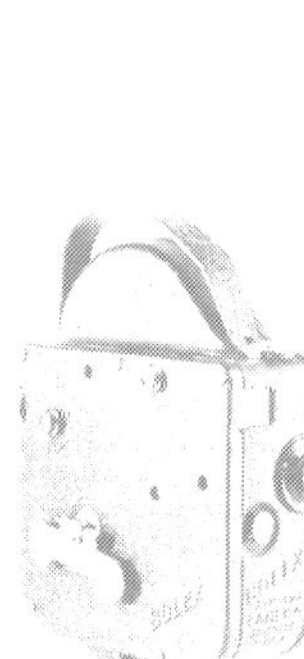

Unaltered to Date: Developing 35mm Film

Paul C. Spehr
Fairfield, Pennsylvania, USA

At the end of the year 1889, I increased the width of the picture from ½ inch to ¾ inch, then, to 1 in. by ¾ in. high. The actual width of the film was 1 ⅜ in. to allow for the perforations now punched on both edges, four holes to the phase or picture, which perforations were a shade smaller than those now in use. This standardized film size of 1889 has remained, with only minor variations unaltered to date.

William Kennedy Laurie Dickson, 1933.[1]

Edison standard film

When Thomas Edison introduced his Kinetoscope in April 1894, it used a film that is almost identical with the 35mm film used today – the same width and with four similar perforations on each side of the image. W.K.L. Dickson's abridged account is accurate except for the date, which he exaggerated in his eagerness to reinforce Edison's claim that his invention preceded all competitors. At the end of 1891 and the beginning of 1892 Dickson made the changes which resulted in a film 35mm wide.

Dickson's pride was justified because it was, arguably, Edison's – and Dickson's – major contribution to the future of the motion picture and, perhaps, the most important technical innovation of the 1890s. In 1895 and 1896, when numerous competitors were designing cameras and projectors, many of them, following Edison's lead, used 35mm film. So many, in fact, that by 1897 it was already called 'standard film' or even 'Edison standard film'. Although competing formats appeared and keep appearing, 35mm was and is so ubiquitous that writers rarely define the gauge or format when they discuss 'film' or 'cinema' or 'movies'. Because the acceptance of 35mm happened so early and became so universal, film historians have regarded it as inevitable and paid very little attention to the factors that influenced what is, in fact, a remarkable phenomenon that was far from inevitable. These historians have also neglected to examine the close relationship

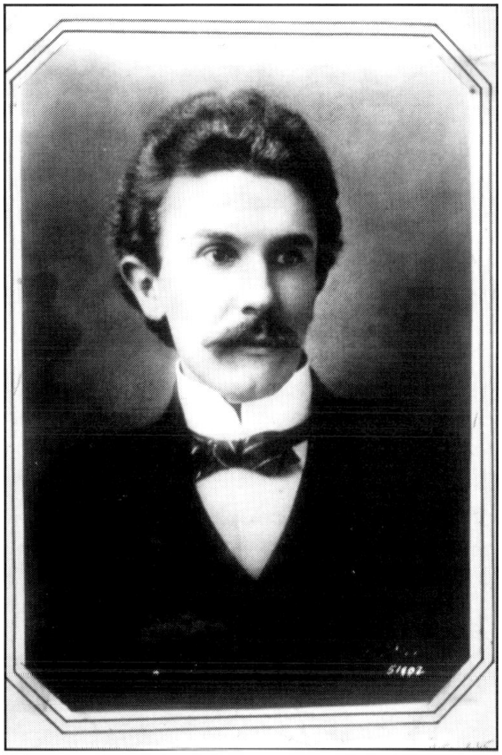

Fig. 1. William Kennedy Laurie Dickson 'as he looked in 1888' according to Eugene Lauste.
[From Eugene Lauste Collection, Photographic History Collection, National Museum of American History, Smithsonian Institution, photo cat. no. 4051.3.45.]

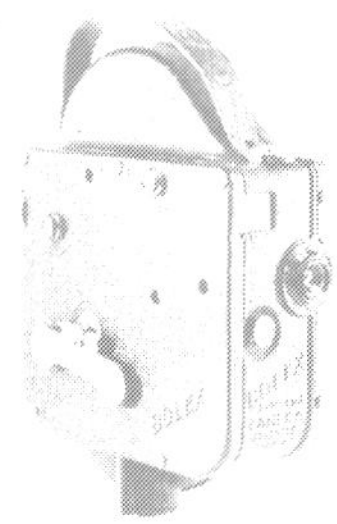

between the movies and other, related industries, particularly the photographic industry and the phonograph.

Far from being a trivial matter, the early acceptance of 35mm as a standard had momentous impact on the development and spread of cinema. The standard gauge made it possible for films to be shown in every country of the world – in cities large and small, to audiences rich and poor. Because it was – and still is – a consistent, predictable technology it gave the motion picture industry coherence as well as stability. It provided a uniform, reliable and predictable format for production, distribution and exhibition of movies, facilitating the rapid spread and acceptance of the movies as a world-wide device for entertainment and communication. 35mm film made it possible for productions made in the US, Britain or France to travel easily across borders and to be shown in small towns as well as big cities. If made in 35mm, there would be projectors that could show the film, which was not true for films made in other gauges. This is still true and, furthermore, images recorded in 1897 can be copied by modern film laboratories and be seen by modern audiences.

This was not inevitable, or even natural. Technical stability is not a normal characteristic of modern invention. Manufacturers usually want exclusive markets and resist developing compatible products that competitors can use. Patents and copyrights as well as the legal entanglements that relate to them are tools supporting exclusiveness rather than consistency. Consider the movies' sister industries, the phonograph and television, where examples of technical incompatibility are frequent: cylinders vs. disks; 33 ⅓ rpm vs. 45 rpm vs. CDs; beta vs. VHS; videotape in ½ in., ¾ in., 1 in., and 2 in. formats; open reels and cassettes. Both industries have been and still are plagued with a never-ending parade of confusing and confounding 'new and improved' technologies, and almost all of it incompatible with what preceded it.

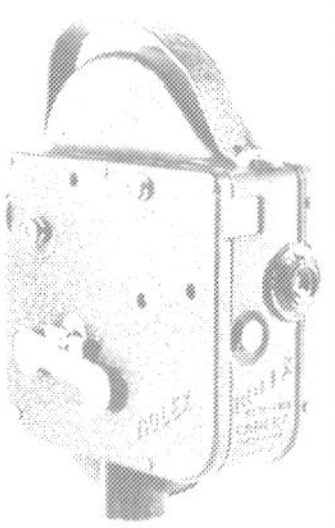

It has been assumed that Dickson arrived at 35mm by cutting a standard roll of Eastman Kodak's transparent roll film in half. John Belton has the best description of this in his section on 35mm film in *Widescreen Cinema* as well as in his paper 'The Origins of 35mm Film as a Standard' in the *SMPTE Journal*, August 1990.[2] Belton's assumption is reasonable enough, but the evidence that survives in various photographic and scientific journals, Edison's papers, and George Eastman's correspondence shows that 35mm evolved in a different way. The story is intricate, involving many important contributors to the introduction of the movies, and it casts an interesting light on the process of invention and the evolution of the industry. The principal players are Thomas Edison, William Kennedy Laurie Dickson (Fig. 1), and George Eastman (Fig. 2), but we will also meet C. Francis Jenkins, Thomas Armat, Robert Paul, Louis and Auguste Lumière, and Thomas Blair as well as a large and varied cast of supporting players. It starts in the 1880s.

A celluloid melodrama, part 1

> *The new film is as thin, light, and flexible as paper, and as transparent as glass … it is wound on spools for roll holders.*
>
> Ad for the Eastman Dry Plate and Film Co., July 1889.
>
> *It is as thin as a blister and as clear as glass.*
>
> *The Philadelphia Photographer*, July 1889.[3]
>
> *An editor of* The Philadelphia Photographer *wondered 'in the name of all that is beautiful' why anyone would patent print photographs on celluloid.*
>
> 6 October 1888.[4]

The narrative begins at The Celluloid Company, a chemical factory in Newark, NJ, and the Eastman Dry Plate Company in Rochester, New York. A revolution was beginning. It was quiet and unobtrusive, but it produced a profound transformation in photography. It began in the early 1880s with the introduction of a celluloid base that could be coated with photographic emulsion. It gathered strength when George Eastman developed a process that linked celluloid film with his recently developed roll holder and put them in an inexpensive camera. Coming on the heels of improved film emulsions that made rapid photography possible, Kodak brought an end to an era when photography was the exclusive domain of professionals and talented amateurs. Photography no longer required time, skill, and where-with-all. Photographers did not have to mix chemicals, operate complex cameras, and work in dark rooms.

George Eastman was the force behind the change. He foresaw a new and completely different photography which put inexpensive cameras in the hands of ordinary people. He also saw a monopoly of the photo industry in the hands of George Eastman. His roll film system, which could fit existing cameras and allowed multiple exposures to be taken, was introduced in 1884. In 1887 the company introduced the Kodak No. 1, the first roll film camera, with an ad campaign featuring the slogan 'You press the button – we do the rest'. Kodak was soon a household name.[5] At first Eastman's system used a rather complicated 'stripping film' with a sensitive emulsion supported by paper but transparent celluloid soon replaced it.[6]

Celluloid was introduced in the 1860s and was used in the manufacture of a variety of products such as combs, collars, cuffs, billiard balls, dolls, etc. Although celluloid coated with photosensitive emulsion was introduced in the early 1880s as a substitute for glass and paper, it was not until 1887 that celluloid began to make an impact on the market. That year Vergara 'Ivory' Film, patented by Francis A. Froedman of Dublin, Ireland, was marketed in England, and John Carbutt, of Philadelphia, introduced a celluloid film based on the pending patent of the Rev. Hannibal Goodwin of Newark, N.J., Allen and Rowell of Boston, and E. & H.T. Anthony, a photographic supply house also began selling transparent 'film'. All of these firms sold sheet film in sizes from 3¼ x 4¼ to 11 x 14.[7]

Carbutt applied his photographic emulsion to a nitro-cellulose base which he purchased from The Celluloid Company of Newark.[8] The Celluloid Company had a virtual monopoly on the manufacture of celluloid products because of their patents, and the methods they developed to handle the volatile, hazardous nitro-cellulose in manufacture and for disposing of the dangerous waste materials which resulted. They produced sheet celluloid by slicing slender sheets from large blocks of celluloid and treating it to produce a very thin, relatively clear film base.[9]

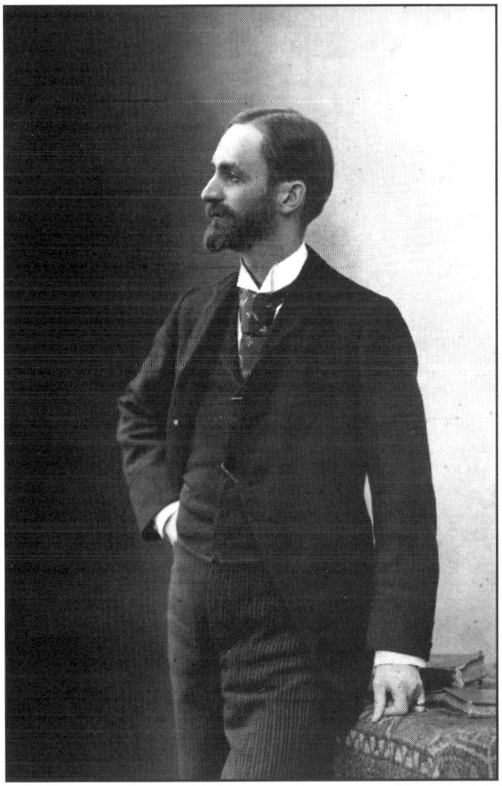

Fig. 2. George Eastman, 1889. A portrait by the Parisian photographer Félix Tournachon Nadar. [Courtesy George Eastman House.]

As the editorial comment from *The Philadelphia Photographer* quoted above indicates, celluloid 'film' was not welcomed by all photographers. Nevertheless, there were those who felt it was an important breakthrough, and it created a major stir in photographic circles. There were demonstrations at camera clubs, and articles in professional journals throughout 1889.

Celluloid was crucial to the creation of the modern motion picture. Experimenters were limited by the possibilities of the substances available to them. Ridged, fragile or opaque substances like metal, glass, or paper, imprisoned moving images in the limited visual cycle of the Zoetrope. Although appealing, the fragments of motion recorded by the chronophotographers like Muybridge, Marey and Anschütz were also limited. Images endlessly leaping over fences, juggling balls or performing exercises intrigued scientists and amused children but there was very little

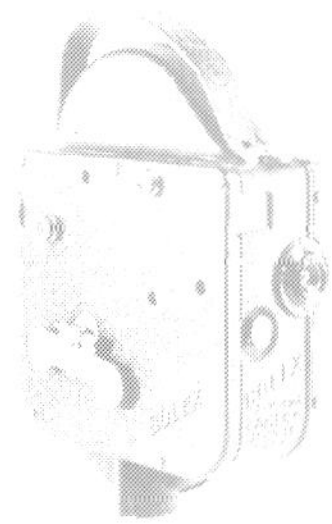

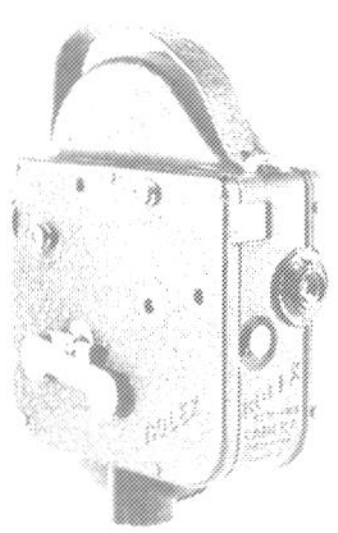

future in them. Pliable, unbreakable celluloid offered new promise.

Eastman's flexible roll film

> *The advantages of these films to the photographer over glass dry-plates, and all other films on the market, briefly summed up are as follows: superior transparency, greater flexibility, lightness, compactness, practicability of printing from either side of the negative, and lack of halation.*
>
> Gustave D. Milburn, 16 August 1889.[10]

The Eastman Dry Plate and Film Co. began experiments to produce a transparent photographic base in the early 1880s, not long after the company was established.[11] Serious experimentation with celluloid began in the spring and summer of 1888 with Eastman's chemist, Henry M. Reichenbach in charge.[12] After testing samples of celluloid from The Celluloid Co. and the Rev. Hannibal Goodwin, Reichenbach and George Eastman devised both chemistry and a system for applying liquid celluloid in thin sheets on long glass-top tables. Eastman applied for two patents in the spring of 1889, and in June 1889 announced that Eastman Dry Plate Co. would begin manufacture of transparent roll film.[13] A new building was constructed for the manufacture, and in the summer Eastman's sales representatives, led by Gustave D. Milburn, began demonstrations and lectures to groups of photographers in major American cities.

After discussions with the Board of Directors, Eastman decided, for safety reasons, they would not manufacture all of the chemicals used to make a celluloid base. Although they made the final blend in their own factory, the company bought a chemical mix, which Eastman called 'dope', from suppliers, then added additional chemicals to produce a celluloid. It was slightly different from the product made by The Celluloid Co., and Eastman's method of applying chemicals to produce the base was unique. Eastman continued to purchase 'dope' from various suppliers through most of the 1890s. They began manufacture of the full chemistry at a date after the period covered by this essay.[14]

Although the company's advertising was effusive and optimistic, Eastman was actually having problems producing transparent film as well as other products. Apologizing for not sending him a supply of the new film, on 16 June 1889 George Eastman wrote to William Walker, now in London: ' ... [for the] past six weeks it has been a succession of petty delays and mishaps ... We have been almost shut down for two weeks on A.M. Films [i.e. stripping film]. The film blisters in spite of everything we can possibly do.'[15] In his otherwise upbeat talk to the Society of Amateur Photographers in New York, Gus Milburn confessed that they were experiencing 'something like a vine or tree' appearing in corners of the photographs.[16] In spite of their early publicity, Eastman's new film was not widely available for sale until the spring of 1890. Even though Eastman had problems, there were photographers anxious to try the new film. One was William Kennedy Laurie Dickson.

Edison: cylinders and celluloid

> *I am experimenting upon an instrument which does for the Eye what the phonograph does for the Ear. Which is the recording and reproduction of things in motion ...*
>
> Thomas A. Edison, Caveat 110, filed 17 October 1888.[17]

The exact date that serious experimentation on the Kinetoscope began is controversial, but by the end of 1888, some work was under way. Edison assigned the project to William Kennedy Laurie Dickson who had been Edison's photographer.[18] In the early experiments Dickson was assisted by Charles Brown, who simultaneously worked on an improved version of the phonograph, a project which occupied much of Edison's time in the first half of 1889. Dickson had access to several machinists, labourers and pattern makers on the staff of the Edison Laboratory. The preliminary work was done in a photographic room in the new laboratory in Orange, NJ.

The Kinetoscope was a secondary assignment for Dickson. In 1887, Edison gave him responsibility for research on a process to separate iron, gold and other valuable metals found in low-grade ore. This was one of Edison's pet projects and during the 1890s it became an obsession. During the entire time that he worked on the Kinetoscope, Dickson was also deeply involved in ore-milling experiments and the large facility that Edison built at an iron mine in Ogden, NJ. The ore-milling project frequently interrupted Dickson's work on the Kinetoscope and Kinetograph, and it was partly responsible for the long gestation period of these machines. By 1898 Edison had spent more than $3,200,000 on ore-milling. By contrast, the company reported that the Kinetoscope experiment had cost $24,118.04 through 1 April 1894.[19] Despite Edison's obsession, the ore-milling project was a dismal failure, while the Kinetoscope proved to be one of Edison's most profitable inventions.

Edison's first Caveat described a machine intended to be an addition to the phonograph (Fig. 3). Tiny images about 1/32nd of an inch wide would be recorded intermittently in a continuous spiral around a cylinder attached to a cylinder phonograph by a common drive shaft. The images would be viewed through a microscope-like viewing device while listening to synchronised sound. Although Edison grandiosely mentioned recording and reproducing Grand Opera, in late 1888, he actually assumed that the visual image would supplement phonograph recordings of official transactions, legislative and judicial proceedings, correspondence and other verbal affairs of business and government. It was a modest device, scarcely capable of lofty operatic ambitions. During the seven-plus years of experimentation on the Kinetoscope, Edison's original conception changed radically as the phonograph evolved into a commercial entertainment device and the difficulties of recording images on a cylinder and synchronizing the sound became more evident.

Dickson's early experiments were made for this cylinder. Surviving purchase records from West Orange as well as Dickson's own accounts of his work show that in October 1888, immediately after Edison sent his initial motion picture Caveat, Dickson began experimenting with a variety of photographic methods, including Daguerreotype, wet collodion, and dry plate.[20] These futile attempts to devise a way of recording microscopic images on the surface of a cylinder gave Dickson a chance to improve his understanding of photography.

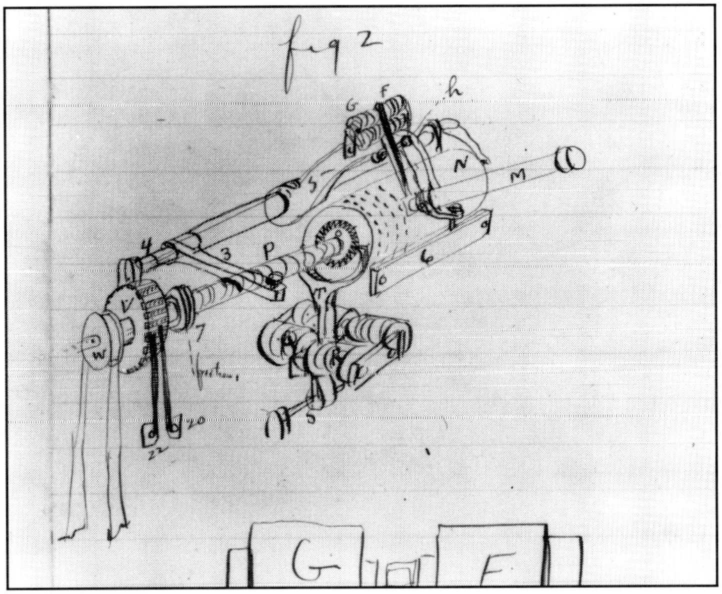

Fig. 3. Drawing from Thomas A. Edison's Caveat no. 110, submitted to the US Patent Office 17 October 1888 showing a device modelled on the cylinder phonograph with consecutive micro-photographs mounted in a continuous sequence around a cylinder. The photos were to be recorded and read with the microscopic tube (letter 'M'). This device was linked with a cylinder phonograph which supposedly recorded and played in synch. [Courtesy US Dept. of Interior, National Park Service, Edison National Historic Site.]

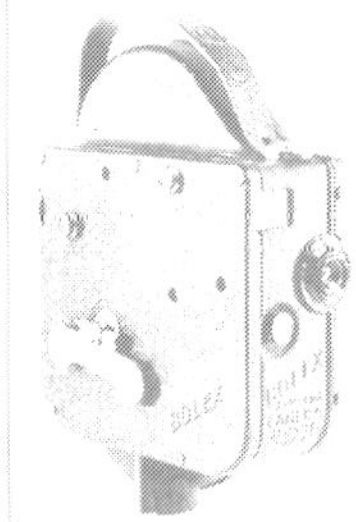

Edison's cylinder scheme was impractical. It is uncertain how long the experiments continued before the cylinder was abandoned, but it is easy to imagine Dickson's delight on learning that strips of flexible, transparent photographically sensitive film were available.[21] Dickson claimed that after some experiments with sheet celluloid, he received his first roll of Eastman transparent film from Eastman's representative at a demonstration in New York City in August 1889, and that it was immediately applied to experiments begun earlier in the summer on a machine using a strip or ribbon of film.[22]

Following the lead of Gordon Hendricks, modern historians have argued that claims by Dickson, Edison and several of Edison's associates that work on a strip machine began in the summer of 1889 are false. Even though it does not prove there was work on a strip-machine by late summer or fall of 1889, there is evidence to support the Dickson–Edison claims.[23] Edison began ordering rolls of Eastman's film in September 1889, and continued to order film in varying quantities for the next three years. The earliest order was 2 September 1889, when Dickson paid for a roll and asked for more with 'your highest sensitometer'.[24] Six rolls, 3/4 in. x 50 ft, were received

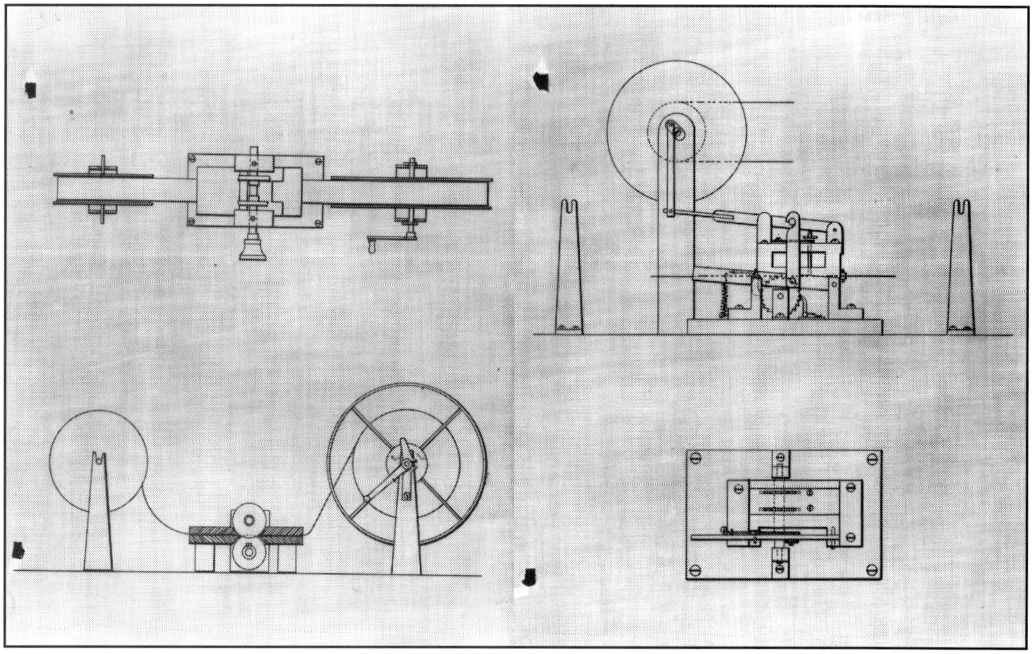

Fig. 4. Drawings of a cutting device (left top and left bottom) and film perforator (right top and right bottom) designed in 1890 or 1891 by W.K.L. Dickson. These drawings are from an application for patent which was prepared and grouped with the applications for the Kinetoscope and Kinetograph. When the applications for the latter were submitted in August 1891, the application for the cutting device and perforator were not sent to the Patent Office. [Courtesy Edison National Historic Site.]

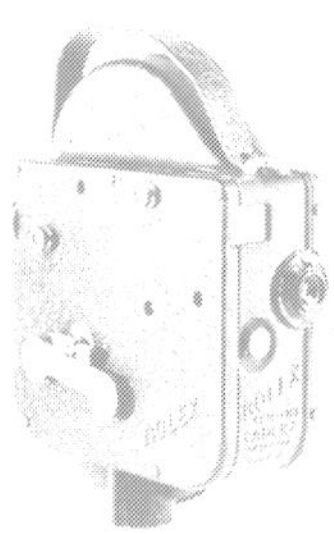

in November and six more of the same size arrived in December.

These orders were placed near the date, 2 November 1889, when Edison sent Caveat No. 114, his fourth and final Caveat about the Kinetoscope, to the Patent Office. This one described a strip machine using perforated transparent film passing from one reel to another, with perforations on either side to guide and drive the film by engaging sprocket wheels. The Caveat specified that the perforations and sprockets were like those used in a Wheatstone automatic telegraph machine.[25] The next orders for film, placed in February 1890, were for rolls of film 1 in. x 50 ft., slightly wider than the ¾ in. film used previously. This order for wider film, a charge made to Edison's kinetoscope for patent drawings and two newspaper articles announcing work on the Kinetoscope – all occurring early in February, 1890 – lend credence to the claims by Edison and Dickson that the experiments with a strip machine were already underway in the fall of 1889 or early in 1890.

The charge for patent drawings was for the work week ending 6 February 1890. These were apparently made for a potential application related to the kinetoscope which was probably initiated by Dickson. These may have been for a perforator and an apparatus for cutting and trimming film strips. Dickson described these devices in his article in the *Journal of the SMPE.*[26] Although there is no record of what the drawings were, the change of width in the order to Kodak indicates that Dickson was now having to prepare the film before using it by trimming a small amount off and then perforating it. A draft patent application for such devices is in the files of the Edison National Historic Site (Fig. 4). The drawings for 'Improvement in Apparatus for Preparing Strips for Kinetoscope' show two devices, one with an adjustable pair of disks for trimming the film, the other for perforating both sides. The apparatuses are shown mounted between a pair of hand-operated rewinds, similar to the ones found in film workrooms today. Such a device would have been necessary to trim film accurately from 1 in. to ¾ in.[27] Cutting film to proper size will crop up again as we explore the evolution of 35mm film.

Apparently Edison felt confident enough about his progress to sanction a public announcement. On 1 February 1890, the *Orange Journal* reported: 'For many months past Mr. Edison has been at work on a series of experiments in instantaneous photography which have been at last successfully concluded.' It reported that from eight to twenty exposures were being taken a second and that commercial development was all that was needed. A similar article appeared in the *New York Herald* on 2 February 1890.[28] If there is evidence that early in 1890 the Kinetoscope was already developed to a state where Dickson seemed to think experiments were 'successfully concluded', why did it take four years before it was presented publicly?

In all probability the mechanism was the least of the problems confronting Dickson. Edison's staff were as familiar with mechanical manipulation of strips or ribbons as they were with cylinders. Edison, a telegrapher by trade, had his earliest commercial success with a machine to record stock price information on a paper strip. His assistants cut their teeth on the mechanisms of ticker tape machines and telegraph devices rapidly recording information on paper tapes advancing through machines. The new laboratory in Orange was equipped with two state-of-the-art machine shops, one for general work and one for precision work. John Ott, the veteran head of Edison's precision room, who was helping Dickson during the summer and autumn of 1889, was very experienced with strip telegraphic devices. It was the nature of the celluloid available to them and difficulty of marrying the photographic material with the machine that presented the most serious challenge to Dickson and his associates. It took them another two and a half years to resolve the problems they encountered.

A celluloid melodrama, part 2

> *The manufacture of transparent film has always been the most difficult part of our business, celluloid have [sic] proved a rather uncertain support for sensitive emulsion. It is only since last July that we have been able to make film that would stand a reasonable time without deteriorating.*
>
> George Eastman to his Board of Directors, 1894.[29]

Although filled with promise for the future, celluloid film had a troubled childhood. Making a product with consistent quality proved to be a vexing problem. Photographic journals published frequent complaints about celluloid film. Vergara film exhibited lumpiness when wet and a 'propensity to assume any shape but a flat one on drying'; 'Ivory' film could not be rolled on a spool because it was too stiff and brittle. There were objections to streaking, excessive grain, uneven thickness of the base, halation (a halo effect in the image) and cockling (wrinkling or puckering of the emulsion). *Anthony's Photographic Bulletin*, 25 October 1890, commented that they 'looked in vain for a professional who uses the rolls ... as a substitute for glass. ... When shall we have a transparent film that ... be of sufficient body to keep it from curling as small as an ordinary lead pencil when it dries?' Most of these complainers were professionals or skilled amateurs.[30]

Eastman produced a conventional product line to satisfy these serious photographers, but roll film was a new enterprise for new customers. Sometime after the beginning of 1890 Eastman's company, now officially reorganised as The Eastman Co.,[31] began regular sale of roll film and they introduced new cameras using the new product. Despite problems with spots on the film and a recurrence of the mysterious vine or tree-shaped flaw, Eastman did a brisk business for the next year and a half. Faulty film was either discarded, or replaced if it was reported. George Eastman had made his reputation by standing behind the quality of his product and he continued the policy. The *American Amateur Photographer*, 2 July 1890, reported that Eastman's celluloid was firmly established on both sides of the Atlantic: '[The] demand for Kodaks and the new film must be something enormous.' In June 1891 Eastman opened three new factory buildings outside Rochester in an area that was to become known as Kodak Park. The film building had 12 tables for pouring the base. They were 200 ft. long and 41 in. wide. The complex had a generating plant and air conditioning.[32]

Eastman dominated the film market, but a rival soon appeared. In 1890, the Blair Camera Co. of Boston began producing and selling transparent roll film. Blair gained technology and skill in celluloid by acquiring the Allen & Rowell Co., and they contracted with The Celluloid Co. to purchase celluloid base. By

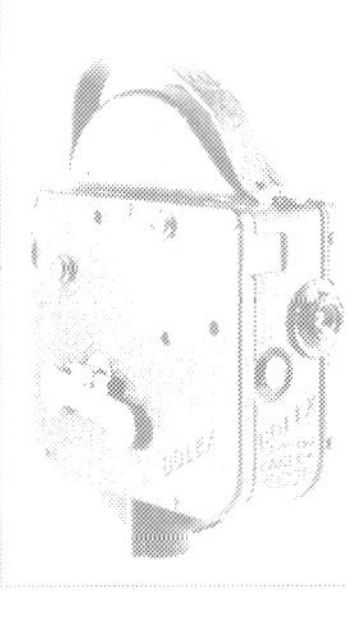

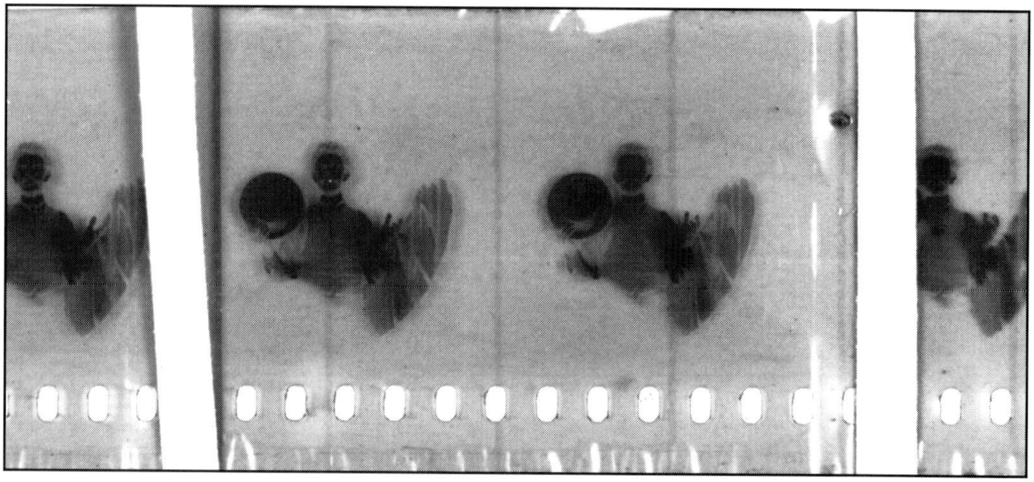

Fig. 5. The 'Dickson Greeting' which was shown to the delegates of the Federation of Women's Clubs on 20 May 1891. Note the circular frame and single row of perforations at the bottom of the film. Enlarged: the original film was

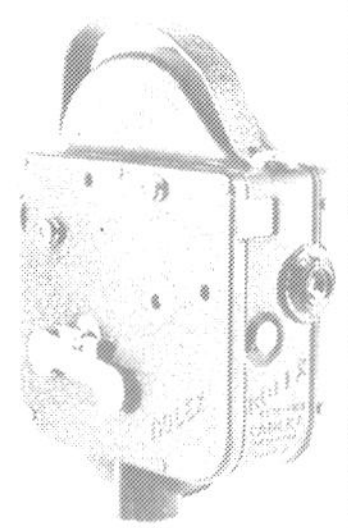

this time, The Celluloid Co. had patented a method for producing continuous strips of celluloid, and could supply Blair with rolls of celluloid that were 20 in. wide and 200–300 ft. long. Their celluloid was translucent rather than transparent, which resulted in a slightly darker tone than Eastman's. Blair coated the celluloid base with sensitive emulsion of their own manufacture, in a process tied to the pending patent of Hannibal Goodwin. Blair's position in the market was strengthened when the venerable firm of E. & H.T. Anthony Co., a stalwart of the American photographic business, purchased an interest.[33] The competition from Blair did not bother Eastman too much; their sales were much better and they were expanding production. Nevertheless, Eastman encountered troublesome obstacles.

Mr. Dickson goes to Rochester and Edison procrastinates

> *Enclosed is a small fragment of film furnished Edison for his phonograph arrangement. He perforates it on both edges and delivers it by means of cog wheels. ... The trouble with the film we have sent him is that the cogs tear the film slightly, as you will see by the enclosed, and gives blurred images.*
>
> George Eastman to Henry M. Reichenbach, Kodak Chemist, 23 July 1891.[34]

Through 1890 into 1891 Dickson experimented irregularly with film strips ¾ in. wide, with his work being interrupted frequently by the ore-milling venture. In May 1891 the experiments had progressed to the point where Edison was ready for a public demonstration. On 20 May the delegates of the Federation of Women's Clubs lunched with Mrs. Edison, and after lunch the ladies went to the laboratory where they were entertained with glimpses of W.K.L. Dickson smiling and tipping his hat to them inside a peep show machine (Fig. 5). The event received generous coverage from the press.[35]

Two months later, on 24 August 1891, Edison's attorneys sent three patent applications for the Kinetoscope/Kinetograph to the Patent Office.[36] They described 'a method of taking and using photographs in such manner as to give a visual impression as of an object in actual motion' by using 'a long gelatine tape film ... coiled on a reel ... pictures 1 in. in diameter between rows of holes into which teeth of wheels engage ... to advance the film'. Each of these applications was challenged by the US Patent Office. One, which attempted to tie together the process of taking with exhibiting, had to be abandoned. The patent for the Kinetoscope, the viewing machine, was granted in 1893, but the application for the camera was disputed and deliberately delayed by Edison until August 1897. Edison's procrastination is important to our story.

On 22 July 1891 Dickson went to Rochester to discuss film with Eastman and to visit two lens manufacturers, Gundlach Optical Co., and Bausch and Lomb. He stayed for two weeks, spending part of the time vacationing in nearby Clifton Springs, NY. On 23 July 1891 he met with George Eastman. Eastman's note to Reichenbach, quoted above, is evidence of the problems Dickson was encountering: torn film, damaged sprocket holes, and poor image quality.[37]

Dickson's visit to Rochester took place between the public exhibition in May and the patent applications. Although the press coverage in May was filled with optimistic statements by Edison, the trip to Rochester must have been because they were revising the machine. Eastman's comment that Edison's film had sprocket holes on both sides is interesting because surviving samples from May–June 1891 have a single set of sprocket holes on the bottom of a film that moved horizontally through the machine. The images taken by this prototype machine were a sequence of circles.

The patent application for Edison's camera, sent a month later, specified two sets of perforations and the 'film ... is preferably of sufficient width to admit the taking of pictures 1 in. in diameter between the rows of holes ... arranged at regular intervals along the two edges of the film'. It added that narrower film with one perforation might be used.[38]

Enlarging the image to 1 in. in diameter was at least the third change in size since the cylinder experiments began. It agrees with Dickson's description in the *Journal of the SMPE* which is quoted at the beginning of this essay: 'I increased the width of the picture from ½ inch to ¾ inch, then, to 1 in. by ¾ in. high. ... ' In August 1891 the Kinetoscope was in transition with a second set of sprockets and larger but, perhaps, still a circular image.

Before summarising his activities during the fall and winter of 1891 let us consider the problems that Dickson encountered that led to this revision. The mechanisms of the camera and the viewing machine had reached a generally acceptable state. The public demonstration and Edison's statements to the press are evidence of this. Film damage and poor image registration resulting from the passage of the film through the machine were the major concern. The fractures around the small perforations were caused by the teeth of the sprockets. The film would ride off the single set of sprockets, producing punctures and tears. The second set of sprockets, larger sprocket holes and tougher film would reduce this problem significantly. The second set of sprockets would also provide a steadier path as the film advanced in the camera – less jitter, and blurring. Increasing the size would improve the sharpness and clarity of the image, and reduce the effect of grain in the film.[39]

Dickson enlarged the image because of unacceptable quality during his early experiments. Dickson described his problems in an article in *Century Magazine*, June 1894. When the microscopic images were enlarged, the grain of the film became excessive and the images recorded on the curved surfaces of cylinders were distorted.[40] A flat surface and larger image area improved resolution, but the blurred images in the sample given to George Eastman show that ¾ in. film had not solved the problem.[41]

The change to 35mm was not immediate. Four rolls of film Dickson ordered while in Rochester were received in August and they were the same measurement as those previously purchased, 1 in. by 50 ft. Two other shipments received in August were also 1 in. by 50 ft. But a change was in the works. On 31 August, Dickson wrote to Gundlach Optical Co., rejecting a lens received from them because it gave an image that was too large. Dickson specified an image 1 in. wide and about ¾ in. high, and drew a sample which was rectangular – a change from the previous circular image. The film images shown in the drawings for the patent application for the camera, mailed a week earlier on 24 August, were curiously ambiguous and seem to reflect this transitional phase. Drawn in the weeks – or months – before they were mailed, the five hand-drawn images on the film strip have no frame line so they might be circular but could be interpreted as square. It appears that the film moved horizontally but the drawings and the text are unclear about this. An article in *Phonogram* in October 1891, reported that the Kinetograph would have images 1 in. square.[42]

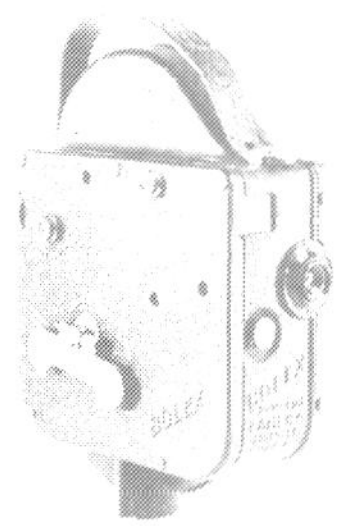

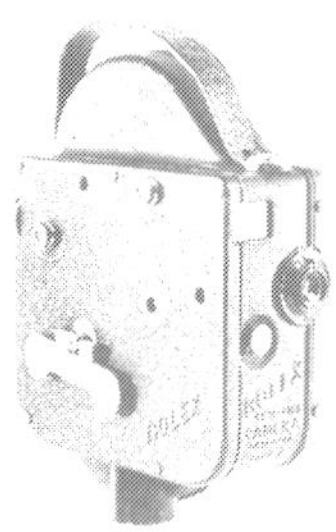

From August through the middle of January 1892, Dickson was active on the Kinetoscope project. He corresponded with Eastman, ordered rubber vats for film developing from Columbia Rubber Co., Akron, Ohio, and lenses from Gundlach Optical Co., Bausch and Lomb, Joseph Lentmeyer in Philadelphia, and William Zeiss. The orders for lenses are particularly interesting because he could not find a lens for the camera that behaved the way he expected. Several lenses were rejected because they did not give a focused image 1 in. wide. Concerned about the depth of field, he asked for one for outdoor work that kept both close and far-away objects in focus. From Zeiss, he requested one lens for close photography and another for more distant work. He also ordered magnifying lenses for use in the viewing machine.[43]

On 22 October 1891, Dickson told Eastman that he was ready to place a trial order for the special film with a thicker base as shown to him by George Eastman in July. Dickson specified that the film be 5/1000 in. thick. Although experiments to produce a film with a thicker base had gone on for some time, Eastman was not selling it yet. In July George Eastman told Dickson that they would make this special film if Edison agreed to purchase a full 'table', i.e. a quantity 41 in. wide by 200 ft. long. They also discussed cutting the film and Dickson speculated that if they received rolls 8 in. wide, they would make a device to cut the film to the required size. The trial order, sent in October 1891, was for a half a table, 41 in. by 100 ft., with one half coated to high sensitometer (a rapid emulsion, probably for camera work) and one half coated to low sensitometer (probably for making prints). Dickson asked Eastman to cut the film $1\frac{1}{2}$ in. wide.

This order generated an exchange of letters and on 2 November 1891, Dickson placed a trial order for 54 rolls of film $1\frac{1}{2}$ in. by 100 ft., half high, half low sensitometer: 'To be as *Tough, clear* & *exactly cut* to $1\frac{1}{2}$ – wide – this we would most earnestly request – we are also very anxious to get as close an adherence of gelatin to backing as possible.' Another exchange of letters took place, discussing whether the film could be exactly 5/1000 in. — Dickson agreed not to hold them to that, but asked that they err on the thick rather than thin side. Then Edison was told to make a cutter and cut the film in Orange. Dickson responded asking Eastman to cut as best they could, adding that their past work had been acceptable.[44] On 5 December, 21 reels were received and on 7 December Dickson wrote to Eastman complaining that the emulsion was not adhering to the base: '[W]hat is to be done? – We cannot of course use it ... Please follow up yr. experiments and send us a new lot. ... The machine is [now] ready to chew up film & should like to make our first trial of the film $1\frac{1}{2}$ in. wide.' Eastman sent nine more rolls on 9 December. In the next order, sent 13 January 1892, Dickson changed the measurement, asking them to add a 32nd of an inch, though he actually wanted $\frac{1}{16}$th of an inch, but assumed they could not furnish it. On 5 March he received a shipment of 6 rolls, $1\frac{9}{16}$ in. by 50 ft.[45] We are now close to $1\frac{3}{8}$ in., the size of 35mm. A little trimming gets us there.

The concern for film, lenses and developing equipment indicates that Dickson was working on the prototype for the final version of the Kinetograph – a camera recording an image 1 in. wide by ca. $\frac{3}{4}$ in. high and feeding the film vertically.[46] It is possible that the brief shift from film 1 in. wide to film $1\frac{1}{2}$ in. wide may have been an experiment with an image 1 in. wide, but on film being fed horizontally. But this is speculation since there is no specific evidence of such a test. Dickson's request for an additional $\frac{1}{16}$th of an inch in January 1893, may have been to make it easier to trim to $1\frac{3}{8}$ in. and still allow room for perforations on either side of a 1 in. image on a film moving vertically in the camera.

All of the film used by this camera and that printed for viewing in the Kinetoscope was probably trimmed before use, a practice Dickson apparently started in 1890. From February, 1890, all film was slightly wider than was used in the camera when it was received. Since all film shipments Edison received during 1892 and 1893 were for film $1\frac{9}{16}$ x 50 ft. – slightly wider than 35mm – it is probable that they continued to cut and trimmed right up until regular film production for the Kinetoscope began in 1894.[47]

Edison received three more shipments from

Eastman during 1892. A note in John Ott's workbook indicates that Edison began making a prototype of the viewing machine in the spring of 1892: 'Making New Model of Kinetiscope [*sic*] Nickel in slot'. During June castings were made for a coin slot device and a prototype cabinet was ordered from John Valentine in Newark, NJ. Construction of the revolving studio, the Black Maria, started in November–December, 1892.[48] Thomas Edison told reporter after reporter that the Kinetoscope would be a featured part of his exhibit at the World's Columbian Exposition. Work on the Kinetoscope seemed on schedule to meet the opening in Chicago on 1 May 1893, but W.K.L. Dickson was not feeling well and George Eastman was having problems. Edison's plans changed.

Eastman, Blair, and Edison: celluloid melodrama, part 3

> *The manufacture of transparent films is yet in its infancy and it is impossible to foresee the exact effect of any slight change of conditions upon a material that is subjected in every day use to tests that are the most delicate known to chemical science.*
>
> George Eastman to the *Chicago Tribune*, 7 November 1892.
>
> *We have not been able to make any good films yet. ... I believe that we are on the track of the trouble but it may take some time to get around it.*
>
> George Eastman to Henry Strong, 7 November 1892.[49]

On Friday, 1 January 1892, the day that The Eastman Co. became The Eastman Kodak Co., George Eastman sent unusual New Year greetings to three of his most valued assistants. His chief chemist, Henry Reichenbach, Gustav Milburn, his leading sales representative, and Dr. S. Carl Passavant, an analytical chemist and assistant to Reichenbach, were told: 'Your services are no longer required by this Company.' Eastman accused the 'Reichenbach gang' of sabotage which spoiled emulsion, ruining $50,000 worth of materials, conspiring with Eastman's competitor, The Celluloid Co., and planning to establish a rival business. Film manufacture came to a halt and Eastman had to search for a new team of chemists to re-establish manufacture of emulsion.[50]

George Monroe, a photographer and emulsion specialist, who had been one of George Eastman's first instructors in photography, was hired. For a while he seemed to have the solution to Eastman's problems. He introduced a new formula which produced thinner film and emulsion. This pleased George Eastman until he received reports of spoiled emulsion. In June 1892, Eastman found a pretext to fire Monroe. In November the company suspended manufacture of transparent film and was effectively out of the transparent roll film market for the rest of 1892 and all of 1893. Finally, in January 1894, Eastman hired William G. Stuber to head the transparency plate department. Stuber introduced improvements and soon was responsible for all transparent film manufacture. By the end of 1894, Eastman was ready to resume leadership in the field.[51]

For a while the transparent roll film market belonged to Eastman's only competitor, the Blair Camera Co., a change that profoundly affected the evolution of the motion picture. Blair, which began selling roll film late in 1890, had made only small inroads into Kodak's market. George Eastman regarded them as minor competition and Blair had problems of their own. Early in 1893 the company underwent a major reorganisation. Darius Goff, Blair's president, announced plans to relocate several of the company's manufacturing plants. Thomas Blair was replaced as general manager and he moved to England to establish a manufacturing branch.[52] The European branch, European Blair Camera Co., Ltd., manufactured film stock using sheet celluloid base supplied by The Celluloid Co. and employing the same process as the American company. They supplied roll film for Kodak's cameras as well as their own. Although Thomas Blair continued as the company's treasurer, the ties between the European and American companies gradually loosened even though they remained officially affiliated.

These events took place while Edison was preparing for the Chicago Fair. At the beginning of February 1893, when the Black Maria was nearing completion, Dickson became ser-

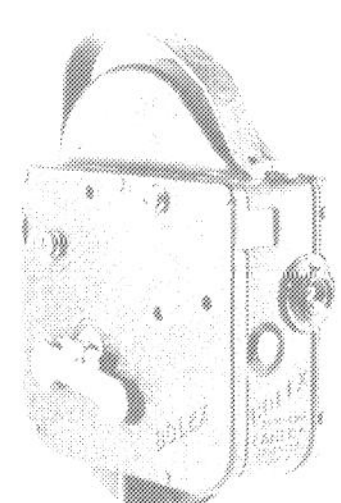

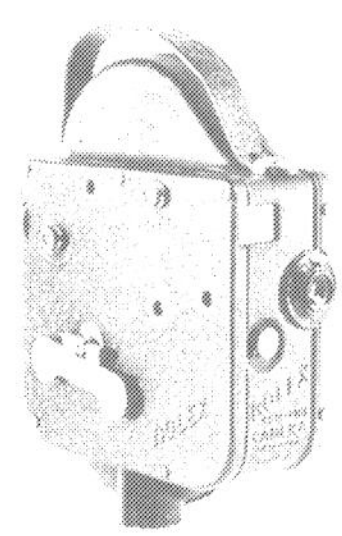

iously ill with what the *Orange Chronicle* called 'brain exhaustion'. Declaring it 'an alarming sickness', Thomas Edison sent Dickson to Florida for two months to recover. Before the fair opened Edison cancelled the space reserved for the Kinetoscope exhibition.[53]

In late February Dickson sent instructions from Florida to ask for a sample test film from The Photo Materials Co., Rochester, a company formed by the 'Reichenbach gang': 'we want it tougher than Eastmans & if they have Experimented further & gotten it more leathery or tough we shd be glad to deal with them.' Dickson's specifications were for 100 rolls, 50 ft. long by $1\frac{9}{10}$ in. wide; 'with an emulsion of proper thickness & speed to take instantaneous impressions … the basis or support to be of an even thickness & also not too brittle'. Dickson added that if it was necessary William Heise, his assistant, could help them communicate with Blair, where similar specifications would apply.[54]

The Photo Materials Co. was not producing roll film. Eastman obtained a court order which prevented them from manufacturing it. Edison had to turn to Blair, Dickson's second choice. On 12 April 1893, an order for two rolls of film was sent to Blair Camera Co., 451–453 Broadway, NYC: 'Emulsion No. 169, similar to sample sent to our Mr. Heiss [*sic*], and of as great a length as you make. If you have a film more sensitive than the one specified, please forward a sample.'[55] Edison's purchase records show that the first film received from Blair was $1\frac{9}{16}$ in. wide, the same width as film purchased from Kodak. Blair kept Edison's business for the next three years, the years during which the Kinetoscope flourished.

There are few surviving records from the Blair company so the information about transactions between Edison and Blair is not as rich as that with Eastman and Edison. The best descriptions of Blair's film are from the records of their competitor, George Eastman, and from testimony by James White who used Blair film at the International Camera Co. and the Edison Co. In 1892, Eastman told William Walker, his manager in London, that The Celluloid Co. was providing 'a very fine support for rollable films. It is made 20 in. wide and of any desired length … smooth on one side and ground on the other like their sheet film. It is very even in thickness, being about $2\frac{1}{2}$ to 3/1000 of an inch thick'. The price was more than Kodak's cost and Blair had problems coating it resulting in 'large losses from imperfect film'.[56] Testifying in the Edison-American Mutoscope Co. Suit, 9 February 1900, James White said:

> [T]he films put out in 1894 were of an opaque or translucent character … [which] differed from the clear, inasmuch as it was not as transparent, and arrested a larger percent of the light rays. …The translucent film is a mat stock, similar to ground glass. …Ground on one side and polished on the other. … A clear film is polished on both sides.[57]

In April 1894 the first Kinetoscope parlour opened in New York and the Kinetoscope operation was transferred to the Edison Manufacturing Co., ending the developmental stage. Between April and September agreements to exhibit Kinetoscope films were reached with the International Novelty Co. which became Raff and Gammon, Maguire and Baucus, and the Kinetoscope Exhibition Co. (Samuel Tilden, Jr., Enoch Rector and the Latham brothers, Otway and Gray). These firms paid for machines and film production, which included use of the Black Maria, the salaries of Dickson and Heise, the cost of talent, and the developing and printing of the films. The processing of film was transferred to the phonograph works which were located adjacent to Edison's lab.[58]

Some time during 1894 or 1895 Blair assumed responsibility for cutting film to the standard size of $1\frac{3}{8}$ in. I have found no information that would establish an exact date, but in 1895, Robert Paul and the Lumières purchased unperforated film cut to $1\frac{3}{8}$ in. or 35mm from Blair. Surviving records of film purchases by Edison's laboratory during 1893, are for film $1\frac{9}{16}$ in. and on thick base.[59] The last recorded order for $1\frac{9}{16}$ in. film was in May 1894. As we shall see, in 1896 Eastman began selling 'Cine' film, cut to 35mm size, matching their competitor, Blair.

By October 1894 the demand for film subjects forced Edison to expand the laboratory operations. It is possible that Blair began furnishing pre-cut film at this time. Cutting and trimming film was a labour intensive, time consuming and risky operation which had to be done in a dark room. It seems likely that in the process of regularising an expanding film manufacture, the Edison Manufacturing Co. or their distributors negotiated with Blair to produce a pre-cut sensitised film.

Blair may have filled Edison's orders by cutting rolls of camera film in half, but the evidence points to it being a special order product rather than off-the-shelf, conventional film. Dickson was very specific in consistently asking for a heavier (thicker) base than the film for still cameras. A tough film was needed for Edison's camera because the intermittent movement exerted a severe strain on film. Although the continuous movement and spool bank in the Kinetoscope was gentler, the film had to be able to stand up to repeated viewing. Thick, tough film was required for positives and negatives. The introduction of the Kinetoscope stimulated a flurry of activity from a host of imitators and innovators. Blair would soon find new customers for their film.

Jenkins, Armat, Paul and the screen machines

> *I have constructed a little instrument which I call a Kinetograph with a nickel & slot attachment & some 25 have been made but I am very doubtful if there is any commercial feature in it & fear that they will not earn their cost. These Zoetropic devices are of too sentimental a character to get the public to invest in.*
>
> Thomas A. Edison to Eadweard Muybridge, 14 February 1894.[60]
>
> *We have made arrangements to handle, and have just purchased for importation, a number of Foreign Films. These are now en route and will probably reach us within the next ten days. They are of Standard width, about 52 ft long and are suitable for use on either Kinetoscopes or Projecting Machines.*
>
> An advertising circular for Maguire & Baucus, Ltd., 44 Pine Street, NY, 25 August 1896.[61]

Edison's initial pessimism about 'Zoetropic devices' was not shared by others. Soon after the Kinetoscope made its debut, a throng of adventurers began working on competing machines. Their success forced Edison to change his mind.

By November 1894 work on motion picture devices was underway in New York, London, Paris, Lyon, Syracuse, NY, and Washington, DC – to name places where experimenters were successful. By the end of 1894 some prototype machines were completed. Although only a few machines were presented publicly, there was feverish work throughout 1895. In 1896 machine after machine premiered in city after city. Without trying to be comprehensive, we will look at the work of several pioneers who influenced the acceptance of 35mm film.

The Phantoscope, a kinetoscope-type viewing device, fashioned by C. Francis Jenkins was one of the first developed. It made its debut at the Pure Food Exposition which opened 12 November 1894 at the Convention Hall, 5th & C Sts., NW, Washington, DC. Jenkins was an energetic and ambitious young inventor who was impatient with the world as he found it. Over the next forty years he registered an astounding number of patents for such diverse devices as projectors, cameras, pocket calculators, launching mechanisms for airplanes, wireless photography and television transmission. Jenkins' Phantoscope was made with backing from E.F. Murphy of the Columbia Phonograph Co. which had the concession for distribution of the Kinetoscope in Washington, neighbouring Baltimore, and Atlantic City, NJ. The Phantoscope showed 35mm Edison films supplied by Murphy. Columbia Phonograph Co. gave Jenkins support because they could not get enough machines and films from Raff & Gammon's Kinetoscope Co.[62]

At this time, November 1894, Jenkins was taking courses at the Bliss School of Electricity in Washington, DC. Prof. Louis D. Bliss introduced Jenkins to Thomas Armat, another student with an active interest in moving pictures. Armat was well-to-do and a partner

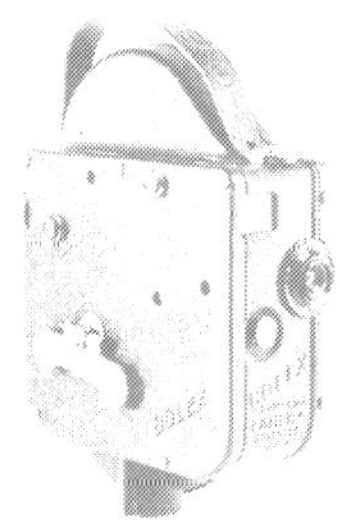

in a family-owned real estate firm. He became interested in experiments with moving images after seeing Ottomar Anschütz's Tachyscope at the Chicago Fair. Jenkins and Armat agreed to work together on a projector.[63]

A workable prototype, designed to show Kinetoscope films, was finished in September 1895, and the Armat-Jenkins Phantoscope projector made its debut at the Cotton States Exposition in Atlanta, Georgia, which opened 18 September 1895. Things did not go well in Atlanta which, by 1895 standards, was awash with moving images: Armat's friend, Henry A. Tabb, was exhibiting the Kinetoscope in downtown Atlanta, and Gray Latham was projecting films with the Latham's projector. To lure the public, the partners offered free admission, encouraging patrons to pay if they enjoyed the show. In mid-October a fire in the neighbouring exhibit, 'The Old Plantation', damaged their exhibit area.[64] Discouraged by the lack of success, the two partners had a serious disagreement that quickly developed into a bitter feud which lasted until Jenkins died in the 1930s. Their squabble had an immediate impact on movies in the United States and ripple effects on the international market.

Henry Tabb told Armat that Edison's agents, Raff & Gammon, were anxious to project films. On his return to Washington, Armat made some minor changes in the machine and contacted them. Their response was lukewarm, so Armat paid the train fare for Frank Gammon who came to Washington on 8 December 1895 for a demonstration. After viewing Annabelle, the Dancer, Gammon returned to New York where he urged his partner to contract for Armat's machine.

Raff & Gammon had tried to persuade Edison to make a projector. Although he allowed experiments on a projector, Edison had very little enthusiasm for what he termed a 'screen machine'. But the peep-show business was declining, and Raff & Gammon persuaded Edison to look at Armat's projector. After a demonstration he agreed to market it, but with the name changed from Phantoscope to Edison's Vitascope. The Vitascope made its New York debut at Koster & Bial's Music Hall, 23 April 1896.[65]

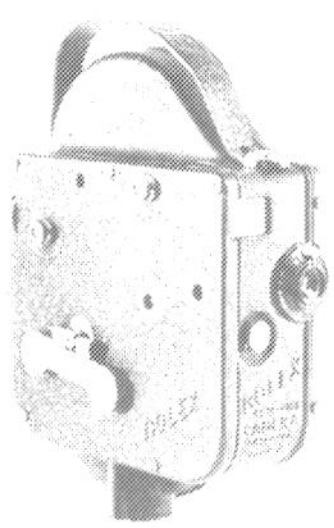

In the meantime, C. Francis Jenkins had also made a deal.[66] He sold his friends at the Columbia Phonograph Co. marketing rights to his version of the Phantoscope, and working with Columbia, they began producing projectors and a variety of other motion picture devices such as cameras, printing machines and perforators. Columbia's Phantoscopes were designed to use Kinetoscope films. Soon Raff & Gammon's clients in the USA and Canada were complaining about unexpected competition from the Phantoscope.[67]

A similar situation occurred in London. Maguire & Baucus, operating as the Continental Commerce Co., premiered the Kinetoscope in London on 17 October 1894. Troubled by a shortage of machines and films, particularly subjects with local interest, in November they wrote to Edison requesting a camera so that they could shoot films in England.[68] By this time Robert W. Paul, an electrician and maker of scientific instruments, had been approached about producing a Kinetoscope machine:

> I was introduced by my friend H.W. Short to two men, George Georgiades and George Tragedes [*sic*] who had installed in a shop in Old Broad St., EC, six Kinetoscopes, bought from Edison agents in New York. ... Additional machines were urgently needed [since there was no patent] ... I was able to construct six before the end of the year.[69]

Paul made about sixty kinetoscopes and some of the first films made in England were shown on these machines.

Edison's agents, Maguire & Baucus, would not sell or lease films for users of Paul's kinetoscopes, so Paul began construction of a camera. He engaged Birt Acres, an experienced photographer to help him. Work on the camera started in February 1895. By March 1895, they had a prototype, and with a few improvements they were ready to shoot films. The first ones were made at the end of March 1895. Construction of a projector followed.

Robert Paul got his first film from European Blair. It was 35mm and on the dark toned, matte celluloid which Blair purchased from The Celluloid Co. Blair apparently provided it pre-cut to Edison's width, but it was without perforations so Paul built his own perforator. Later Paul purchased negative stock with clear base from Eastman but made prints for the kinetoscope on Blair's dark-toned stock. By the end of May 1895, Birt Acres had struck out on his own and constructed a camera-projector, also designed to use 35mm film. Acres was soon selling films too. In 1897 Acres began to manufacture cinema film.[70]

The Kinetoscope opened in Paris in late October or early November 1894.[71] By the end of the year Louis and Auguste Lumière were at work on their camera-projector which would become famous as the *cinématographe*. They applied for a patent on 13 February 1895, and on 22 March 1895, gave a demonstration at Société d'Encouragement pour l'Industrie Nationale in Paris where *La Sortie des usines Lumière* was shown. The Lumière's film was 35mm wide, but with only one round perforation on each side of the frame. This was a unique perforation system, and restricted the use of Lumière films to the widely-patented Lumière camera-projector.

The Lumières were proprietors of a prominent photographic firm which manufactured photographic plates and equipment.[72] The earliest experiments with the Cinematograph used photographic paper, but they soon switched to celluloid and made plans to manufacture film themselves. Although some celluloid sheet film was produced in France and prominent experimenters like Marey and Demenÿ had used celluloid, no celluloid roll film was manufactured there. The Lumières sent a representative to the USA to arrange the purchase of base from The Celluloid Co. The Lumières had trouble coating samples of the film and looked elsewhere for a supply of film stock. They may have made an overture to European Blair for a concession to manufacture Blair's film in France, but ultimately they turned to Victor Planchon of Boulogne-sur-Mer who was producing celluloid sheet film. It took Planchon a while to produce a product that met the Lumière's standards: 'It will need to be very strong to guard against snapping and the emulsion will need to be laid in a very even coat.'[73] By the end of 1895, Planchon had satisfied the Lumières well enough that they backed a new company, Société des Celluloses Planchon, to make celluloid. Planchon's factory was established in Lyon. This agreement was reached in January 1896, and it took several months for Planchon to begin manufacturing film: meanwhile, they purchased unperforated film from European Blair. They had used Blair's unperforated 35mm film during 1895, and this may have influenced their choice of that width for their camera-projector. During the early months of 1896, when the Cinematograph was being first introduced, Lumière films were made on Blair's stock.

The Lumières' plan to produce film stock was well publicised. In January 1896 the Lumières received a letter from Edison asking if they would be interested in setting-up a plant in the USA to manufacture film. Edison would furnish the building and material. This intriguing offer was not accepted. In March 1896 a letter from George Eastman to William Walker in London mentions that Eastman saw a description of the Lumière's coating machine in *Scientific American*.[74]

Thomas Edison, King of Patents

> *... such machinery as was employed, was machinery which I have no reason to believe Mr. Edison, or any other one man in the world, had the exclusive right to use. Mr. Edison did not invent the sprocket wheel, nor gear wheels, nor shutters, nor belts.*
>
> Woodville Latham, 8 December 1897.[75]

Not all of cinema's pioneer inventors took advantage of the pre-cut film stock available from Blair. Several of them used film with different formats. The Latham's Eidoloscope, projecting the Griffo-Barnett fight, was shown in New York City, 20 May 1895. The Eidoloscope, called the Pantoptikon at first, used film that was 2 in. wide, with a frame the same height as Edison's, and with four sprocket holes per frame on either side. The Mutoscope, patented 21 November 1894, by Herman Casler, was a flip-card, coin-in-the-slot machine. The Mutoscope and the camera and projector that Casler subsequently designed, used film that was $2\frac{23}{32}$ in. wide and almost 2 in. high. Georges Demenÿ's Chronophotographe, using 60mm film, was modified

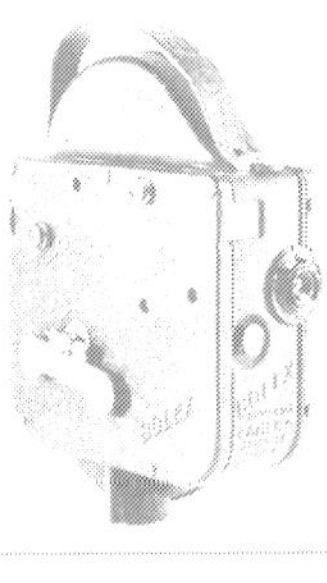

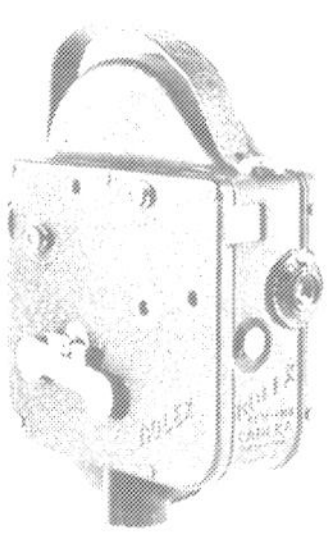

into a successful projector in 1895 and was publicly exhibited in 1896. The successful exhibition of these machines stimulated Edison and Lumière who made projectors that used film larger than 35mm. Each of these machines was unique, and it was impossible to use them with films made for another machine.

From the 1920s, when Terry Ramsaye first wrote about it, historians have accepted the explanation that the Latham and Casler used a wider film on the advice of W.K.L. Dickson as a way of avoiding Edison's patents. This explanation may have validity. Though uncredited, the source of Ramsaye's information seems to have been Dickson's friend and partner in the American Mutoscope Co., Henry Marvin. Much has also been made of Edison's failure to take out patents in Europe which seemed to open the door to competitors.

Our view of this period, however, has been coloured by the effect of the decade of 'patent wars' that came after 1897. In 1895 and 1896, while various apparatuses were being introduced, Edison's patent situation was anaemic. He had only two confirmed patents, for the Kinetoscope and for a stop-motion device to provide intermittence. His patent for the camera was still pending and was the subject of serious challenges by the Patent Office. It was not accepted until 31 August 1897.[76] Not only did Edison have no patent for a motion picture projector, he did not even have one pending. He had not applied for a patent on the specifications for film. The Kinetoscope patent applied only to viewing machines, and Edison's machine ran continuously so it could not be used against a machine with intermittent movement. Moreover, his stop-motion device was not adopted by others since there were other un-patented movements that worked better.

Even if Edison had patented his machines in Europe, it is doubtful that this would have prevented the rapid spread of competing machines. As Woodville Latham observed, Edison could not claim ownership of most of the basic components used in cameras and projectors.[77] The flood of machines that appeared within a year of the public exhibition of the Kinetoscope is the most compelling argument against the notion that Edison could have reserved the film market to himself. Nor could anyone else. There was an open and comparatively unrestricted environment which fostered the rapid, international circulation of the newborn motion picture. Before 1896 this had little effect on film manufacturers, but in 1896 'Cine' film suddenly developed importance.

Eastman and cine film

> *It seems to me wise for you to use every endeavor to supply this trade because if there is a demand for it somebody will, as you state, certainly make it and the making of it will establish them in the market as competitors in camera films.*
>
> George Eastman to George Dickman, Eastman Kodak, London, 14 May 1896.[78]
>
> *There is certainly lots of unworked territory in Europe.*
>
> George Eastman to Harris H. Hayden, NYC, 27 November 1896.[79]

Like Edison, George Eastman was sceptical about the future of the motion picture. He was much more interested in photographic plates, the expanding market for Kodak cameras, and the rolls of film that fed the cameras. By 1896 the camera business was booming. In May 1896 the always-expanding Eastman Kodak Co. employed 700 workers.

On 21 March 1896 George Eastman wrote to Wm. H. Walker, in London:

> We have been supplying film to the Eidoloscope people but not to the Kinetoscope Co. [Raff & Gammon], the latter requiring a heavier film than we have been able to make heretofore. We have recently been making some experiments, with a view to supplying them, but have been obliged to discontinue them because the prospect is that we can sell more film than we can make the coming summer to users of our cameras.[80]

Apparently Walker, former head of Eastman's European branch, reported an increase in orders from motion picture companies for special film. George Eastman remained reluctant. On 28 May 1896, he wrote George Dickman, now manager in London: 'I hope

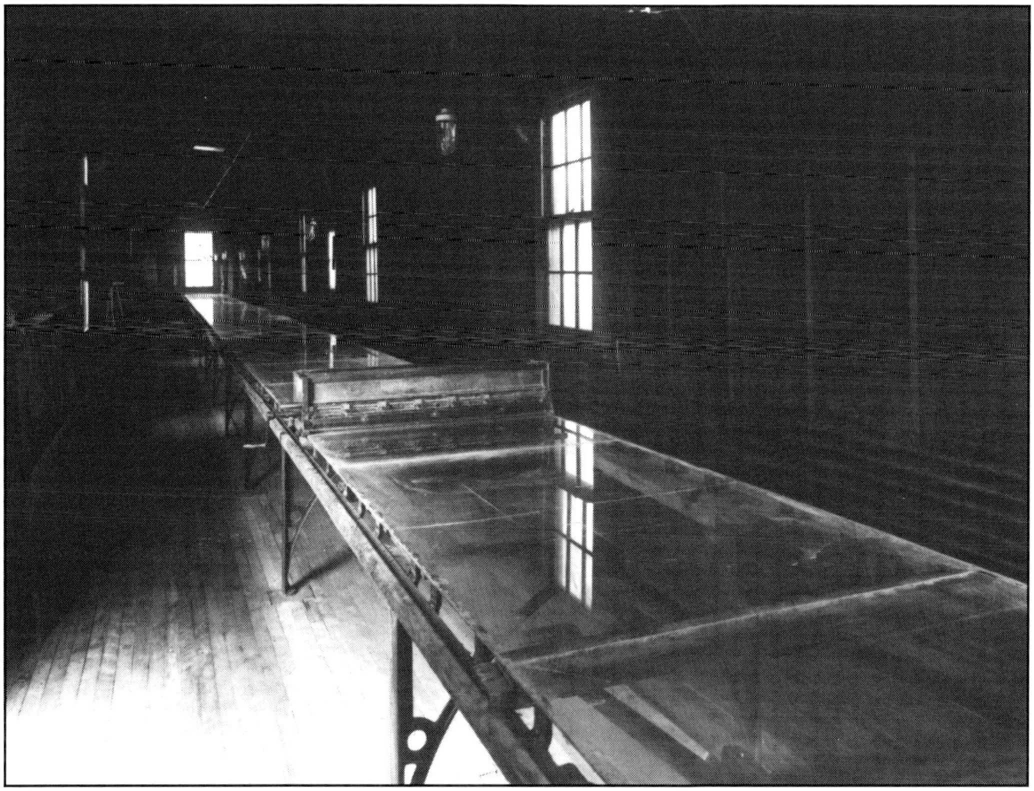

Fig. 6. The glass coating tables for celluloid film at Eastman Kodak's plant in Rochester, NY. The tables were 200 ft. long and about 42 in. wide. The liquid celluloid was spread on the glass from movable hoppers (centre) that had valves to control the thickness of the solution on the glass. [Courtesy of George Eastman House.]

you will realise your anticipations in regard to the Cinematagraph film. I think the demand for it will be only temporary.'[81] Eastman was wrong.

Dickman came to the USA in June 1896 to visit Kodak's facilities, and the topic of cine film for Europe continued to be an issue. At the end of June, Eastman gave a go ahead to manufacturing film for Europe. The film was made in Rochester because the tables for pouring celluloid base in London were only 80 ft long and Rochester's were 200 ft. (Fig. 6). Rochester also had more experience in making thicker film. To meet the demand for tougher film, they experimented with samples 4½/1000 in. thick (5/1000 with emulsion added) and a perforating machine. There were discussions of adding a new building to manufacture cine film. Finally, a base with a thickness of 5/1000 in. was settled on (7/1000 in. when coated with emulsion). On 9 July a large shipment was on the way, destined for France where much of the European demand had originated.

The day the shipment left, 9 July 1896, George Eastman sailed for Europe accompanied by George Dickman and their families. Eastman spent three months touring Europe, combining his vacation with business. To manage affairs during his absence, Eastman brought the company's President, Henry Strong, to Rochester from his home in Tacoma, Washington. Strong supported the 'Cine' trade as a special project during Eastman's absence. He reported several sizeable orders in the regular reports which he sent to Eastman via the London office. On 24 August 1896 he reported that Edison had switched their now sizeable business to Kodak. By November when George Eastman returned to Rochester, cine film had become a regular Kodak product.

The market for cine film provided some surprises. Cinema producers ordered much more positive than negative film while buyers of still film usually bought equal amounts of negative and positive. Some companies resisted buying the thicker, more expensive cine stock. The Eidoloscope Co. had bought film of regular thickness and the American Mutoscope Co., which started ordering film

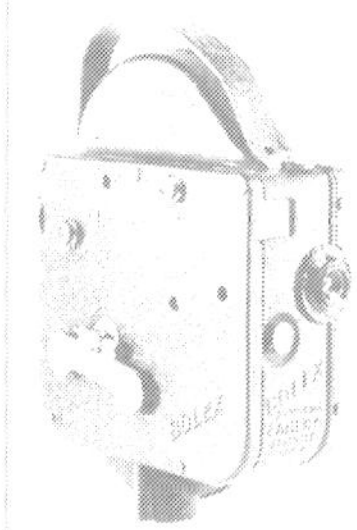

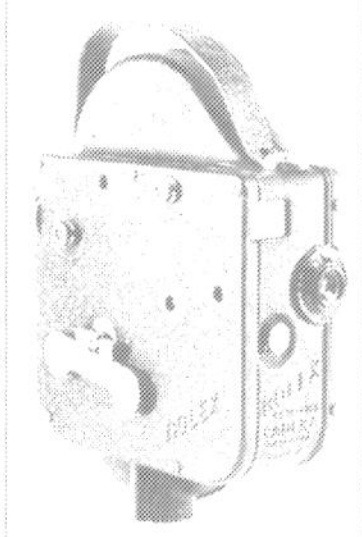

early in 1896, ordered standard Kodak roll film. Their camera and the Mutoscope had been designed to use film made for still cameras and they refused to switch to the thicker film base. The Mutoscope company ordered some bromide paper for printing Mutoscope rolls from Eastman and some from rival companies.[82]

Astute observers will notice that prior to 1896 the term '35mm' was not used for the 'standard' film. Until the Lumières began manufacture of their film, all 'cine' film was made in the United States or England. In both countries, the size of film was measured in feet, inches or fractions of inches. The universal acceptances of the designation '35mm' as well as George Eastman's acceptance of the term 'Cine Film' reflects the early impact of France on the development cinema.[83]

Movies for the masses

> [*T*]*he Cinématographe* [*is*] *the main attraction of the season.* ... [*It will bring to the screen*] *bullfights, beheadings in China, atrocities in Armenia, and lynchings in Texas, some people are going to make barrels of money.*
>
> *Phonoscope*, November 1896.[84]

In August 1896, only four months after the premiere of the Vitascope, the 1⅜ in. film devised by Dickson for Edison's Kinetoscope was already being called 'standard' width by the trade. That month Maguire & Baucus advertised a number of 'Foreign Films' of 'standard width' including scenes of the Coronation of the Czar. Blair Camera, European Blair, and Eastman Kodak were selling photographic roll film cut to this measurement to producers, and it was also being made by Société des Celluloses Planchon for Lumière.

But not everyone wanted film in a standardised format. Several of the largest producers sought to limit the market for their films by using a specialised film size, often combined with a unique sprocket. A controlled market was what Edison originally sought, though he counted on restrictive contracts to a limited number of distributors to curb competition, a pattern borrowed from the phonograph which had been commercialised just before the Kinetoscope was introduced. Edison's major American competitor, the American Mutoscope Company, kept an even tighter monopoly of their films and machines. Already circumscribed by a unique format, they regulated exhibition by supplying machines, films and operators to a limited number of theatres who contracted to show their films. They did not sell or lease their films or their machines and they only exhibited in large theatres in major cities. Their Mutoscopes were also carefully controlled.

In 1896 and 1897, it was conceivable that cinema might have developed as a succession of fragmented, competing formats similar to those which have plagued the phonograph and video industries. In this introductory period, the novelty of the machine had almost as much value as the images being shown, and competition between formats was intense. Exclusive formats like the Cinematograph, the Eidoloscope, the Biograph and the Veriscope competed with 'standard' formats like the Vitascope, Paul's Animatographe, and Birt Acres' Kineopticon. The unique formats usually had superior image quality, but 35mm had an ultimate advantage – variety, availability and, ultimately, quality.

In the May 1897 issue of *Phonoscope*, Maguire & Baucus advertised that in response to the demand for foreign film subjects:

> [*W*]*e have entered into an agreement with A. Lumiere & Sons, Lyons, France*, by which we *control* the sale of these celebrated films in the *United States and Great Britain*. They are accurately perforated for any Standard Gauge Machine.

In the same advertisement they announced that they had taken over the remaining stock of Lumière films in the USA and were selling the films, Cinematograph machines, as well as films produced by International Film Company, and a stock of used Edison film subjects.[85] The Lumières were among the first of the exclusive formats to accept the 'standard' format.

In the US, Britain and other countries, dealers like Maguire & Baucus were selling machines and films. Ambitious showmen could purchase Jenkins' Phantoscope, Paul's Animatographe, Acres' Kineopticon or other machines along with stocks of film. Although projectors were expensive at first, a flood of competing machines soon brought prices down. In November 1896, Edison, unhappy about distributing a machine that was not of his own design, announced that his Projectoscope could be bought, and he began selling his films, putting him in direct competition with Raff & Gammon's Vitascope. Raff & Gammon were flooded with complaints from their clients about competition from Phantoscopes, Fantascopes, Zooscopes, Cinematographes, Centographs, Magniscopes, etc. After a fleeting attempt at going into production, Norman Raff and Frank Gammon closed shop and returned to Raff's hometown of Canton, Ohio.[86]

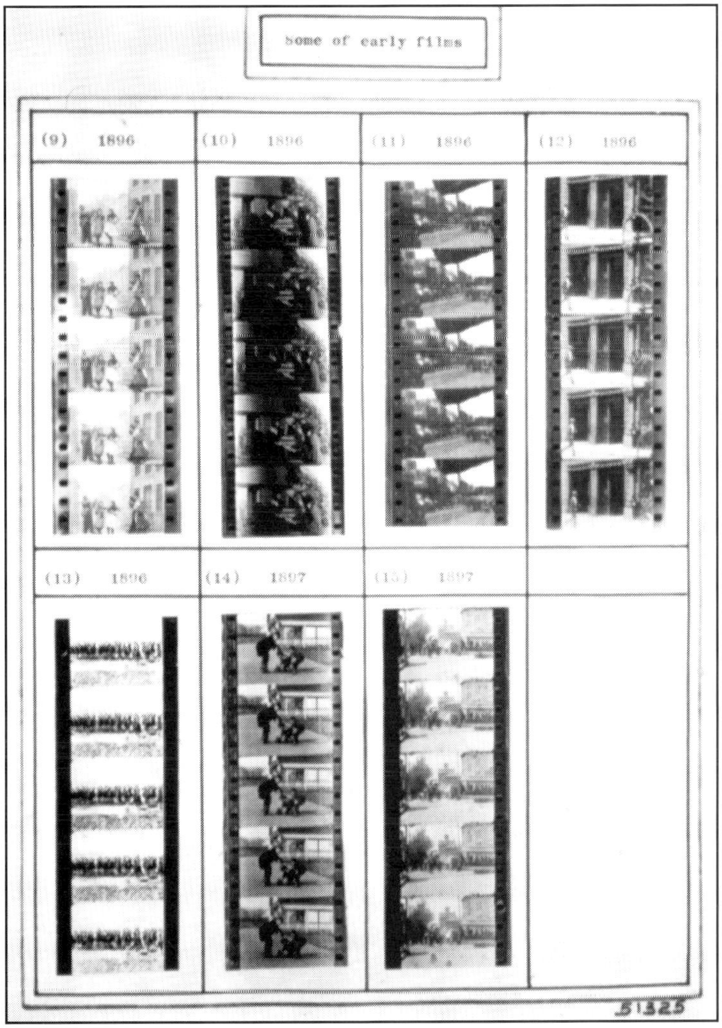

Fig. 7. Examples of early 35mm films from Eugene Lauste's scrap-books which show the standardised film width and consistent (but not identical) frame size. All have four perforations on each side of each frame. Lauste identified these as: (9) Schooldays *(1896); (10)* The Gossips *(1896); (11)* Elevated Railroad, NY *(1896); (12)* Fire Department, Orange NJ *(1896) [actually, in Newark, NJ]; (13)* Charge of the French Dragoons *(Lumière, 1896); (14)* Boot Blacks *(Vitagraph, 1897) [He says the man is William E. 'Pop' Rock]; and (15)* Street Scene in Brooklyn *(Vitagraph, 1897). [Courtesy National Museum of American History, Smithsonian Institution, photo cat. no. 4051.3.57.]*

For a brief time, during the last months of 1896 through 1897, an open market for films and projectors flourished in the USA (Fig. 7). It came to an end in 1898, when Edison unleashed his barrage of lawsuits to clear the field of unwanted competitors.

It was during this brief window of open competition that 35mm film became established as the 'standard'. It was the preferred gauge of the flood of 'scope' and 'graph' machines that appeared in country after country, city after city, town after town, and village after village. While theatre patrons in large cities still enjoyed spectacular projections by large-format machines like the Biograph and the Chronophotographe, audiences in neighbourhoods and smaller towns flocked to shows projected by 'standard' machines. It was the democratic format, and by 1898 moving images made by Pathé, Edison, Paul and Méliès had been seen by audiences in most parts of the world – and there was a continuing demand for more. One by one the specialised formats disappeared. Spectacle could not compete with the variety – and, eventually, quality – of subjects available on 35mm. The mass audience was supplied by films made on 35mm stock in 35mm cameras and shown on 35mm projectors – a situation that continues today. It is the format developed out of W.K.L. Dickson's search for an image 1 in. wide. As Dickson said: 'This standardized film size ... has remained, with only minor variations unaltered to date.' Sixty years after Dickson's death, this is still true.

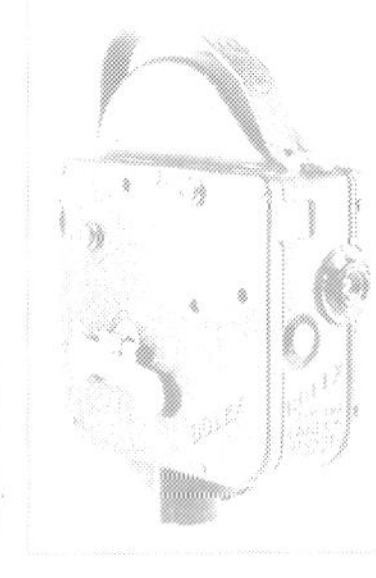

Notes

1. 'A Brief History of the Kinetograph, the Kinetoscope and the Kineto-phonograph' by William Kennedy Laurie Dickson, *Journal of the Society of Motion Picture Engineers* (December 1933): 14.
2. John Belton, *Widescreen Cinema* (Cambridge, Mass.: Harvard University Press, 1992), 18–19 and *SMPTE Journal* 99, 8 (August 1990): 652–661.
3. An ad for The Eastman Dry Plate & Film Company's 'Transparent Film for Roll Holders', *American Amateur Photographer* (July, 1889). Editorial comment from *The Philadelphia Photographer* (July 1889). Gordon Hendricks Collection, Archives Center, National Museum of American History, Smithsonian Institution (hereafter Hendricks Collection).
4. An editorial comment in *The Philadelphia Photographer* (6 October 1888). Hendricks Collection.
5. Reese V. Jenkins, *Images and Enterprise* (Baltimore, London: The Johns Hopkins University Press, 1975), 96–120 and the correspondence of George Eastman in the Library of the International Museum of Photography at the George Eastman House (hereafter Eastman Correspondence). The roll film system and the Kodak No. 1 camera were designed by William H. Walker, a skilled camera designer hired by George Eastman. Walker became the manager of Kodak's European operations based in London. The advertising campaign for the Kodak camera was created by the J. Walter Thompson agency who placed ads in major national magazines.
6. Eastman Correspondence and Jenkins, 100–102. After exposure, the negative was developed and the emulsion was 'stripped' from the paper by dissolving the gelatine binder. The emulsion was then transferred to glass from which prints could be made. Although complicated, it was lighter and less fragile than glass, and it allowed photographers to go into the field without heavy glass plates and cumbersome chemical apparatus.
7. Jenkins, 122–127 and Hendricks Collection.
8. Hendricks Collection. Only a few companies manufactured celluloid. In an article, 'Celluloid', by Samuel P. Sadtler, *Scientific American* (29 January 1887): 69, Sadtler related that celluloid was manufactured by The Celluloid Company, Newark, NJ; American Zylonite Co., Adams, Mass.; British Zylonite Co., with patents similar to American Zylonite; a company at Staines-on-the-Seine working under license of Celluloid Manufacture Co. and a company in Hanover, Germany, which started but shut down because of the hazzard of explosion. The Celluloid Co. sued American Zylonite for patent infringement. Nitrocellulose, chemically related to TNT, is very flammable and hence dangerous to manufacture, store, and handle.
9. Hendricks Collection. John Carbutt describes the manufacture of celluloid: 'It may interest you to know, as far as I am able to describe it, the way this celluloid is made. It is this: They buy the very finest of tissue paper, bleached as white as possible. That is nitrogenized. It is then ground up with camphor and pressed out into a large slab. After some evaporation has taken place it is shaved off in thin slices and put into frames and stretched down and put into a press through which air is forced, and it takes about six weeks to cur it, as they term it. Then the finishing process is brought on. And my object in selecting and inducing them to manufacture it with a fine mat surface was to prevent halation, so much complained of in glass.' Carbutt (at the convention of a photographic society in Boston), 7 August 1889, *Wilson's Photographic Magazine* (September 1889): 529–532.
10. Hendricks Collection. Gustave D. Milburn, Eastman Dry Plate Co., in a paper read before the Society of Amateur Photographers of New York, 16 August 1889, quoted in *The Photographic Times and American Photographer*, 414.
11. Eastman Correspondence, Jenkins, 71–75, *Kodak Milestones* (Rochester, N. Y.: Eastman Kodak Co., c. 1959) and Elizabeth Brayer, *George Eastman: A Biography* (Baltimore, London: The Johns Hopkins University Press, 1996). George Eastman began making film emulsions in 1878, invented an emulsion coating machine in 1879, and began making and selling dry-plates in 1880. In 1881, with backing by his friend Henry A. Strong, they established the Eastman Dry Plate Co. with Strong as President and George Eastman as Treasurer. In September 1881, Eastman quit his position at a Rochester bank to devote full time to the company. Eastman was an aggressive businessman who invested his profits to expand the company, to purchase competing photographic companies and acquire patents that he felt offered serious competition or could enhance the growth of the company.

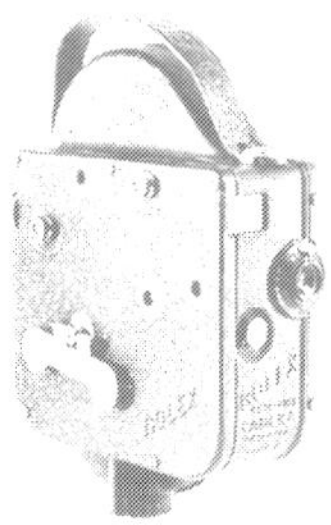

12. Eastman Correspondence.
13. Eastman Correspondence. The patent for the chemical process was in Henry Reichenbach's name,

and the patent for the mechanical process was in George Eastman's name. Both were assigned to the company.

14. Eastman Correspondence. In a letter to Fred Crane, a chemical manufacturer in New York City, 18 February 1892, Eastman gave the following formula for making film: ' ... the dope purchased from Cooper [chemical company]; cotton, 22 lbs; wood alcohol, 74 lbs. and add: fusel oil, 22 lbs.; amyl acetate, 9 lbs.; camphor, 13 lbs; wood alcohol, 17 lbs.'
15. Eastman Correspondence.
16. Eastman Correspondence. By 1890 they identified the cause of these streaks as static electricity and made an effort to get rid of it, but with only mixed success. Changes in chemistry and drying methods sometimes caused additional problems. Static electricity continued to plague the company through the 1890s.
17. Edison Papers, Edison National Historic Site, West Orange, NJ (hereafter EHS). The caveat was a formal notice to the US Patent Office of experiments intended to result in a patent, submitted in order to establish priority ahead of other potential applicants.
18. EHS. A card in the files at the Edison National Historic Site lists Dickson as Edison's photographer as early as 1884, the year after Dickson joined Edison. 'Mr. Edison asked me if I understood photography. I replied I did, also the chemistry of photography. He then proposed to start a department in which I was to have exclusive charge developing his ideas in moving photography' (Dickson's testimony, Motion Picture Patents Co. vs. IMP 1911). A letter from A.O. Tate, dated 15 September 1884, said Dickson did Edison's photography. At that time he was working at Edison's Goerk St. facility in New York City. He joined Edison's research laboratory in Newark/Harrison, New Jersey in 1885. The laboratory in Orange (now West Orange), New Jersey, was built in 1887, opening near the end of November 1887.
19. Robert Conot, *Thomas A. Edison, A Streak of Luck* (New York: Da Capo Press, 1979), 345 and EHS.
20. EHS and Hendricks Collection.
21. Dickson, *Journal of the SMPE*: 'I was glad to get away from drums, disks, *etc.*, and a hopelessly limited number of pictures, looking forward some day to getting decent lengths of strips of film.'
22. EHS, Dickson, *Journal of the SMPE* and Hendricks Collection. Edison's surviving records show that in March 1889, Edison asked The Celluloid Varnish Co., Newark, NJ about supplying sheets of celluloid. They replied that they could not supply any at the time, but would supply sheets up to 30 ft. long if 'demand arises'. The earliest celluloid film receipt recorded was sheet Zylonite, received 11 May 1889. On 25 June 1889, a dozen 8 x 10 sheets of Allen & Rowell's Ivory film and a dozen 11 x 14 sheets of Carbutt's celluloid were received. Dickson said Carbutt film was used on the cylinder and some were cut into strips for an early experiment with strip film.
23. Gordon Hendricks, *The Edison Motion Picture Myth* (Berkeley, Los Angeles: University of California Press, 1961); rpt. in Gordon Hendricks, *Origins of the American Film* (New York: Arno Press and The New York Times, 1972). Gordon Hendricks' book is an exhaustively researched study of Edison and Dickson, and it is an early revisionist study of Edison's work. Unfortunately, it is coloured by Hendricks' passionate dislike for Edison and Edison's methods. In his effort to de-mystify Edison, he omitted information supporting Edison's claims, imposed unsubstantiated interpretation on evidence and misinterpreted other information. The result is a work that focuses on what Edison did not do and tells little about what Edison did. These flaws seriously colour the value of a work that has been widely accepted as the definitive work in the field. While this essay is not intended to be a discourse about Hendricks' work, it tries to establish what Edison and Dickson did, and offers some different interpretations on Edison and Dickson's work. Hendricks' extensive notes, now at the Smithsonian, were an invaluable source of information in this essay.
24. This letter, in Dickson's hand, is reproduced in Terry Ramsaye, *A Million and One Nights* (New York: Simon & Schuster (1926, 1964), 70. Hendricks argues that this was used for cylinder work because of the width, though Dickson's letter specifically refers to length and straightness: 'I never succeeded in getting this substance in such straight long pieces.' Length would have been a lesser consideration for cylinder work.
25. EHS and Hendricks Collection. The texts of Edison's four Caveats can be found in Appendix B of Hendricks' *The Edison Motion Picture Myth*, 158–163.
26. The drawings were part of a patent application prepared in the office of Dyer & Seely, Edison's patent attorneys and the Edison application number, 928, is the same as that for the Kinetograph

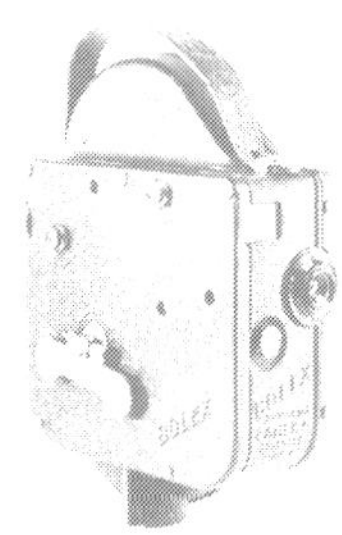

which was submitted in August, 1891. The exact date that these drawings were made is not clear, but the pay roll time sheets for Edison's lab charged George Mayer's time for patent drawings to the kinetoscope account on 6 February 1890. This is the only charge for patent drawings made against the kinetoscope account during the period from 1889 through 1893 when the kinetoscope patent was issued.

27. Dickson, *Journal of the SMPE*, 12: 'we had to devise certain essentials, such as a circular film cutter or trimmer, a perforator, a clamp with steady pins to fit the punch holes, to use in joining the films with a thin paste of the base dissolved in amyl acetate.'
28. Orange Public Library and Hendricks Collection. It is possible that W. K. L. Dickson was the source of both articles. Details about the Kinetoscope and the construction of a new photographic building constructed in August and September 1889 while Edison was in Europe attending the Paris Exposition probably came from Dickson who designed the building.
29. Eastman Correspondence.
30. Hendricks Collection. John Carbutt in 'A Perfect Substitute for Glass . . .' read before the Franklin Institute, 21 November 1888, published in *Journal of the Franklin Institute* (December 1888), *Photographic Times* (November, 1888).
31. Eastman Correspondence. It became the Eastman Company on 24 December 1889, and it was reorganised again as The Eastman Kodak Co. on 1 January 1892.
32. Eastman Correspondence and Hendricks Collection. Eastman announced that roll film was available for sale in August 1889 but they had trouble producing film during 1889 so the film was not widely available until early in 1890. In April 1890, *American Amateur Photographer* reported that there were eight Kodak cameras making negatives from a 2 3/8 inch circle to 5 x 7 inch oblongs.
33. Eastman Correspondence and Jenkins, 136–143, 190–191. There were occasional complaints about flaws in Eastman's film caused by the joints in the glass plates on Eastman's tables. Blair's film made on The Celluloid Company's continuous strips lacked these flaws. The business relations between these companies was very complicated. Eastman was involved in suits and patent interferences with Hannibal Goodwin, Blair Camera and Celluloid Co. At the same time, Eastman was negotiating to purchase Blair through most of the 1890s. Eastman acquired Blair in 1899.
34. Eastman Correspondence.
35. Hendricks Collection and EHS. It was reported by *New York Herald*, *New York Sun*, *Orange Journal*, and *Orange Chronicle*. Most newspaper accounts indicate that the images were viewed through a hole in the top of a box, but the *Journal* said it was projected. Edison's Charles Batchelor mounted several strips from films shot at this time in a book which survives at the Edison National Historic Site.
36. National Archives. Application no. 403,534, for 'Improvement in Kinetoscopes, Method of Taking & Subsequently [showing] ... ' became US Patent no. 589,168, 31 August 1897, for a camera, Application no. 403,535, 'Method & Apparatus for Taking the Pictures' was abandoned, 4 June 1896. Number 403,536, 'Method & Apparatus for Exhibiting Photographs of Moving Objects' (the Kinetoscope) became US Patent No. 493,426, kinetoscope, 14 March 1893. Application no. 403,534, was the subject of a lengthy contest before being awarded in 1897. After it was awarded it was the basis for Edison's famous patent suits.
37. EHS and Hendricks Collection.
38. National Archives. The specification for an image one inch wide with perforations on either side was in Edison's application for the camera patent but not in his other applications. The specification survived the long period of rejections, delays, and appeals the camera application went through before becoming, No. 589,168, 31 August 1897. Edison had to surrender this patent in 1902, as a result of an appeals decision in his suit with the American Mutoscope Co. It was then reissued as Patent no. 12,037, 30 September 1902. The same description of an image one inch wide with two sets of perforations is in the reissued patent.
39. Terry Ramsaye Collection, Georgetown University Library. In an undated chronology which he prepared for Terry Ramsaye in the 1920s, Dickson confirmed that the double set of perforations was introduced to improve the steadiness of the image.
40. Antonia & W.K.L. Dickson, 'Edison's Invention of the Kineto-Phonograph', *The Century Magazine*, Vol. XLVIII (May 1894): 206–214. His co-author, Antonia Dickson, was his older sister. During the 1920s and until his death in 1935, Dickson was questioned about his work with Edison by a number

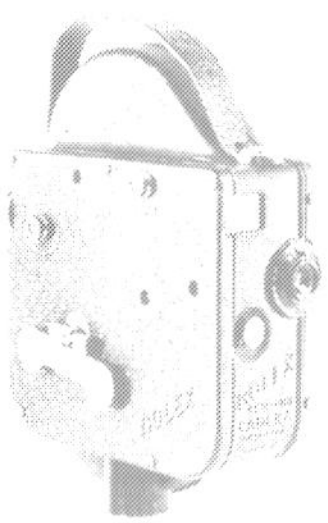

of prominent film historians including Terry Ramsaye, Will Day, Merritt Crawford as well as Earl Theisen and Glen Matthews of the historical committee of the Society of Motion Picture Engineers. Dickson's accounts are a mixture of valuable, authentic information embellished by blatant exaggerations, particularly about dates. By reading carefully to identify misleading information, a great deal of first-hand information about his work for Edison can be found. Comments about film and image quality recur in these stories.

41. Dickson's concern about image quality and his belief that better images could be obtained from larger original photographs may have influence two other projects that Dickson was connected with, Latham's Eidoloscope and the Mutoscope. The Latham camera used a film two inches wide, though about the same height as Edison's, and the Biograph camera used 68mm film that was both wider and taller. The Biograph's programmes were praised for the exceptional quality of their projected images. Conventional wisdom, dating from Terry Ramsaye's *A Million and One Nights*, is that Dickson advised a larger format to avoid Edison's patents. This may have been a consideration, but, as we shall see, Edison had no control over the size of the image, so quality may have been a primary consideration.
42. EHS, National Archives and Library of Congress. The text of the patent specifies: 'An unbroken transparent or translucent tape-like photographic film provided with perforated edges and having thereon equidistant photographs of successive positions of an object in motion, all taken from the same point of view, such photographs being arranged in a continuous straight-line sequence, unlimited in number save by the length of the film.' Edison National Historic Site, National Archives and Library of Congress.
43. EHS. Dickson's correspondence indicates that he had trouble with Gundlach's lenses, and the record of receipts at the lab for 1892 and 1893 lists lenses received from Bausch and Lomb but not from Gundlach. Dickson's most detailed letter was to Zeiss, though there seems to be no record that lenses were ever received from Zeiss.
44. The debate with Eastman about trimming the film may reflect a reluctance on Edison's part to assume the onerous and potentially risky task of trimming each roll of film. Even though Dickson had a device to trim the film, it was designed to trim small edges off pre-cut rolls and it would not have been capable of cutting the wider rolls Eastman wanted to furnish now. Although no specific evidence exists, Edison may have been unwilling to assume permanent responsibility for trimming all the film used. Trimming film during the experimental phase was one thing, doing it in a commercial situation where quantities of film might be involved was a different matter. This may account for Edison's reluctance to patent a cutting/trimming device.
45. EHS. All correspondence cited here are from the Edison National Historic Site.
46. There is a natural confusion about the terms 'Kinetoscope' and 'Kinetograph'. Before 1894, Edison and his associates used both terms somewhat indiscriminately. In 1891 and 1892, Kinetograph was used most often and it described both the camera and the viewing machine. When Edison began commercial distribution in 1894, the peep-show machine was called the Kinetoscope and the camera was the Kinetograph – a distinction that has generally been followed since 1894.
47. EHS and Hendricks Collection.
48. EHS and Hendricks Collection. Eastman shipped 6 rolls on 5 March 1892, 12 rolls on 9 May 1892, and 13 rolls on 5 August 1892. John Ott's notebook covered a period from May–June 1892.
49. Eastman Correspondence. The first quote is from a not-for-publication letter George Eastman wrote to the *Chicago Tribune* in response to an article accusing Eastman of producing defective film. The second is from a letter written the same day to Henry Strong, President of the Eastman Co.
50. Eastman Correspondence; Brayer, 89.
51. This is a simplified condensation of a complicated situation. For more detailed descriptions, see Brayer, 82–100; Jenkins, 152–156.
52. Eastman Correspondence. See also Jenkins, 144–145. Darius Goff wrote to George Eastman, 10 February 1893, informing him that Thomas Blair was going to England. Goff suggested that discussions about a merger with Eastman Kodak continue, even though the two companies were also involved in legal disputes.
53. EHS and Hendricks Collection. *Orange Chronicle* (11 February 1893) and Thomas Edison to H. E. Dick, 4 February 1893. H.E. Dick was the brother of A.B. Dick, a Midwestern lumber manufacturer, who took over Edison's mimeograph and developed it into a successful business machine company.

Edison's decision to delay the Kinetoscope may also have been influenced by the 'Panic of 1893', one of the worst depressions to affect the US economy during the nineteenth century. Arthur M. Schlesinger, Jr. (ed.), *The Almanac of American History*, (New York: Barnes & Noble Books, 1993), 376.

54. EHS.
55. Hendricks Collection, from material at Edison National Historic Site.
56. Eastman Correspondence.
57. Hendricks Collection.
58. EHS and Hendricks Collection. On 1 April 1894, the day of the transfer, Wm. Gilmore, a former assistant to Samuel Insull, began work as Edison's business manager. He was charged with improving Edison's business activities, particularly the North American Phonograph Co. which had operational problems. Gilmore supervised most of the Kinetoscope arrangements which were made through the Edison Manufacturing Co.
59. Hendricks Collections from EHS.
60. EHS.
61. Hendricks Collection.
62. There are documents from C. Francis Jenkins at the Franklin Institute, Philadelphia, Pa. and the Wayne County Historical Museum, Richmond, Indiana. Jenkins was a busy self-promoter who wrote numerous articles about his work on movies as well as the work of other persons. These appeared in several journals and in three books that he published. There is testimony from Armat and others involved in his work in various hearings and suits related to Jenkins' inventions. Jenkins was a founding member of the Society of Motion Picture Engineers, and contributed to a series of articles by and about pioneers of the motion picture which appeared in the *Journal of S.M.P.E.* during the 1920s and 1930s.
63. Thomas Armat donated some of his papers to the Library at Georgetown University. There are detailed depositions by Armat in several patent cases, particularly in Patent Interference No. 18,461, Amet vs. Latham vs. Casler vs. Armat, 1897–1898 which involved conflicting patent applications for their motion picture projectors. Additional testimony was taken in Patent Interference No. 18,032, Jenkins vs. Armat & Jenkins. Armat also wrote several letters to the Smithsonian protesting their exhibits on motion pictures. An account of his work appeared in the *Journal of the S.M.P.E.* (March 1935): 17–22.
64. National Archives and Hendricks Collection. The fire was reported in the *Atlanta Journal* (15 October 1895).
65. Armat Papers, Georgetown; Raff & Gammon Papers, Baker Library, Harvard Business School; EHS and National Archives.
66. Franklin Institute, National Archives and *Phonoscope*. Jenkins' contact with Columbia Phonograph Co. came through his work as a stenographer in office of the US Lifesaving Service, a predecessor of the US Coast Guard. Because it was originally viewed as a device for recording business correspondence, trials, legislative sessions, etc., many of the recently-established phonograph companies were founded by and staffed with stenographers. Although started by a group of stenographers, Columbia Phonograph Co. was pioneering new markets, particularly coin-operated record parlours where patrons could listen to music. John Phillip Sousa was an early favourite in Columbia's music parlours. Their Kinetoscopes were placed in such parlours.
67. Raff & Gammon Collection, Baker Library and *Phonoscope*.
68. EHS.
69. Hendricks Collection; Robert W. Paul, *Journal of the S.M.P.E.* (November 1936) and *Proceedings of the British Kinematograph Society* no. 38: 2.
70. Hendricks Collection; Robert W. Paul, *Journal of the S.M.P.E.* (November 1935) and *Proceedings of the British Kinematograph Society*; Deac Rossell, 'A Chronology of Cinema,1889–1896', *Film History* (Summer 1995); John Barnes, *The Beginnings of the British Cinema*, I, 189 and 220, II, 23.
71. There seems to be some confusion as to the exact date. Several French sources say it was late October, but the shipping records at the Edison National Historic Site show the shipment of kinetoscopes consigned to France left New York at the beginning of November 1894. See Gordon Hendricks, *The Kinetoscope*, (New York: Beginnings of the American Film, 1966); rpt. in *Origins of the American Film*, 112.

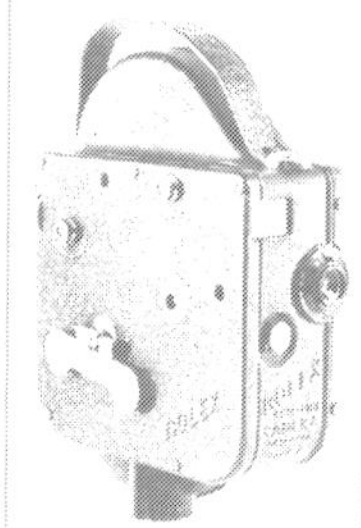

72. As early as 1889, the Lumières had an agreement with Balagny to produce celluloid flat film, but at the time the Cinematograph was being developed and marketed, the Lumières were best known in photographic circles for their recent experiments in colour photography.
73. Auguste and Louis Lumière, *Letters: Inventing the Cinema*, (ed.) Jacques Rittaud-Hutine, (London, Boston: Faber and Faber, 1995), 37–40, hereafter *Lumière Letters*. Louis Lumière to Planchon, 5 November 1895. Following the premiere of the Cinematograph in December 1895, Georges Balagny complained about the Lumière's agreement with Planchon and also demanded a percentage of Lumière's film sales. His claim was apparently based on an agreement between Lumière and Balagny. On 2 January 1896, Auguste Lumière wrote a terse letter denying Balagny's demands and proposing an end to their agreement. *Lumière Letters*, 88–89.
74. *Lumière Letters*, 96 and Eastman Correspondence. Eastman to Wm. Walker, 21 March 1896.
75. National Archives. Testimony in Patent Interference 18,461, Latham vs. Casler vs. Armat.
76. National Archives. US patent no. 589,168, Thomas A. Edison, Kinetographic Camera, applied 24 August 1891 as 403,534.
77. See epigraph at the head of this section of the essay.
78. Eastman Correspondence.
79. Eastman Correspondence. Harris Hayden was an early employer of Eastman, and he was Eastman's financial adviser and stock broker for many years.
80. Eastman Correspondence.
81. Ibid.
82. Ibid.
83. I did not find the designation '35mm' in any contemporary correspondence by George Eastman or the staff at Edison's laboratory, nor was it used in contemporary photographic journals in the US or England. The term 'film' was commonly used to describe pliable transparent photographic material. Various people, including George Eastman and Thomas Edison, claimed to have been the first to use it. It appears in photographic literature as early as 1888, and by 1891, it was being generally used to describe celluloid film.
84. 'Chic Paree', *Phonoscope* (November 1896).
85. Hendricks Collection and *Phonoscope* (May 1897), 4.
86. Edison National Historic Site, Hendricks Collection, Raff & Gammon Collection; *Phonoscope*, 9 March 1897. Raff & Gammon purchased the assets of the Latham's bankrupt Lambda Company including production facilities in New York City but their plan to produce films did not develop. They sold the Latham's assets, including the Latham's patent, to E. & H. T. Anthony who later sold them to the American Mutoscope & Biograph Co. The Latham patent included the 'Latham Loop' which figured prominently in the patent wars.

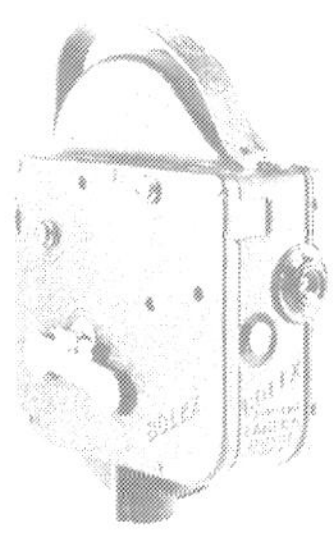

Seeing Seeing: Hermann von Helmholtz and the Invention of the Ophthalmoscope

Oliver Gaycken
Department of English, University of Chicago, USA

> [W]ith the relative autonomy of a technical apparatus, indeed that of a machine and of a prosthetic body, this artifact that is the university has *reflected society* only in giving it the chance for reflection, that is, also, *dissociation*. ... The time for reflection is also the chance for turning back on the very conditions of reflection, in all the senses of that word, as if with the help of a new optical device one could finally see sight, could not only view the natural landscape, the city, the bridge, but could view viewing.[1]

In 1850, Hermann von Helmholtz published a short treatise entitled *Description of an Ophthalmoscope*. Near its beginning, he describes his objective in the following words: 'The present treatise contains the description of an optical instrument by which it is possible to see and recognize exactly in the living eye the retina itself and the images of luminous objects which are cast upon it.'[2] Helmholtz's invention frequently is described as an event of foundational importance in the field of ophthalmology, an event that, indeed, creates that field in its modern sense.[3] While such claims are not exaggerated, the ophthalmoscope nonetheless should be situated within a long tradition of reflection upon the eye.

In the translator's opening footnote to Helmholtz's *Description*, we read that in 1807, 'Jean Méry, of Paris, performed his famous experiment with a cat. Having immersed the animal in water, he first observed that the pupil dilated (as a result of suspended respiration) and then he beheld in all its glory the fundus of the animal's eye – the entrance of the optic nerve and all the colours and vessels of the choroid.'[4] In Méry's experiment, the force required to make the eye's interior visible could not be more obvious. What we shall see, so to speak, in the experiments and technologies that follow Méry's is that instead of being able to remove this incidence of force from the act of examination, the technical refinements for exposing the body to vision are always haunted by the violence of this early scene of observation. The visual examination of the living body involves, at an irreducible level, a vision that causes the observed to become, in a certain sense, a living corpse.

The ophthalmoscope is a significant episode in a progressive valuation of vision as the supreme sense for medical examination, a process that Stanley Reiser calls 'the anatomisation of the living'.[5] Whereas the physicians of Méry's era had primarily listened for disease by asking questions and listening for symptoms, at the beginning of the nineteenth century, physicians began to prefer visual methods of diagnosis, a preference that would reach a zenith with Röntgen's discovery of X-rays in 1895. Physicians and other researchers paid an increasing amount of attention to inventing and implementing devices that could penetrate the living body's exterior, and make parts of the body visible that until then had only been accessible after

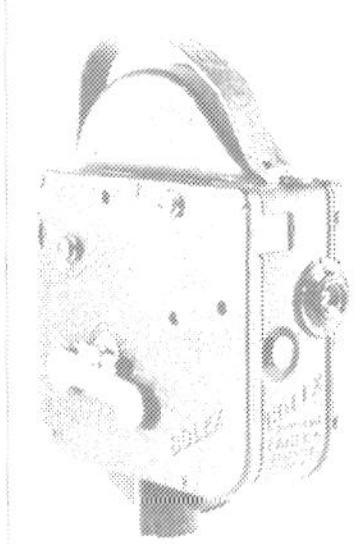

death, on the pathologist's slab. The speculum in all its different forms, from the ophthalmological to the gynaecological to the laryngological, began to assume a privileged position in medical diagnosis.

Helmholtz's invention of the ophthalmoscope came about when, during his preparations for classes he was teaching in general pathology and physiology at the university in Königsberg, he sought to demonstrate contemporary work on how under certain conditions the eye took on a luminous appearance. The peculiar sight of an eye aflame with light, which is particularly noticeable in certain animals under low light conditions and which the ancients believed was generated by the eye itself, had occasioned a host of experiments that attempted to answer the question of this light's origin. During the first part of the nineteenth century, the experiments of Bénédict Prévost and Johannes Müller, among others, proved that the light in the eye was reflected, not self-generated, since it never appeared in total darkness. William Cumming and Ernst Brücke had performed experiments in which they were able to reproduce the puzzling luminosity, but they did not devise a method for focusing the light produced by their indirect illumination of the eye's interior. These experiments occasioned Helmholtz's invention of the ophthalmoscope, which he describes in the following passage:

> I was endeavouring to explain to my pupils the emission of reflected light from the eye, a discovery made by Brücke. Brücke was only a hair's breadth away from the invention of the ophthalmoscope. He only neglected to ask himself to which optical image the rays of light returning from the illuminated eye belonged. It was not necessary for him to pose this question for his purposes at that point. Had he posed it, he would have been just the man to answer it as quickly as I did. I turned the problem this way and that a bit to see how I would be able to demonstrate it most simply to my listeners and then ran into the defining question. I was well acquainted from my medical studies with the need of eye doctors in cases that at that time were grouped under the name of black cataract. I hurried to glue together the instrument from pasteboard, eyeglass lenses, and cover glasses used for microscopic objects. At first it was difficult to use. Without the certain theoretical conviction that it must work, I may not have persevered. But after eight days I had the great joy of being the first who saw a living human retina lying clearly before him.[6]

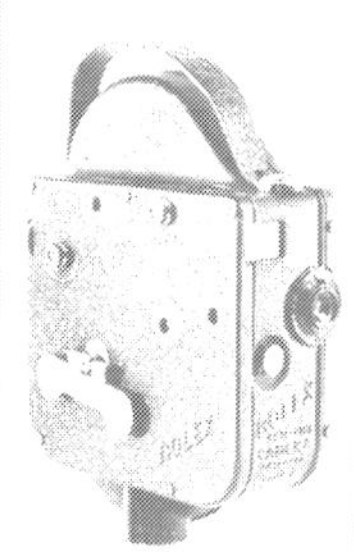

This reminiscence is interesting for a number of reasons. In the first place, we can note the interdependence of the practical and the theoretical, a hallmark of Helmholtz's methodology, which allows the ideality of his conviction in the theoretical principles of geometrical optics to sustain his research in the face of practical difficulties. Another detail worth noting at this point is how the final sentence conveys a sense of discovery, and specifically, a powerful emotion ('great joy') that is evoked by seeing something that no one had seen before. Elsewhere Helmholtz glosses the term 'black cataract' with the phrase 'terra incognita', a wording that suggests not only something we might term medical imperialism (Helmholtz as Columbus), but also that the fundamental struggle in which the ophthalmoscope engages is the struggle of light against darkness.[7] We will return to the sense of wonder and amazement that the ophthalmoscope engendered.

While the *Description* does not make any explicitly broad claims, Helmholtz envisions that the ophthalmoscope's introduction will have considerable consequences for medical examination in general. In his popular scientific lectures on the physiology of vision, he proposes nothing less than total attention to the methodology of ophthalmic medicine by all other branches of medicine: 'As once astronomy was the pattern from which the other sciences learned how the right method will lead to success, so does ophthalmic medicine now display how much may be accomplished in the treatment of disease by extended application of well-understood methods of investigation and accurate insight into the causal connection of phenomena.'[8] In the turn from observing heavenly bodies to human ones, the truthfulness of the new methods of investigation, 'the right method', asserts itself by analogy to the previous methods.

My attention to Helmholtz's rhetoric stems from the conviction that the *Description* should not be read only from the vantage point of medical history, but also from those of philosophy and the prehistory of cinema. When we read the *Description* rhetorically, what emerges is a discourse of direct observation marked and indeed made possible by the appearance of its opposite, indirection, or mediation. When the eye reflects upon itself, we might think that it would be possible to see seeing, to unlock the enigma of another's gaze. The eye of the observed, however, resists the advance of vision; the *terra incognita* prevents or problematises direct vision, and specularity gives way to spectrality. When we juxtapose accounts of the early cinema period with accounts of the discovery and subsequent use of the ophthalmoscope, a number of features of these accounts are strongly resonant, suggesting that these two rarely-connected technologies of vision have more in common than previously has been thought.

'At first everything appears … absolutely dark.'[9]

To begin to read Helmholtz's reflection, we must understand what it is about the eye, or more precisely, *in* the eye, that denies ordinary vision access past the pupil. As Helmholtz explains, the light-refracting [*lichtbrechend*] media of the eye 'under ordinary circumstances, hinder us from seeing illuminated parts of the retina behind the pupil'.[10] In other words, the eye's structures of mediation hinder the observer's eye from getting a picture of the retina.

Under normal conditions the only thing we can see when we look closely at a pupil is our own reflection, so the question of how to reconnoitre the eye's interior, of how to short-circuit the loop of self-reference, becomes one of the *illumination* of the retina through the frame of the pupil. What Helmholtz must devise is a technique, a method, and finally an apparatus that enables the observer to follow the path of light into and out of the eye; the observer's eye must, in other words, follow a path of light that it had hitherto never been able to emulate. The difficulties of this transformation are several, and in order to situate his task, Helmholtz outlines the particularities of the eye's optics that make it necessary in the first place. This tracing of the path of light into the eye establishes the conditions of the ophthalmoscope's possibility:

> [W]ithout special expedients, we can see nothing of the illuminated portion of the retina, because we cannot bring our eye into the direction of the returning light without at the same time cutting off the incident [*einfallende*] light absolutely. To our pupil no light from the depths of the other's eye can return which has not proceeded from it [i.e. our pupil]. And as, in general, none at all has proceeded from our pupil, it sees in the darkness of the other's eye merely the reflection [*Wiederschein*] of its own blackness; only those portions of the retina become visible to it on which its own dark image is copied [*sich abbildet*].[11]

This originary darkness is the eye's lesson of mediation, the fascinating enigma at the origin of the image that forms the point of departure for Helmholtz's inquiry. Locked in the fidelity of the reproduction of an image of darkness, the image of its own failure to illuminate, the scientific eye must invent a device to dissemble or circumvent this unenlightened situation.[12]

Helmholtz's aim is to look into the eye directly, since the route of indirection is insufficient: 'Some other method is necessary for the attainment of our object, a method which makes it possible to look into the eye not merely somewhat, but exactly in the direction of the incident light.'[13] We should keep the order of insights in mind, however. It is only because the eye does not provide 'absolutely precise images', that researchers were first able to remark upon its reflective characteristics. Some portion of the eye's light goes beyond the point of emission and results in the partial illumination beheld by Jan Purkynje, Cumming, Brücke, Charles Babbage, and others. The light reflected in the shine of the eyes of animals at night, the eyes of the Ethiopian albino, and the eyes of people caught in the flash of a photograph is what

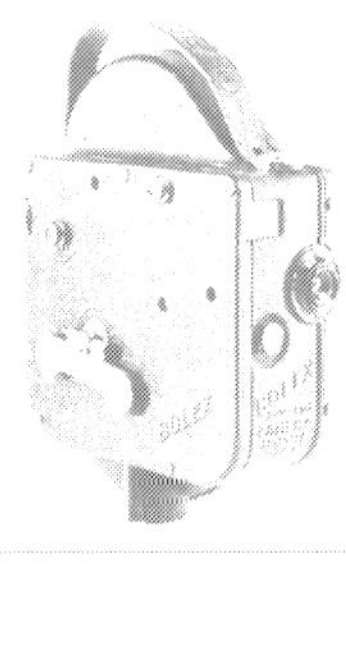

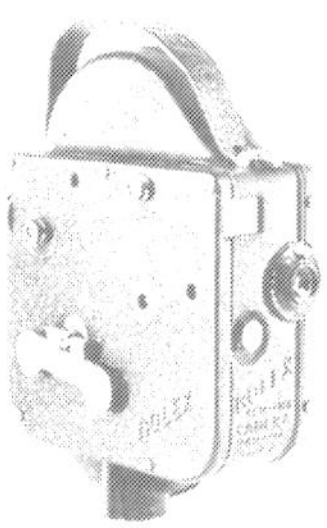

makes Helmholtz's inquiry possible. Without the aberrations and imprecisions of light's encounters with the eye, Helmholtz's method of analysis would be locked into the reflection of blackness with which it begins. Helmholtz's direct, precise approach can only follow from the recognition of indirection.

Helmholtz traces the path of light into and out of the eye according to the laws of geometrical optics. The geometrical theory of light, based on the concept of rays of light that travel in straight lines, was being replaced in the minds of scientists of optics by Fresnel's theory of the undulatory or wave motion of light. Although Helmholtz was aware of Fresnel's work, the ray-based optics of the previous centuries were all that was necessary for his work on the ophthalmoscope.[14] In other words, although he is one of the key figures in the nineteenth-century shift to physiologically based theories of vision (especially in Jonathan Crary's account), for the work on the ophthalmoscope, Helmholtz utilises the ray-based, point-to-point epistemology linked to the camera obscura.

His tracing of light's path is to be an ideal case: 'Let light fall from a luminous point upon a fittingly accommodated eye, concerning which we assume that it is formed with absolute accuracy, that is, that all the incident rays from the point in question concentrate upon a single point of the retina'.[15] In this situation, the path of light must provide complete accuracy with reference to the original which, we should remember, is the observer himself. Helmholtz's invocations of the laws of reflection that are a part of the geometrical theory of light are not simply expedient. Although this theory accurately describes what happens when light encounters the eye, it nevertheless also supplies the rhetorical surplus value of an assurance about the identity of the light and its image: 'If now *a* is a luminous point outside the eye, and *b* its image [*Bild*], a point on the retina, then the ocular media will concentrate the returning light precisely at *a* into an image at *b*. The image of the illuminated retinal point will coincide exactly with the original point of luminosity.'[16] The emphasis on reflection's precision is continual.

That light cannot deviate is a well-known formulation of the ray theory of light. Even its breaking conforms to strict mathematical laws: Snell's Law gives the formula for the computation of refraction. This computability, which allows for the chance of correction, does not mean, however, that the light returning from the observed's retina (which cannot, of course, be seen without the help of a special instrument) is *precise* in the sense of being *the same as* the original. What the observer sees, after all, is an image. Helmholtz's illustration of the path of light into the eye is instructive in this regard:

> If we now replace our sheet of white paper by a prepared photographic plate, each point of its surface will be altered by the light which is concentrated on it. This light is derived from the corresponding point in the object, and answers to it in intensity. Hence the changes which take place on the plate will correspond to the chemical intensity of the rays which fall upon it. This is exactly what takes place in the eye.[17]

Within the exactitude of this comparison of the retina to a photographic plate something else happens. In the conflation of the specularity of geometric optics with the spectral residues of photochemistry, a dislocation that resembles photography itself takes place. Helmholtz's analogy leads us to think how the eye has always been like a photographic camera and that, within the eye's structure, images have always been technically reproduced. Instead of assuring the identity of an object with its image, the desire to overcome the eye's darkness with the ophthalmoscope's light of reflection can only reproduce exactly the essential disjunction that separates the object from its image, the continual *Augenblick* or blink of an eye in which the eye registers the passage of light into its interior. In the same way as the eye has always been seen, but never in an identical manner, the non-identity that guarantees the formation of the image shall have marked every apparatus of vision.

Having explained the ophthalmoscope's necessity and having acknowledged its debts to previous investigations, Helmholtz proceeds to describe the scene in which it will be used:

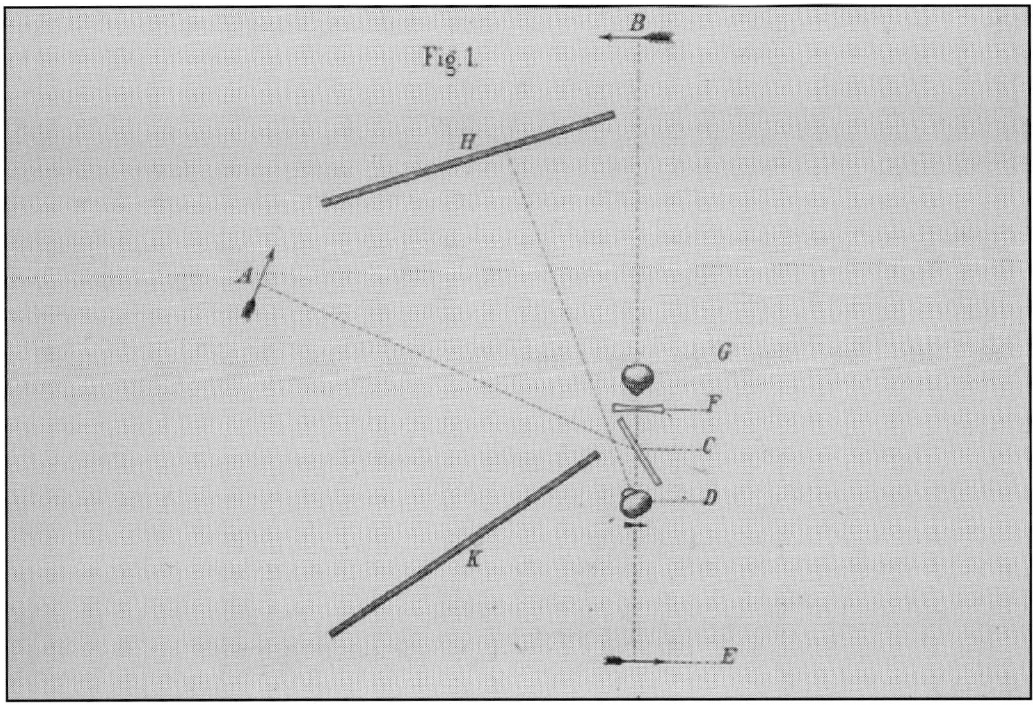

Fig. 1. Hermann von Helmholtz, Description of an Ophthalmoscope *(1851).*

> In a darkened room where only a single source of light, a well burning lamp or an opening in a window shutter for the sunlight, is present, let one set a small plane glass plate in such a way that the observed eye may perceive therein the mirrored image [*Spiegelbild*] of the light, without, however, its necessarily gazing at this mirrored image directly (Fig. 1).[18]

The piece of glass at C functions as both a screen and a mirror because it has been set obliquely in the line of the observer's direct vision. As a result of this obliquity, the observed sees an image of the candlelight, a reflected representation of the physician's power to illuminate, and without this figuration or relay of power, his privileged look into the other's eye would not exist. The observer must be separated by a reflective transparency from the image of the flame, the illuminating effect. In so far as the ophthalmoscope represents the eye of the observer, it effectively splits the observer's gaze – the front of the gaze, the mirroring aspect of the glass, sends light into the other's eye, while behind this protective/projective medium his eye looks through the glass, and the lens at F, and perceives the image of the retina. While screening his power for the observed, the observer screens himself from the illuminating and dazzling effects of the light. He does not have to expose himself to its truth, he does not have to take it for truth; indeed, doing so would constitute an annulment of his carefully constructed position. This peculiar dynamic of indirection achieves the effect Helmholtz has been setting up so precisely:

> One sees that, in this way, it becomes possible to look into the subject's eye in precisely the same direction as that in which light falls upon it. Under these circumstances the eye of the observer in fact receives light from the depths of the other eye, and sees its pupil apparently grow luminous [*scheinbar leuchten*].[19]

The ophthalmoscope, like any window, can be both a mirror and a transparent screen. For the observed it offers an image of dazzlement but for the observer it offers a view into an illuminated interior.

The observer must be the active agent in establishing this constellation for observation, a necessity that arises from the positioning of the plane glass plate. In order to be examined, the observer's eye must be

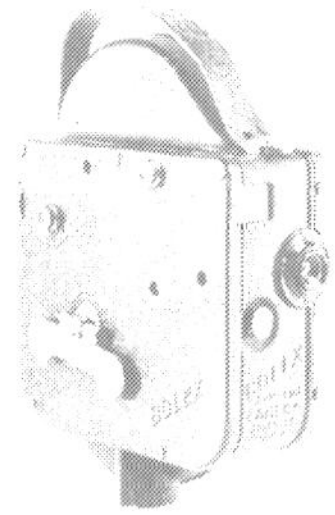

made to see only light, as opposed to the image that appears for the observer. The implication of this configuration is that the observed gives up the capacity to look, at least insofar as looking is normally understood. As Helmholtz explains:

> To *look* at anything means to place the eye in such a position that the image of the object falls on the small region of perfectly clear vision [i.e. the *fovea centralis* or the yellow spot]. This we may call *direct* vision, applying the term *indirect* to that exercised with the lateral parts of the retina – indeed with all except the yellow spot.[20]

Helmholtz describes how at first he allowed the observed to position the plane glass plate, a method that was not successful in providing a view for the observer. It is better not to involve the observed's vision in any active way, to let the observed rest in the peace of nonrecognition during the examination:

> It is better ... to perform the experiment in another way, whereby the observer holds the glass himself. One must, by this method, shade the face that is being observed, and make the reflecting plate so small that it is barely large enough to see through. The light reflected from it then produces on the shaded face of the observed a small, bright spot, which has about the form of the reflecting glass. This point [*Schein*] should be so managed by the observer that its centre falls upon the observed eye, while he himself looks through the glass. In this way the glass may easily be placed correctly, and the observed eye may, without the slightest difficulty, be turned toward all sides in order to cause the image of the flame to fall on different parts of the retina.[21]

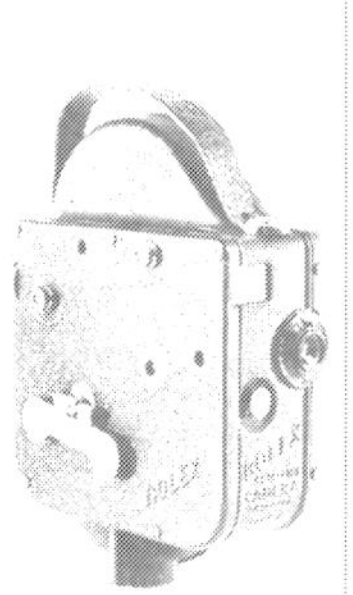

The observer, in other words, should draw a bead on the observed: it is not surprising that Stanley Reiser uses a martial metaphor to talk about the ophthalmoscope: 'General practitioners ... began to incorporate the ophthalmoscope into their small but increasing armoury of medical instruments.'[22]

We can now turn to some contemporary reactions to the new device. Reiser gives an account of the early reactions to the sight provided by the ophthalmoscope: 'The delicacy and beauty of the eye's interior, "this spectacle of the red vessels on the transparent white ground", fascinated physicians.'[23] The terms 'spectacle' and 'fascination' bring some of this imaging technology's resonances with early cinema into relief, as does the following report about the reaction of one of the pioneers of ophthalmic medicine to the sight of the retina: 'When von Graefe first saw the fundus of the living human eye, with its optic disc and blood vessels, his face flushed with excitement, and he cried, "Helmholtz has unfolded to us a new world! What remains to be discovered?"'[24]

Another similarity to the early cinema is the anxiety the new diagnostic procedure produced: '[T]he reaction of some patients hampered the adoption of the instrument; a number became alarmed during the examination, first at being ushered into a darkened room, then at having reflectors and lenses project a powerful light into their eyes.'[25] These fears about exposure to light recall the reactions attributed to early cinema audiences at the first instances of projection as well as the widespread discourse of cinema's possible deleterious effects on the health of its patrons. This anxiety was met with professional reassurances that, as Reiser observes, employed a twofold strategy: 'the instrument's supporters denied that apprehensiveness or trauma would occur if the apparatus for the examination were not prominently displayed and if the procedure was conducted carefully.'[26] In other words, a certain amount of expertise was necessary for the user of the ophthalmoscope to see the retina without causing pain or distress. The need for this expertise became evident when lay-purchasers of the ophthalmoscope were disappointed by their inability to produce the images that had caused so much excitement in the medical community.[27] Equally important and intertwined with the expert's methodology was the elimination of the apparatus's visibility from the scene of its employment. A similar development would take place during the first decade of cinema exhibition, when the operational aesthetic was replaced by the preference for the concealment of the apparatus from the site of exhibition.

The shift from initial wonder at the image itself to a need to have that image tell a story is also familiar from the early history of the cinema. As the cinema of attractions gave way to a cinema of narrative integration, so too a certain *mental* attitude that is similar to the narrativisation of the camera's gaze supplants the initial *visual* fascination with the ophthalmoscope's images. It should come as no surprise that this narrativisation in the case of the ophthalmoscope is concomitant with an increased capacity for rationalisation and control. In a practical sense, the ophthalmoscope allowed for the apprehension, in both senses of the word, of the irregularities of the human eye. Recruits feigning nearsightedness or defendants in court cases who claimed to be unable to read because of faulty eyesight were among the first to come into contact with its incisive insight. The ophthalmoscope was one of the first instruments that allowed for an examination without the patient's full approval.

The question of what happens when the observed looks back is, however, of some interest:

> Moreover, the place of direct vision (the yellow spot) is essentially distinguished in appearance from the parts which lie immediately about it. In order to get this point before oneself, one causes the eye which is being observed to look directly at the mirrored image of the flame. ... Then, too, one is greatly annoyed while gazing on the yellow spot, by the tiny image from the cornea, which obtrudes itself precisely in the center of the visual field, while, during the observation of the lateral part of the retina, it lies to one side.[28]

The direct look of the observed is annoying to the observer precisely because it obtrudes an image into the observer's central field of vision. And since the image of the observed's cornea corresponds to the front of the piece of plane glass on which the image of the flame appears, what the observer sees when the observed looks directly back is exactly that image that the observed is made to see. Here, the impossibility of a direct exchange of looks reappears. In the *fovea* the mediative structures that underlie the rest of the retina are lacking; in other words, the site of clearest vision manifests itself as a site of absence. If, to return to the epigraph, reflection requires a space for dissociation, that space – which in the eye is responsible for discriminating distances, establishing the subject's place in space, and its distance from other objects – presents a view for the observer that is a nonview. The view of the observed's most accurate place of reflection is the spot that offers an 'unfavourable' view for the observer. Even though the ophthalmoscope has allowed access past the pupil, it faces the same frustrated desire with which we started, which is to say that the possibility of seeing seeing is not possible because of the reflection of an other.

That we are involved with more than a logic of specularity, no matter to what extent it is repetitively invoked, is clear when we consider that an ophthalmic examination is not a procedure that someone could do alone: there can be no autodiagnosis. Although anyone can make their eye shine by stepping before a mirror with a piece of plane glass and moving the image of a flame or another light source over the image of their eye in the mirror, the illumination is indirect because the line of vision brings the flame to bear on a point on the outside of the retina. The 'direct' line of sight can only be possible where there is more than one person involved, when one reflects upon an other.

The ophthalmoscope is not, finally, a device for seeing the self seeing. It is not even a device for seeing the other's gaze. Instead it stages the impossibility of the occurrence of such an event. What we can say about Helmholtz's treatise, however, is that it teaches us that the fall of the light of thought into the eye gets medical history underway in a moment of illumination and blindness. The position of aligning oneself through an apparatus with the sun makes the person who looks through that apparatus into a ray of light; taking the place of the sun, the self is displaced according to the logic of the prosthesis that extends the self and at the same time marks its limits. By positioning himself as the light, as the force

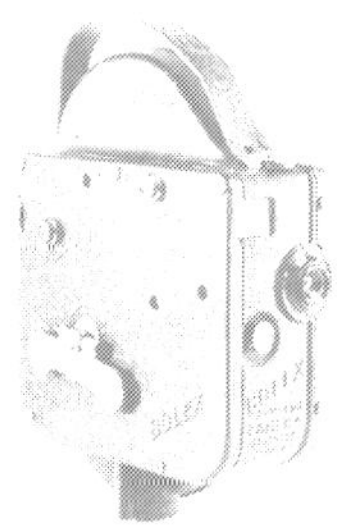

that falls into the eye of the observed, Helmholtz takes part in the falling into the eye that is also the fall into metaphor, the fall into history, the fall into virtuality and prosthetics, a fall, in other words, into reflection.

Notes

1. Jacques Derrida, 'The Principle of Reason: The University in the Eyes of Its Pupils', *diacritics* 13, 3 (fall 1983): 19, emphasis in original.
2. Hermann Ludwig Ferdinand von Helmholtz, *Helmholtz's Description of an Opthalmoscope*, trans. Thomas Hall Shastid (Chicago: Cleveland Press, 1916), 7; translation modified; originally published as *Beschreibung eines Augen-Spiegels zur Untersuchung der Netzhaut im lebenden Auge* (Berlin: A. Förstner'sche Verlagsbuchhandlung, 1851); reprinted in Ernest Engelking (ed.), *Dokumente zur Erfindung des Augenspiegels durch Hermann von Helmholtz im Jahre 1850* (Munich: J.F. Bergmann, 1950).
3. 'In fact, there are just two kinds of ophthalmology, that which came before and that which followed after Helmholtz's *Beschreibung eines Augenspiegels*' (translator's note, Helmholtz, *Description*, 29).
4. Helmholtz, *Description*, 7.
5. Stanley Joel Reiser, *Medicine and the Rise of Technology* (New York: Cambridge University Press), 45–68, esp. 45–50.
6. Hermann von Helmholtz, 'Erinnerungen', in *Vorträge und Reden* (Braunschweig, Vieweg, 1903), 12, my translation.
7. 'Up until now a series of the most important diseases of the eye was grouped together under the name "black cataract", a Terra incognita, because one did not discover anything about the changes in the eye in either the living or even after death.' Letter from Helmholtz to his father, 17 December 1850, in Engelking (ed.), *Dokumente zur Erfindung des Augenspiegels*, 6, my translation. A contemporary of Helmholtz's wrote about this group of conditions that 'the patient sees nothing, but neither does the physician', a statement that underlines how vision and knowledge are linked in the discourse surrounding the ophthalmoscope. (Richard Greef, 'Historisches zur Erfindung des Augenspiegels', *Berliner klinische Wochenschrift* 38:48 [1901]: 1201, quoted in Arleen Tuchman, 'Helmholtz and the German Medical Community', in David Cahan (ed.), *Hermann von Helmholtz and the Foundations of Nineteenth-Century Science* (Berkeley: University of California Press, 1993), 34). For a discussion of how 'the entire history of our philosophy is a photology, the name given to a history of, or treatise on, light', see Jacques Derrida, 'Force and Signification', in *Writing and Difference*, trans. Alan Bass (Chicago: University of Chicago Press, 1978), 27.
8. Hermann von Helmholtz, 'The Recent Progress of the Theory of Vision', in *Popular Scientific Lectures* (New York: Appelton, 1873), reprinted in Richard and Roslyn Warren (eds.), *Helmholtz on Perception: Its Physiology and Development* (New York: John Wiley and Sons, 1968), 63.
9. Helmholtz, *Description*, 8.
10. Helmholtz, *Description*, 8.
11. Helmholtz, *Description*, 10.
12. Michel Foucault's writing on the historical shifts leading to the modern medical observer are suggestive at this point, particularly in his use of the image that conflates the eye and the flame: 'For Descartes and Malebranche, to see was to perceive …; but, without stripping perception of its sensitive body, it was a matter of rendering it transparent for the exercise of the mind: light, anterior to every gaze, was the element of ideality – the unassignable place of origin where things were adequate to their essence – and the form by which things reached it through the geometry of bodies; according to them, the act of seeing, having attained perfection, was absorbed back into the unbending, unending figure of light. At the end of the eighteenth century, however, seeing consists in leaving to experience its greatest corporeal opacity; the solidity, the obscurity, the density of things closed in upon themselves, [they] have powers of truth that they owe not to light, but to the slowness of the gaze that passes over them, around them, and gradually into them, bringing them nothing more than its own light. All light has passed over into the thin flame of the eye, which now flickers around solid objects and, in so doing, establishes their place and form.' (Michel Foucault, *The Birth of the Clinic: An Archaeology of Medical Perception*, trans. A.M. Sheridan Smith (New York: Vintage, 1975), xiii–xiv).

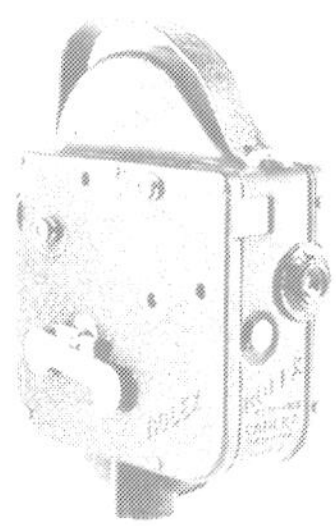

13. Helmholtz, *Description*, 10.
14. In the letter to his father, Helmholtz mentions that the discovery of the ophthalmoscope required no further knowledge of optics than what he had learned in the *Gymnasium* (*'erforderte weiter keine Kenntnisse, als was ich auf dem Gymnasium Optik gelernt hatte'*).
15. Hermann von Helmholtz, 'The Recent Progress of the Theory of Vision', in *Popular Scientific Lectures* (New York: Appelton, 1873), reprinted in Richard and Roslyn Warren (eds.), *Helmholtz on Perception*, 69.
16. Helmholtz, *Description*, 10.
17. Helmholtz, 'The Recent Progress of the Theory of Vision', in Richard and Roslyn Warren (eds.), *Helmholtz on Perception*, 70.
18. Helmholtz, *Description*, 11.
19. Helmholtz, *Description*, 11.
20. Helmholtz, 'The Recent Progress of the Theory of Vision', in Richard and Roslyn Warren (eds.), *Helmholtz on Perception*, 71–72, emphasis in original.
21. Helmholtz, *Description*, 13.
22. Stanley Joel Reiser, *Medicine and the Rise of Technology*, 51.
23. Ibid., 47.
24. John Gray McKendrick, *Hermann Ludwig Ferdinand von Helmholtz* (London: T. Fisher Unwin, 1899), 83.
25. Reiser, op. cit., 50.
26. Reiser, op. cit., 50.
27. Helmholtz published a second article on the ophthalmoscope, 'Über eine neue einfachste Form des Augenspiegels', in *Archiv für physiologische Heilkunde* (1852), reprinted in Engelking (ed.), *Dokumente*, 45–61. This article responds to the invention of another ophthalmoscope, devised by Reute, professor in Göttingen. Helmholtz found a way to improve and simplify Reute's device. He also gives a shortened version of his original treatise and outlines the differences between the two approaches. Reute's method of illumination is based on Brücke's method that involves looking just past the light that is being projected into the eye. It yields a stronger illumination when the observed's eye is least accommodated for the light source, but there is a central dark spot where the observer's eye casts its shadow. When utilising this method, it is better for the observed to have as unclear an image of the light source as possible, so lenses are placed before normal eyes to make them either nearsighted or farsighted because the illumination is then greater. Helmholtz's device is based on von Erlach's method (a glass plate that projects the image of the candle into the observed's eye). The illumination is not as strong as is possible in Reute's method, but the image offered to the observer is better in some ways (i.e. no central dark spot). Also, this method makes the corneal reflection, which appears when the observer looks into the observed's eye along the observed's axis of vision, much less prominent – Helmholtz reports that he was not able to see the yellow spot using Reute's method because of the corneal reflection (53). Helmholtz sums up the differences between the two devices: 'The essential differences of the two methods are that both Reute's original and the simplified versions of the ophthalmoscope allow for a greater field of vision at lower magnification and greater brightness', 59. This second publication underscores how the ease of the ophthalmoscope's use for the general practitioner was a great concern to Helmholtz.
28. Helmholtz, *Description*, 28–29.

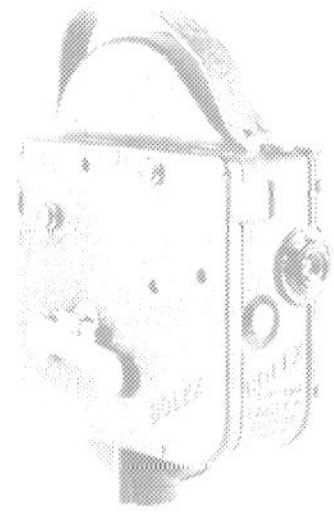

On Fairies and Technologies

Frank Kessler

Department of Theatre, Film and Television Studies, Utrecht University, The Netherlands

– J'ai vu parfois, au fond d'un théâtre banal
Qu'enflammait l'orchestre sonore,
Une fée allumer dans un ciel infernal
Une miraculeuse aurore;
J'ai vu parfois au fond d'un théâtre banal

Un être, qui n'était que lumière, or et gaze,
Terrasser l'énorme Satan;
Mais mon cœur, que jamais ne visite l'extase,
Est un théâtre où l'on attend
Toujours, toujours en vain, l'Etre aux ailes de gaze!
Charles Baudelaire, 'L'Irréparable' (excerpt)

When Georges Méliès published his *causerie* on 'Les vues cinématographiques' in 1907, he told his famous anecdote about how he discovered the so-called 'substitution stop-trick' as a result of a jam in his camera. Putting this effect to creative use, he consequently filmed a couple of metamorphoses changing men into women, as well as a number of other pictures where people suddenly vanished or appeared. 'It's thanks to this simple trick', Méliès observed, 'that I made the first *féeries*: *Le Manoir du Diable, Le Diable au couvent, Cendrillon*, etc.'[1] According to Méliès, the origin of cinematographic *féeries* is thus directly linked to the discovery of a specific trick technique, which is possible only through motion picture technology. This, however, leads to the question why an almost brand-new machine – the first of the films Méliès mentions is from 1896, the other two are from 1899 – and a newly discovered technique were used to recreate a type of stage show which, already more than a century old, dealt with subject matter which explicitely referred to the world of the fantastic.

This question, in fact, opens up a broad range of different issues that one has to explore step by step in order to understand the complex relationship between modern technology and the *féerie* genre. To begin with, all sorts of stage and camera technology are used to create the spectacular effects that are integral to both *féerie* plays and films. This includes also the presentation of films within *féeries* shown on stage. Another important aspect, then, is the tension between the rational world of modern technology and the magical universe of the *féeries*. At this point one has to consider the place of fairies in the imaginary of the late nineteenth century. And finally there is the representation of modern technology in *féeries*, its role both as an element within the narrative and as a visual motif in the image.

Leaving aside the problem of giving a precise definition of *féerie* as a theatrical or cinematographic genre,[2] I shall restrict myself here to a brief and rather general characterisation. A *féerie* is a play or a film – which, by the way, does not necessarily require a fairy as a character – presenting a fantastic subject where the spectacular elements of the *mise-en-scène*, the sets, and the costumes are foregrounded, whereas the narrative serves largely as a pretext to introduce all sorts of marvellous effects.[3] For the stage this implies a number of intricate devices allowing the execution of complex tricks, including the explosion of vehicles and the scattering of body parts.[4] As Georges Sadoul puts it: 'The authors of the *féeries* were probably rather the painters and the stage technicians, and only to a lesser extent those who wrote the libretti.'[5] Thus the main attractions for audiences are,

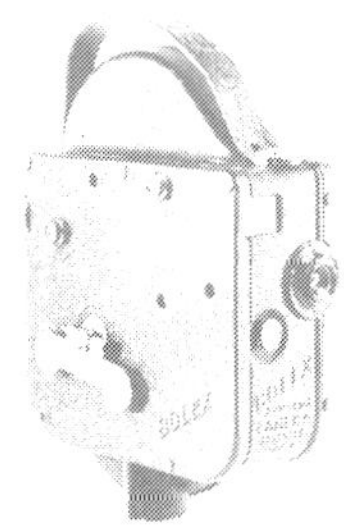

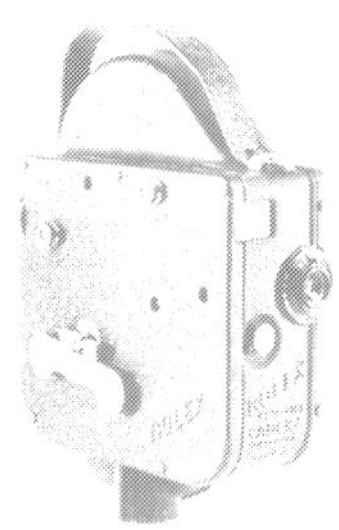

primarily, special effects and spectacular tricks. However, one should not forget that this exhibition of the possibilities offered by an intricate backstage machinery is not an end in itself: its main task is to contribute to the presentation of a fantastic and marvellous universe. Baudelaire's evocation, quoted above, of a common stage with a fairy illuminating an infernal sky with a miraculous sunrise could actually be understood on a quite literal level, since the poem was inspired by the *féerie* called *La Belle aux cheveux d'or* where the heroine was played by a woman Baudelaire loved.[6]

Given the importance of these spectacular elements and the constant demand for new sensational attractions, it is hardly surprising that the stage *féerie* also turned to the new technology of the moving image in order to amaze the audience. According to a recent discovery at the Cinémathèque française, it appears that as early as 1896, a hand-coloured 58mm film of approximately one minute depicting a fantastic dance is included as a special '*clou*' in a *féerie* called *La biche au bois* which was given at the Théâtre du Châtelet in Paris.[7] Another example is the film that Georges Méliès made in 1905 for a presentation of the play, *Les 400 coups du diable*, and which he later integrated into his 'grande féerie fantastique en 35 tableaux', *Les quat' cents farces du diable* (1906).[8] The film appears here in a way as a double attraction: as a film being part of a stage show and also through the marvellous events that are depicted in it.

As far as the cinematographic genre of *féerie* is concerned, the remark by Méliès quoted earlier establishes, as we have seen, an explicit link between the discovery of a trick effect made possible by the technology of the moving picture camera, and the genre. A man with considerable stage experience, Méliès – besides using, of course, also the possibilities offered by theatrical machinery in his Montreuil studio – undoubtedly looked for an application for this new device in an area with which he was familiar. Beginning with filmic versions of magical illusions, the next step took him, apparently quite automatically, to the *féerie* where the substitution stop-trick could be used to create effects similar to those presented on the stage with the aid of complex machinery. It would, however, be too simplistic to see the filmic *féerie* as a continuation of or even successor to the stage show then already in decline. The fact that the technology of the moving pictures accomplishes the same spectacular effects as the theatrical *féerie*, or even surpasses them in an easier and more economic way does not mean that the latter had become obsolete. The films follow the stage form to a certain extent of course, and benefit from the existence of a generic model with which audiences were familiar. But both address their audiences in specific – though undoubtedly related – ways and function in different contexts of presentation and reception.

Both share, however, the same fundamental characteristic, namely that the magical universe and the wondrous things they present to the audience depend on the respective technologies that have to remain undisclosed in the background. This is the paradoxical role of the trick effects in both theatrical and cinematographic *féeries*: they have both to be invisible (in order to give maximum credibility to the magic) and, at the same time, the audience's attention has to be drawn to them in such a way that their virtuosity can be admired, since trick effects count among the specific attractions the genre has to offer. At least for the stage *féerie*, the balance between the fantastic and the technology which makes possible its very existence seems to be a rather precarious one. In his historical study of 1910, Paul Ginisty complained that 'the *féerie* which could have been sovereign, rested content with being but a pretext to mechanical artifices'.[9] With regard to the filmic genre, one could say that the effects are even more crucial as this kind of an aesthetics of amazement is central to the cinema of attractions to which the films, in spite of their obvious narrative component, undoubtedly belong.

Seen in a broader perspective, the relationship between modern technology and craft on the one hand, and the magical or supernatural on the other, seems to be part of a more general phenomenon. The apparent rationalism of the late nineteenth century produces, as it were, its own irrational Other. This can of course take many different forms. Tom Gunning, for instance, has discussed how photography became involved in the spiritist movement as a seemingly objective recording instrument for all sorts of unexplainable manifestations.[10] Gunning also points to the link with Méliès' use of cinematic trick technique in the 1903 film *Le portrait spirite* (The Spiritualist Photographer), where Méliès explicitely announces the trick effect through which the 'spiritualistic photograph' is obtained.

With regard to fairies, it is quite surprising to discover that the *Grand dictionnaire universel du XIXe siècle* by Pierre Larousse, which, in a way, can be considered a sum of nineteenth-century science and knowledge and as such a profoundly rationalistic enterprise, dedicates two whole pages, i.e. eight columns, to the lemma *fée*! The article explains the different classes of fairies in popular legends, the hierarchy of the fairy kingdom, their role in literature and plays, and also song excerpts from *opéras-comiques*. The fairy is seen, of course, as a fictive, fairy-tale creature, but it is still quite amazing that the entry is given in such detailed manner. Without drawing any far-reaching conclusions, one can still assert that this is at least an interesting indication of the presence of fairies in the imaginary of the late nineteenth century.

One can also observe that the fairy was quite often invoked as a metaphor to talk about the wonders of modernity. Thus, at the 1851 Great Exhibition in London, a German visitor to the Crystal Palace, erected there as a monument to the possibilities of modern iron and glass architecture, concluded: 'It is by sober economy of language that I call the sight of this room incomparably fairy-like.'[11] And in Clairville's and Jules Cordier's show, 'Palais de Cristal ou les Parisiens à Londres' (1851) where the Great Exhibition was almost immediately turned into stage spectacle, a verse ran:

> Chaque industrie, exposant ses trophées
> Dans ce bazar du progrès général,
> Semble avoir pris la baguette des fées
> pour enrichir le Palais de Cristal.[12]

Walter Benjamin himself seems to have adopted this metaphor when, in his *Passagen-Werk*, he compared the Paris *passages* to 'fairies' caves'.[13] Furthermore, in letters to Scholem and Adorno, he remarked that in an early stage, his *Passagenarbeit* was supposed to have a subtitle: *Eine dialektische Feerie*.[14] And with regard to the Paris *Exposition universelle* of 1900, a contemporary author, Jules Trousset, celebrated the reign of the *fée électricité*, the Fairy of Electricity: 'The true sovereign of the exposition of 1900 will be Electricity, the young and brilliant fairy, who endows contemporary industry with its two chief factors: movement and light.'[15]

In all of these examples the metaphor of the fairy serves to describe the marvellous and astonishing qualities of phenomena which in themselves could be seen as almost emblematic of modernity: iron and glass architecture, the products of modern industry, the *passages*, or electricity. It is as if the authors could only convey their own amazement at these wonders or trigger their readers' imagination properly by situating them in the realm of the fantastic and the supernatural. Trousset's description of electricity as a fairy is particularly interesting as this trope links the forces that move modern industry with a figure from popular legends that one should picture as a female and rather ethereal being. The semantic field that the author opens up seems hardly appropriate to the rationalistic sphere of industrial production. In a way this choice of metaphor could be seen as an attempt not so much to defamiliarise its object, but to *familiarise* the powers driving modernity. Since the Fairy of Electricity is doubtlessly meant to be a good fairy, and not an evil one, everything that could potentially appear disturbing or threatening in modern technology vanishes behind the image of a lovely creature, whose brilliance and swiftness so effectively evoke the more prosaic ways in which light and movement occur in the world of industry. In this context it may be useful to emphasise the fact that the trope of the fairy should not be confused with those well-known allegorical female figures of the period that were used to represent Science, Craft, Industry, or Modernity, because these do not function in the same manner. Here the female bodies carry the attributes of whatever abstract entity they stand for, and the allegory can be read only through them. At the same time, there is absolutely nothing fantastic or supernatural

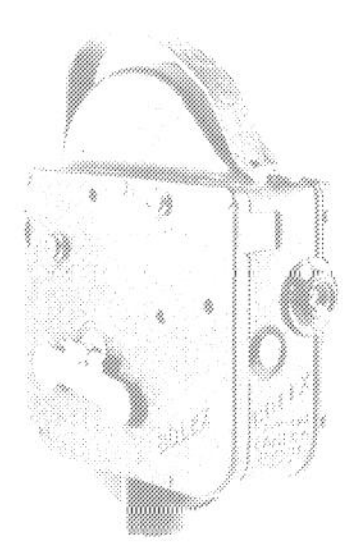

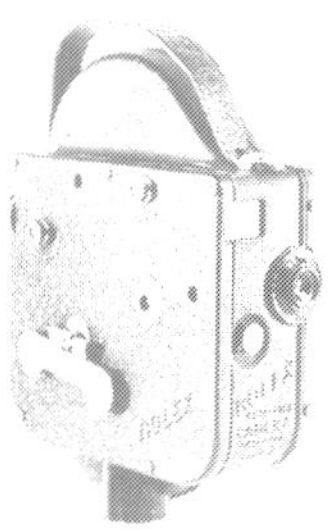

about them. The Fairy of Electricity, on the other hand, is not an allegory. The trope functions differently: it is not through attributes signifying an abstract term that the association with the female figure is motivated, but through the fairy-like nature ascribed to electricity.

In this regard one could say that metaphors like *la fée électricité* present an interesting analogy with the *féerie* as spectacle. In both cases the visibility of the technological reality behind the phenomenon is seen as disenchanting, whereas the image of the fairy provides the invisible forces with a fantastic, albeit familiar and unthreatening, materialisation. This image functions in a way as the dialectical Other of technology. If the latter participates in the process of registering a disenchantment with reality, the fairy permits situating it within the realm of enchantment. This dialectical relation between technology and the image of the fairy also offers an interesting new perspective on the *féerie* as spectacle. It is in fact anything but a naive genre: essentially based on technology, it simultaneously disguises and displays the forces that enable it to create a charming magical world, where the cold rationalistic aspects of its very foundations are rigorously excluded.[16]

Another dimension is added to this already complex relationship when the *féerie* includes modern technology in its subject matter. In 1875, *Le voyage dans la lune*, very loosely based on Jules Verne's novel in two volumes first published ten years earlier, opened at the Théâtre Gaité in Paris. The music was composed by Jacques Offenbach whom Siegfried Kracauer credits with renewing the genre by 'mixing fairy tales with science and fusing modern utopias with the torpid patterns of the *féerie*'.[17] Paul Ginisty also considered this show as the beginning of an evolution, but he declared:

> The scientific element is introduced into the *féerie* with *Le voyage à travers l'impossible* by d'Ennery, based on Jules Verne (Porte-Saint-Martin, 25 November 1882), something like a stage summary of the *Voyages extraordinaires* where a couple of characters of the inventive story-teller, Doctor Ox, the son of Captain Hatteras, Captain Nemo etc., are put together and are probably rather surprised to meet in this way.[18]

Ginisty went on to add that d'Ennery, although exploiting Verne, in fact presented a very negative image of science, and spoke despisingly of '*oeuvre infâme*'.[19] Interestingly enough, the two stage *féeries* just mentioned have the same titles as two of Georges Méliès' best known films, even though the intrigues of the respective theatrical and filmic works seem to bear hardly any resemblance to each other (nor to Jules Verne's writings to which they apparently refer). The way in which the name of Jules Verne appears in this context is quite peculiar. His influence on the theatrical and cinematographic *féeries* seems to be limited to a few characters, motives, or elements, like the gigantic canon which propels the voyagers towards the moon. And in the case of d'Ennery's adaptation for the stage, the scientifically-minded author was even bluntly betrayed.

Georges Méliès, on the other hand, has been called the 'Jules Verne of cinema' by Merritt Crawford in 1930, and 'Father of cinema, son of Jules Verne' by Lo Duca in 1952.[20] Even though this kind of genealogical affiliation linking the filmmaker to the famous writer of fantastic scientific novels may seem plausible at first sight, a closer look reveals that considerable differences separate them. Nevertheless, the association of the two is interesting, and a closer study will help elucidate the complex relationship between fairies and technologies.

In an illuminating article on the question, Guy Gauthier points out that the difference between Verne and Méliès is in fact both a generation gap and a contrast in temperament.[21] Verne represents the mid-nineteenth century and its belief in science and progress, and his novels, the fantastic elements notwithstanding, are firmly rooted in the scientific knowledge of his time. Méliès, on the other hand, is indeed the creator of cinematographic *féeries*, where he can unleash his imagination without having to respect any norm or degree of plausibility.[22] This is quite obvious in the two films mentioned previously, *Voyage dans la lune* (1902) and *Voyage à travers l'impossible* (1904). Both the sun and the moon are depicted as anthropomorphic figures with faces, and the landing of the vehicles is represented as a direct bodily penetration. The projectile in the first film hits the eye of the moon in one of Méliès' most famous shots, whereas the flying train in the second is swallowed and chewed by the sun. And although the launching of the rocket in *Voyage dans la lune* does show at least a superficial similarity to the novel, questions like the scientific calculation of the trajectory, and the forces of attraction and gravity which are discussed at length by Verne's space travellers, are of course completely absent from Méliès' films (except for the rapid sketch on the blackboard in the opening scene). In both of these, the capsules just fall down from the celestial bodies (an expression which is to be taken quite literally here, as we have seen) when they are pushed over a cliff.

In Jules Verne's books, the potentially destructive or otherwise dangerous dimensions of modern inventions are often a central theme, most prominently so with Captain Nemo's 'Nautilus', but also with the uncontrollability of the rocket which turns the space travellers into powerless and passive prisoners inside the projectile. In Méliès' films this aspect is not excluded, but treated in a playful and humorous way. The accidents and crashes of the different vehicles in the two fantastic travel pictures do not have any serious consequences, since they are spectacular moments allowing for pyrotechnics or other trick effects. This aspect is highlighted specifically in the catalogue description of *Voyage à travers l'impossible*:

> 21st *tableau* – A Bitter Pill
>
> The train arrives full-steam and runs into the sun's mouth. After a series of comic grimaces, the latter starts to fret and fume as a result of the indigestion caused by this unforeseen bitter pill.
>
> 22nd *tableau* – A Formidable Crash
>
> Fantastic solar landscape providing a most striking effect. The train falls on the sun. The locomotive, the tender, and the carriages pile up upon one another in an indescribable chaos. This catastrophe produces on the solar surface a volcanic outburst with blazing fire and the emission of sparks giving a superb decorative effect. (This trick is an absolute novelty.)[23]

The dangers of technology are thus transformed into spectacular trick effects where threatening elements like fire and flames appear as decorative elements.[24] Science and modern inventions, together with then still utopian machines like rockets, are depicted in a manner which turns them into the attributes of a fantastic and almost dreamlike universe where sun and moon are gigantic heads, and the stars and planets are represented by young women or bearded men. Even the most disturbing and disquieting aspects of modern technology are turned into an enchanting and charming spectacle along with the other components which are part of the world of a *féerie*. In Méliès' films this integration of different elements is achieved on a visual level through the use of artificial backdrops and props. Landscapes, buildings, vehicles and machinery are all made of painted cardboard, and thus create a homogeneous whole. The stylisation of all the elements facilitates the integration of modern technology into the magical universe of the *féerie*.

This strategy is visible in almost all *féeries* which, regardless of the type of narrative they present, are exclusively shot in visibly artificial studio sets. There is, however, at least one exception to this rule, namely the 1906 Pathé film *Le fils du Diable*.[25] This *féerie* is remarkable in more than one way (not least because of the ending, where a young woman commits suicide in order to be reunited in Hell with the Devil's son), but in the context of this essay, it is especially interesting because it mixes the artificial stage-like world with shots taken outdoors.

The film's narrative is about the Devil's son being bored, and his parents worrying about his health. They call a physician – who is none other than Molière's *médecin malgré lui*, Sganarelle – who recommends sending him to Paris. They leave together in a motor-car (a real one), driving first through the subterranean tunnels of Hell and then arriving in Paris. Since it is carnival, they and their bird-headed chauffeur do not seem odd to the others (even

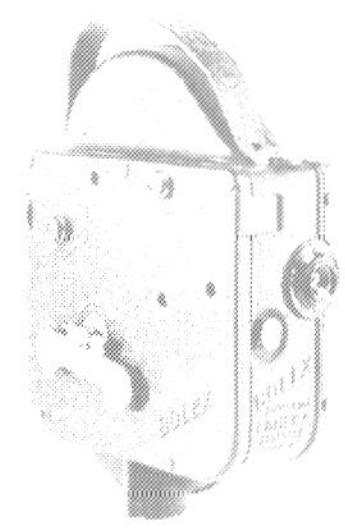

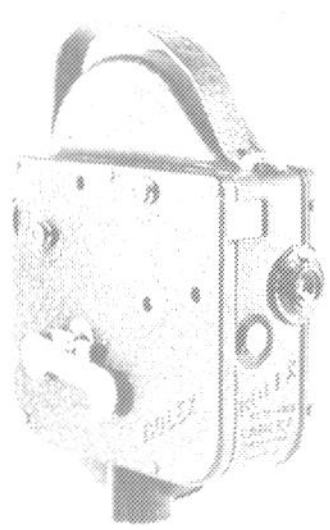

though, in a shot taken on a crowded boulevard, real-life passers-by flock around the car and its strange-looking passengers). A woman steals the purse of young Satan, and when he is unable to pay the bill, Sganarelle telephones Hell, and Satan sends a diabolic messenger with money. Looking for a plain and simple woman, the Devil's son picks a peasant girl from a selection of women whose life-size images are shown to him by Sganarelle. They leave the city and have to change into ordinary-looking people wearing contemporary costumes in order not to frighten the population. The car has a providential accident through which young Satan meets the girl he is looking for. They want to marry, but he cannot stand being in church, and flees. So she has to commit the deadly sin of suicide to be reunited with him.[26]

In *Le fils du Diable*, the fantastic universe of the *féerie* and modernity belong in fact to two separate, yet somehow connected, worlds. Satan's subterreanean empire is filled with bizarre creatures which are his subjects serving him in his palace. The diabolic family wears costumes from an undetermined past, and looks like an almost ordinary fairy-tale royalty. To call his servants, the Devil blows his horn, but to communicate with the world above, he uses a telephone. Having arrived in Paris, the opposition between the fantastic costumes and masks of the travellers coming from the underworld produces a visual clash with the natural landscape, and later with the street-life in Paris. Here the genre is indeed taken to its representational limits, yet this injection of images taken from contemporary reality does have a particular charm. However, the visual unity of the diegetic world is broken, and especially the scenes shot in the streets of Paris give an almost documentary record of how these strange filmic characters all of a sudden turn into a spectacle for the passers-by.

One might consider *Le fils du Diable* an attempt to renew the *féerie* as a *cinematographic* genre by exploring the possibilites of combining artificial and real settings as was commonly done in other types of films. But breaking up the closed universe of *féerie* also introduces a new dimension which in the end works against the traditional functioning of the genre. Creating a closed, magical and fantastic universe allowed filmmakers to integrate elements of modernity in a playful manner. In *Le fils du Diable* this is still the case when Sganarelle sticks his giant syringe into the ground and puts headphones on it in order to telephone Satan. But the car, and especially the car accident, are no longer just spectacular elements (in fact, in the print I have analysed, the crash itself is not shown). Here the accident has a distinctive narrative and dramatic function which differs profoundly from the mainly attractional role such incidents have in traditional *féeries*.

Thus, in conclusion, the complex relationship between *féeries* and modern technology appears to function within a rather rigid framework of representational strategies. Cinematographic (or theatrical) techniques make possible the creation of trick effects, and the attraction lies both in their fantastic or magical quality, and in the technical or mechanical virtuosity through which they are achieved. The same principles also govern the introduction of modern technology as subject matter. Again, spectacular and fantastic dimensions are highlighted; threats and dangers are present only in so far as they can motivate even more amazing trick effects. In this respect, the *féerie* seems to participate in that peculiar representation of technology through the trope of the fairy, which one can find in various forms in late nineteenth-century culture. Through this trope technology appears as something which is unattainable and almost inexplicable, yet unthreatening and in a way familiar. Undoubtedly, in this lies also its main ideological function, namely a reassurance with regard to the fears that are caused by the rationalistic powers of modernity. In *féerie*, the trope is turned into a spectacle which, however, functions only within a closed and homogeneous diegetic universe. When images of social and technological reality are introduced, the genre gets thrown off balance. Soon audiences demand something else – around 1910, the *féerie* starts disappearing from the catalogues. The fairy's spell may have been able to turn technology into magic – but not for long.

Acknowledgement: This essay is part of the Utrecht Media Research Programme 'Historical and Cultural Construction of Media and Media Forms'. Its presentation at the Stockholm conference has been supported by a grant from the Utrecht Onderzoekinstituut voor Geschiedenis en Cultuur (OGC). Many thanks also to William Uricchio for suggestions and corrections.

Notes

1. Georges Méliès, 'Les vues cinématographiques' in Georges Sadoul, *Georges Méliès* (Paris: Seghers, 1970), 107 (all translations from French or German sources are by the author).
2. I have discussed this problem in two papers: 'Die Feerie als filmisches Genre' delivered at the 1995 Marburg conference 'Der Film im Ensemble der Künste' (publication forthcoming) and 'La féerie – un genre des origines' delivered at the 1998 Udine conference, published in Leonardo Quaresima, Alessandra Raengo, Laura Vichi (eds.), *La nascita dei generi cinematografici* (Udine: Forum, 1999), 229–238.
3. A definition of the stage *féerie* is given in Arthur Pougin, *Dictionnaire historique et pittoresque du théâtre et des arts qui s'y rattachent*, 2 vols. (Paris: Firmin Didot, 1885; rpt. Plan-de-la-Tour: Eds. d'Aujourd'hui, 1985), 360–364; an important historical source is Paul Ginisty, *La Féerie* (Paris: Louis-Michaud, n.d. [ca. 1910]). For the filmic *féerie,* see Jean Giraud, *Le lexique français du cinéma des origines à 1930* (Paris: CNRS, 1958), 119.
4. See the seminal essay on the subject, Katherine Singer Kovács, 'Georges Méliès and the *Féerie*', in John L. Fell (ed.), *Film Before Griffith* (Berkeley, Los Angeles, London: University of California Press, 1983), 244–257.
5. Sadoul, op. cit., 34.
6. See Singer Kovács, op. cit, 244, note #2.
7. See Laurant Mannoni, 'Une féerie de 1896. *La biche au bois*', *Cinémathèque* 10 (automne 1996): 117–123.
8. For a description, see Jacques Malthête, *Méliès. Images et Illusions* (Paris: Exporégie, 1996), 231–233.
9. Ginisty, op. cit., 9.
10. Tom Gunning, 'Phantom Images and Modern Manifestations: Spirit Photography, Magic Theater, Trick Films, and Photography's Uncanny', in Patrice Petro (ed.), *Fugitive Images: From Photography to Video* (Bloomington, Indianapolis: Indiana University Press, 1995), 42–71.
11. Quoted in Walter Benjamin, *Das Passagen-Werk* (Frankfurt-am-Main: Suhrkamp, 1982), 248.
12. Quoted ibid., 256. (Every industry exposing its trophies / at that bazaar of general progress / seems to have used the fairy's magic wand / to enrich the Crystal Palace).
13. Ibid., 700 (the German expression Benjamin uses is '*Feengrotten*').
14. See Benjamin's letters to Scholem, 30 January 1928, and to Adorno, 31 May 1935 in Walter Benjamin, *Briefe* (Frankfurt-am-Main: Suhrkamp, 1978), 455 and 663. Susan Buck-Morss, in her important study, *The Dialectics of Seeing: Walter Benjamin and the Arcades Project* (Cambridge, Mass., London: The MIT Press, 1989), misreads the German expression as '*dialektische Feen*' [dialectical fairies], but translates, correctly, as 'dialectical fairy scene'. She does not, however, refer to *féerie* as a stage or film genre.
15. Jules Trousset, *Les Merveilles de l'Exposition de 1900* (Paris: Montgrédien, 1899), II, quoted in Emmanuelle Toulet, 'Le cinéma à l'Exposition Universelle de 1900', *Revue d'histoire moderne et contemporaine* 33 (avril–juin 1986): 180. In the English version of this article in *Persistence of Vision*, 9 (1991), Tom Gunning translates '*fée*' by 'sprite'.
16. When Benjamin in an early phase wanted to call the *Passagenarbeit* a 'dialectical *féerie*', his intention may have been to transform 'a world of mystical enchantment into one of both metaphysical and political illumination', to quote a wonderful phrasing by Susan Buck-Morss, op. cit., 23 (where it appears in a somewhat different, yet related context).
17. Siegfried Kracauer, *Jacques Offenbach und das Paris seiner Zeit* (Frankfurt-am-Main: Suhrkamp, 1976), 323.
18. Ginisty, op. cit., 214.
19. Ibid., 215.
20. Merrit Crawford, 'Georges Méliès, le Jules Verne du cinéma', *Le Nouvel Art Cinématographique*

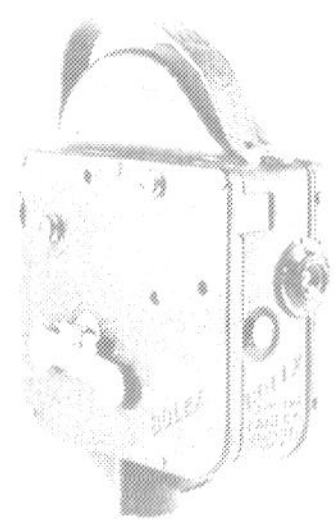

(2ème série), 6 (avril 1930): 36–46; Lo Duca, 'Méliès, père du cinéma, fils de Jules Verne', *Cahiers du Cinéma*, 10 (mars 1952): 52–53.

21. See Guy Gauthier, 'Von Jules Verne zu Méliès oder Von der Gravur zur Leinwand', *KINtop. Jahrbuch zur Erforschung des frühen Films*, 2 (1993): 53–58.
22. Ibid., 53.
23. See the catalogue description in Malthête, op. cit., 228.
24. And again this example shows the intimate relation between the *féerie* and innovation on the level of trick techniques.
25. I have seen a 35mm print of this film (which was screened during the 1996 Domitor conference at the Cinémathèque française) and a 16mm print held by the Cinémathèque française (CF 5363 PMU7, 134 m) which I saw on a viewing table. Neither of the prints is complete. In the 35mm print the ending misses, whereas the 16mm print does not feature a very interesting scene where many of the Pathé actors come out of a cinema in a cameo-like appearance. I should like to thank Dominique Païni and the Cinémathèque française, and specifically Claudine Kaufman and Rodolphe Lussiana who made the 16mm print available to me.
26. For a description, see the Pathé catalogue of 1907, 234–236.

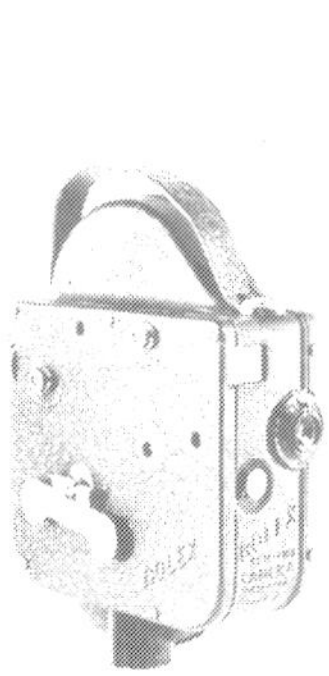

Seeing in the Dark: Early X-ray Imaging and Cinema

Solveig Jülich
Department of Technology and Social Change, Linköping University, Sweden

The phenomenon of temporary blindness to dim light after exposure to intense illumination is an experience we have come to accept as, in one way or another, common to all people. Yet, despite the familiarity and apparent simplicity of dark adaption, physiologists of vision have shown it to be a highly complex affair. And what is more interesting, the uses and meanings of scotopic vision are so rich that it could well be said to have a cultural history of its own. This paper focuses on dark adaption as a site for early X-ray imaging and cinema.

Both the science of radiology and popular cinema were 'born' in 1895. That year the German physicist Wilhelm Conrad Röntgen discovered the 'new light' and the Lumière brothers gave their first public demonstration of the *cinématographe*. However, historians have not paid much attention to this coincidence. Rather, they have tended to treat the careers of these two technologies as separate stories. But as recent publications have begun to show, the histories of X-ray imaging and cinema touch upon one another in a number of concrete ways. For instance, as early as 1897, the Scottish physician John Macintyre took a series of X-ray photographs of the movements of a frog's leg, and combined them to make a short animated filmstrip which was screened for a group of scientific and lay viewers. Subsequently, X-ray motion pictures of the physiology of the heart, the stomach, and the lungs were produced. This topic has been explored by Lisa Cartwright in her pioneering study of the development of science film (Fig. 1).[1]

My article provides an account of how seeing in the dark became a source of both pleasure and anxiety in the emerging institutions of radiology and cinema in Sweden. First, I show how both radiology and the cinema drew upon the conventions and practices of the darkened auditorium, common to the worlds of theatre and popular illusion, in order to produce new kinds of visual experience since at the turn of the century, cinematographic images as well as moving X-ray shadows were displayed to audiences in unlit rooms at exhibitions and fairgrounds. Secondly, I argue that the issue of scotopic vision in radiology as well as in the cinema was raised in relation to a concern for the dilemma of the subjectivity of the observer.

Darkened rooms

The darkened auditorium has a long history. Since the Renaissance theoreticians of the theatre knew that the more brightly a picture is lit and the darker the position from which it is viewed, the more distinct it appears. This insight, however, had no practical effect on the Renaissance and Baroque theatre, as the social function of the theatre continued to demand an illuminated auditorium. The fact that the auditorium gradually got darker as the stage grew lighter throughout the eighteenth century can be interpreted as a change in the social, aesthetic and moral role of the theatre. The audience that assembled in the auditorium now concentrated all its attention on the drama on the stage.[2]

In the eighteenth century there also arose a popular entertainment industry for produc-

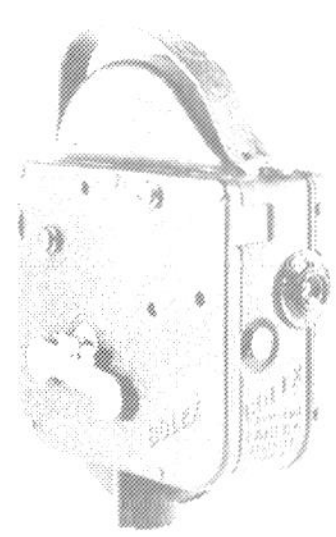

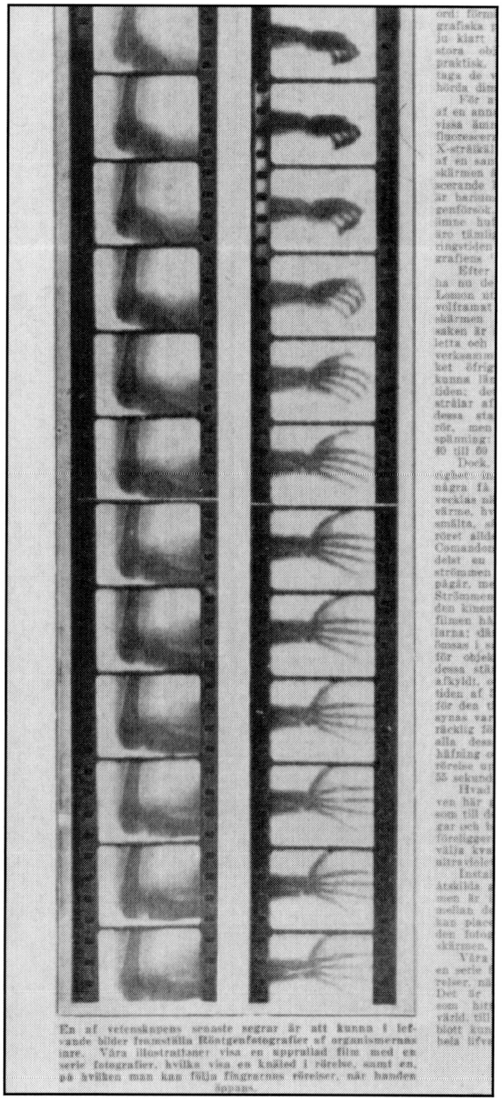

Fig. 1. In the 1910s moving X-ray pictures of the human body were presented as 'one of the latest victories of science'. [Source: Allers Familj-Journal *36, no. 9 (1912).]*

ing images meant to be seen in the dark: *vues d'optique*, shadow-plays, laterna magica, and *théâtre mécanique*.[3] Around 1800 several new media developed in which light was the most important component in the production of illusion. Especially, the panorama and the diorama were great public successes. Although they functioned differently they had one major thing in common: the viewer sat in the dark and watched an illuminated scene.[4]

Traditional and spectacular new forms of popular entertainment came together at the World Expositions of the middle and late nineteenth century. As Tom Gunning has pointed out, these events provide 'one of the richest instances of the visual and technological culture that emerged in industrialised countries' at this time.[5] The Scandinavian Art and Industry Exhibition in Stockholm in 1897, modelled on national and international forerunners, served as another site where not only the products of modernity were marketed, but also modern forms of spectatorship within the context of a new consumerist society.[6]

The artificial staging of light and darkness played a central role in the Stockholm Exhibition. Electricity was one of the major attractions. The illumination of the walks by electric light served as a manifestation of the power of the new energy.[7] And electricity was, of course, the vital ingredient for another popular event at the exhibition: the demonstration of X-rays. The experiments took place in a tower room of one of the historically reconstructed buildings of 'Old Stockholm'. On the dome, the letter 'X' surrounded by thunderbolts had been set up.[8] Just around the corner, the Lumière *cinématographe* had been installed in a modest room on the ground-floor. A journalist covering the exhibition reported on the screenings. The audience was seated, and when the film finally started, the light went out and their attention was drawn to moving shadows on a white screen.[9]

The popular demonstrations of X-rays were performed by the physician Thor Stenbeck and his assistant. Stenbeck also participated in a medical section of the exhibition located in the Nordiska museet (the national museum of cultural history). A number of X-ray photographs taken by him were displayed to illustrate the progress of science and technology. Among these were several of patients who had been shot in the head. The bullets were said to have been successfully removed during surgery.[10] These were static images that had been taken through exposing a photographic plate to X-ray light. No camera was needed. In contrast, the technique Stenbeck used in 'Old Stockholm', fluoroscopy, was used to produce moving X-ray pictures. A screen covered with barium platinocyanide, a kind of platinosalt, fluoresced if X-rays were present. An opaque substance placed between

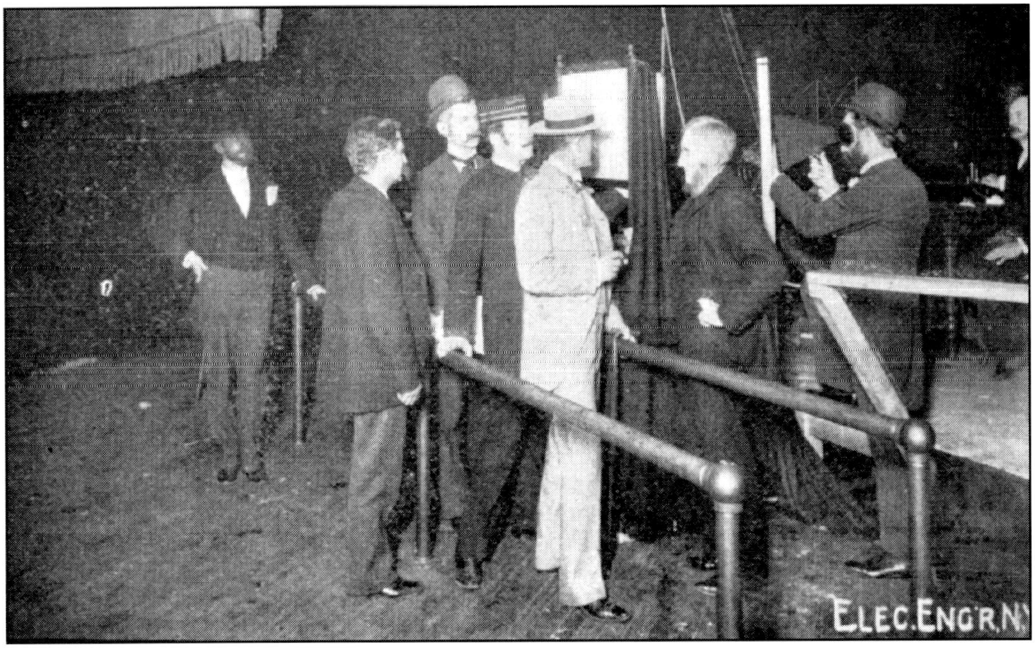

Fig. 2. Fluoroscopic demonstrations were a common international phenomenon at the turn of the century. One of the most spectacular and well-attended was Edison's X-ray exhibition in New York, May 1896. Visitors entered a darkened room and were instructed to examine the anatomy of their own hands on the interactive screen. [Source: The Electrical Engineer *21, no. 422 (1896).]*

the evacuated tube and the fluorescent screen produced a dark shadow. There was no permanent image left on the screen. As soon as the generation of X-rays ceased, the fluorescent material stopped glowing and the screen went dark.[11] Stenbeck's demonstrations attracted a lot of visitors. They could experience how the bones of their own hands appeared like dark shadows within the lighter shadow of the surrounding flesh. Darkness was, of course, a crucial condition for seeing the images on the screen (Fig. 2).[12]

The Stockholm Exhibition was not the first occasion when X-rays and cinema were demonstrated at the same location in Sweden. As has been well documented by film historians, when the first motion picture screening took place it was at the industry exhibition in Malmö 1896. During the summer of that year, Harald Limkilde, from Copenhagen, projected his series of films in a theatre which had been temporarily constructed at a fairground.[13] Less known, however, is that the newly discovered X-rays were also an attraction at the exhibition. Ole Olsen, another Danish showman who earlier had toured in Sweden with a troupe of African tribes-people, had invested in an X-ray apparatus on a trip to Germany, and brought it with him to Malmö. Olsen later became film producer and founder of Nordisk Films Kompagni, an international motion picture company with stars like Asta Nielsen.[14]

Judging from reports in the daily press, Ole Olsen had no further ambition with his X-ray installation than attracting an audience to the Malmö exhibition's fairground.[15] In contrast, Stenbeck also wished visitors to receive some educational instruction. Before the demonstrations started he always held a short lecture about the history and nature of X-rays, and talked about their applications in medicine and customs examination.[16] This was wholly in line with the educational aims of the planners of the Stockholm Exhibition. Their expressed ambition was to design the events so as to educate and create curiosity. This programme had a moral component.[17] But at the same time there was an ambiguity in Stenbeck's demonstrations. Though the visitors were provided with a scientific explanation for the phenomenon, he skilfully manipulated their senses to manufacture an astounding and bewildering experience. Darkness was a necessary condition for seeing the shadow pictures, but it was also a highly suggestive medium for creating a spectacular and lucrative show. And even if the event in itself was designed primarily to educate, the effect of the visit was perhaps not so much to

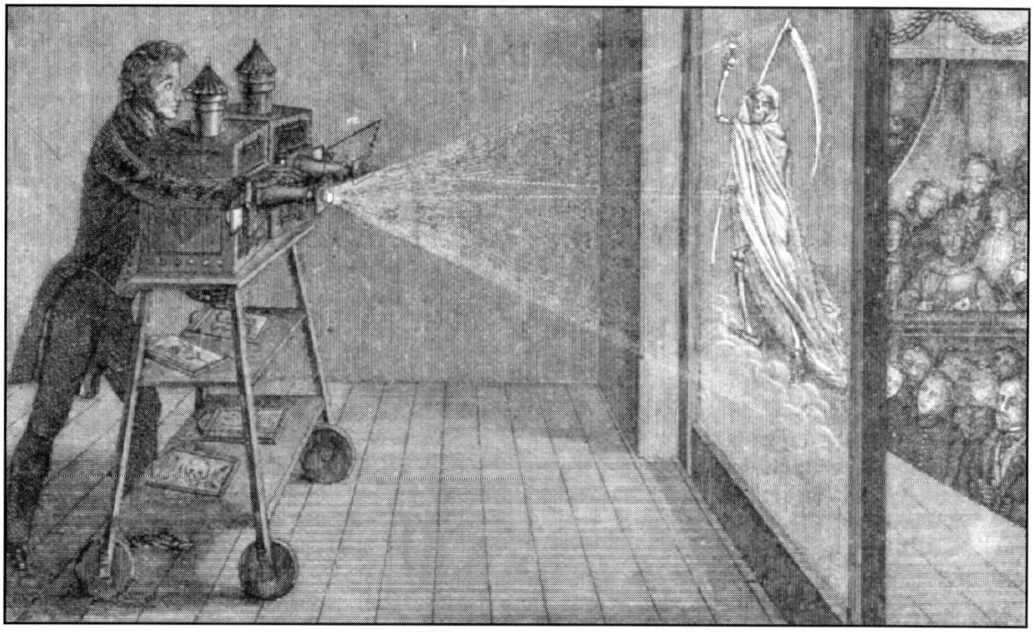

Fig. 3. Fantasmagorical shows displaying ghostly pictures of skeletons to audiences seated in darkened auditoriums can be seen as a nineteenth century forerunner to the popular X-ray demonstrations of human anatomy that took place at the turn of the century. [Source: Svenska Familj-Journalen 6 *(1867).]*

produce knowledge but, rather, to shock and dazzle. Visitors went along to be thrilled and entertained by a new kind of visual spectacle.[18]

This tension between education and visual entertainment was likewise apparent at a popular demonstration held by Stenbeck at the Royal Academy of Sciences in November 1896. The tickets had been sold out long in advance, but the entrance hall to the lecture room was still crowded with people wanting to get in. At the start of the evening, Stenbeck stood on the platform facing the audience and with his back towards a long table filled with different scientific instruments, 'electricity machines', glass tubes, mirrors and clocks. As usual he first lectured on the physics of X-rays, but according to an account in the daily press, not many in the audience paid much attention to his speech. Then, suddenly, the auditorium went pitch black and the spectators were alerted that the demonstration had begun. On the table the electricity machine made humming sounds. A yellow-green light was emitted from the fluorescent screen. One of the assistants placed his head behind the screen, but when no one in the back row could see his cranium, Stenbeck had to assure them that it was clearly visible if inspected from closer range. After this item it seems as if the light was scheduled to be turned on again. But, apparently, someone who had been assigned to manage the gaslight chandelier failed to fulfil his task. Or was this part of the programme? The physician was invisible to everybody, but his voice was heard saying that, in reality, the eye's ability to adapt to the dark was very useful since this made it easier for the audience to see X-rays. Trickery or not, spectators were reported to have been delighted and enthusiastic over the arrangement that let them experience such 'magic'.[19]

As Tom Gunning has pointed out in his essay on the cinema of attractions, both visitors to the first film screenings and demonstrations of X-rays came primarily to witness the marvels of technology. The machine was an attraction in itself. The film lecturer, who presented the views to the audience, focused attention on the novelty of the event and heightened viewer curiosity.[20] This role of mediator between the audience and the unique event is similar to Stenbeck's performance. The film lecturer continually reminded the spectator about the nature of their visual experience. Gunning relates early cinema's particular aesthetics of display, of showmanship, to the tradition of the magic theatre.[21] Early X-ray demonstrations, too, borrowed

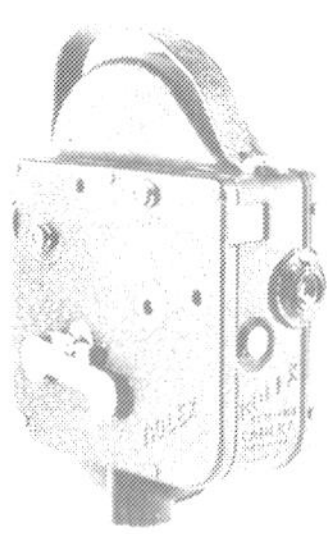

from existing forms of popular entertainment (Fig. 3).

Similar to the 'myth' of early cinema's agitated audience, the traditional account of popular X-ray demonstrations describes the first spectators as horrified by the view of the interior of the body.[22] But though some of them definitively expressed fear and revulsion there was not really 'terror in the aisles'. X-rays were often associated with death. For instance, one of the observers at the Stockholm Exhibition described how X-rays transformed a young, beautiful woman into 'a ghastly image of a skull that moves the jaw, nods, and so on'. Science in this form, he lamented, exposed viewers to the 'most uncanny picture of the decay of everthing'.[23] But, similar to the satisfaction supplied by the cinema of attractions, there was also an element of pleasure in viewing the repulsive and macabre. This is evident in many descriptions from the Stockholm Exhibition.[24] The attraction of the repulsive was, then, rationalised by Stenbeck by appealing to intellectual curiosity.

As recent studies have pointed out, theatrical forms of public science are nothing new. Skilfully manipulated demonstrations have long been used to communicate laboratory-made knowledge. But these demonstrations have often been surrounded by controversy. Lecturers have been accused of deceiving the audience by cheap tricks.[25] Stenbeck may have reflected on the dangers of associating oneself with dramatic spectacle. By locating the experiments in a reconstruction of medieval Stockholm, he voluntarily invoked the image of wizards and alchemists playing with supernatural powers in dark chambers. Initially, this may have been thought to be a successful way of promoting the new science, but soon it became more important to present an image of the radiologist devoted to work that emphasised precision and exactitude. If darkness was a prerequisite for seeing X-ray images, at least it had to be shown that the radiologists were in full control and that they had nothing to hide. In connection with planning the new radiological department of the Serafimerlasarettet, the university hospital in Stockholm, its head, Gösta Forssell, initiated a series of experiments on the problems of dark adaption during fluoroscopic examination.[26]

The tension between education and entertainment was also evident in the attitudes to screenings of film at the Stockholm Exhibition and elsewhere. What happened in the dark? Though the critics emphasised the possibilities of using the medium for educational and moral purposes, they more often complained about cheap amusement and depravation. The suggestive powers of moving pictures were considered to be harmful to spectators, especially young audiences. These critics were successful in their agitation for laws restricting the exhibition of films. In 1911 a central government authority for the censorship of films was established in Stockholm. One of the first actions of this new institution was to conduct experiments on dark adaption, an issue to which I shall return.[27]

Dark adaption and the creation of a new type of scientific observer

So far, I have mainly discussed the pleasures spectators found in the new visual attractions that were introduced in the public sphere at the turn of the century. But I am also interested in the ways in which subjectivity was shaped in scientific culture. As a number of recent studies have indicated, there was a new concern with the skills and discipline of the scientific observer in the nineteenth century. According to Jonathan Crary, beginning in the 1820s there was 'a complex remaking of the individual as observer into something calculable and regularisable and of human vision into something measurable and thus exchangeable'.[28] Vision became formalised and abstract and the human observer became increasingly part of a technical apparatus. Simon Schaffer has traced systematic attempts to discipline laboratory personnel up to the beginning of the nineteenth century. He has studied the use of the 'personal equation' to standardise and mechanise observations by those hired as astronomical assistants in the period.[29] Finally, Lorraine Daston and Peter Galison view the adoption of what they call 'mechanical objectivity' as an expression of what they claim was the growing fear of subjectivity in the natural

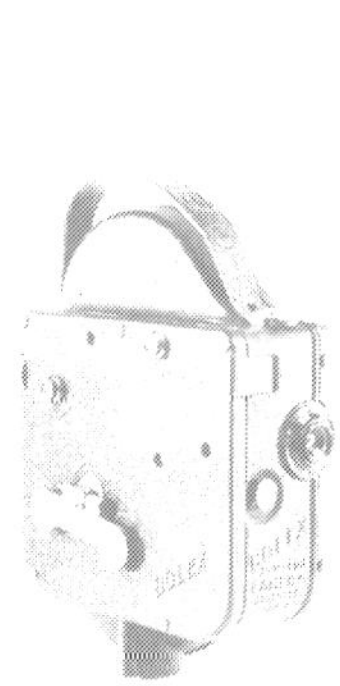

sciences, and of the moral necessity of 'censuring some aspects of the personal'.[30] One solution to the problem of dark adaption in radiology and cinema, I will argue, was to discipline the body of the observer and to rely on prosthetic vision. At the same time, another representational strategy was employed which depended more on the art of judgment than moral self-restraint and technical support.[31]

The phenomenon of dark adaption had been discussed in antiquity, but it was not until the nineteenth century that it became an object for scientific experimentation. The concept of 'adaption' was first introduced and distinguished from accommodation by Herman Aubert, who together with Helmholtz and Hering was one of the prominent figures in physiological optics at the time. In *Physiologie der Netzhaut* (1865), he wanted to reserve the concept of adaption to describe the adjustment of the eye to the intensity of light, whereas accommodation should refer to the adjustment of the eye to focus on objects at different distances. Aubert undertook a series of experiments with himself as a test subject. He systematically plotted the time courses of dark adaption, calibrating the intensity of the light source with an instrument of his own invention, the episcotister. In this way he determined the 'absolute threshold' for illumination after periods of dark adaption that ranged from no time at all up to more than two hours. The resultant curve showed that there was more adaption in the first five minutes than in the next 115 minutes, and that adaption was still not complete after two hours. These experiments became a model for later investigators of dark adaption such as W. Nagel and H. Piper.[32]

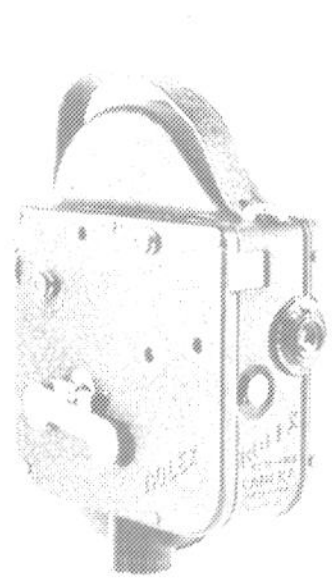

Aubert's work was important for the formulation of what later became known as the duplicity theory. In *Physiologie der Netzhaut*, Aubert demonstrated that there were differences in function between the receptor's mechanisms in the eye. Soon anatomists related these differences in function to the distribution of rods and cones in the retina. This concept was developed particularly by the physiologist Johannes von Kries. According to him, the cones were colour-sensitive and used in high illumination. The rods were achromatic and used in low illumination. Because the rods existed primarily in the periphery of the central fovea, which he believed was the area of clearest vision, the perception of details was poor. Seeing with the cones was later called photopic vision, while seeing with the rods was referred to as scotopic vision. Though the duplicity theory was considered a progress when it was made public at the beginning of the 1890s, it was not generally accepted until the second decade of the twentieth century.[33]

At the turn of the century, the scientific literature on quantitative investigations of dark adaption rose substantially.[34] The German physicist Otto Lummer commented on this tendency in a lecture that later was reviewed in a Swedish journal. According to him, research on 'the ability to see in the dark' had been intensified through the great scientific discoveries of the time: X-rays and radioactivity.[35]

Radiology and the problem of scotopic vision

While Lummer's interest in scotopic vision was mainly theoretical, it posed an acute problem to early X-ray workers. Many physicians testified to difficulties in observing fluoroscopic images. It was hard or almost impossible to discern details in the moving picture. Intuitively or by empirical observations, these workers soon found that the shadows on the screen were easier to perceive if they first made the room as dark as possible. In spite of this early empirical insight about the importance of dark adaption in fluoroscopic examination, it took several decades before radiologists began to look for physiological explanations of the phenomenon.[36] An exception to the rule was the experimental study made by the French physician, Antoine Béclère. In 1899 he published the first systematic attempt to explain the visual problems encountered in fluoroscopy in relation to the function of the human eye. By explicitly posing the problem in terms of physiology rather than physics he opened up questions about the subjectivity of the observer. Exactly how long did radiologists have to stay in the dark before they started an examination? How did one person's ability to perceive shapes in the dark differ from another? Could one really trust what the radiologist saw in the dark?[37]

Following Béclère, the Swedish physician Gösta Forssell (1876–1950) undertook a series of experiments on scotopic vision during fluoroscopic examinations. Forssell was the first full-time hospital radiologist in Sweden. In 1906 he had been appointed director of the X-ray department of the Serafimerlasarettet, the clinical hospital of the medical university in Stockholm, the Karolinska Institute. It was here, a couple of years later, in a small room used as a diagnostic and therapeutic laboratory as well as an office, archive, and waiting room that he began his investigations (Fig. 4).[38]

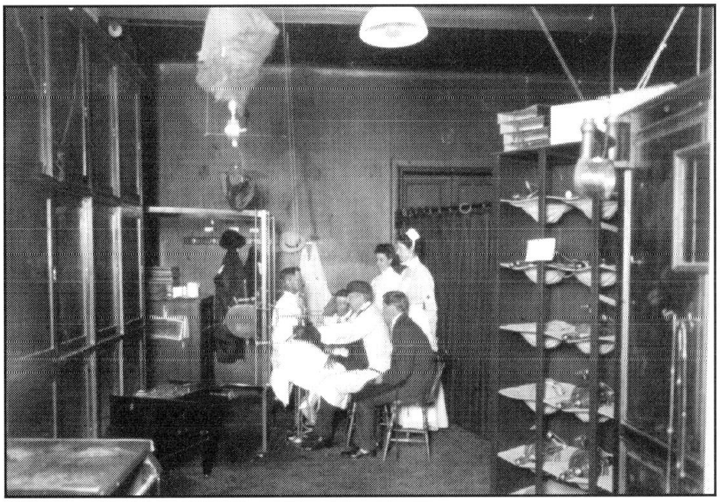

Fig. 4. The physician Gösta Forssell performing a fluoroscopic examination in the X-ray laboratory at Serafimerlasarettet circa 1911. The windows are covered to exclude light and enable the observer to see the images on the screen more clearly. [Source: Museum of Medical History, Stockholm.]

To conduct his experiments Forssell invented crude instruments. Two of them were devices to exclude either cone vision or rod vision, the former (cone or photopic vision) being considered synonymous with central vision. Forssell also used the term 'direct' vision since he thought this kind of seeing was responsible for the perception of fine details. On the other hand, rod vision or scotopic vision was responsible for visual perception in the dark. Forssell often describes dark adaption as peripheric vision or indirect vision.

Thus to exclude peripheric vision, Forssell used a device consisting of paper tubes in different dimensions. Looking through one of these tubes, only direct vision was possible. To exclude central vision, he put in round plates of lead in a pair of empty spectacle frames. This arrangement initially caused him some trouble since he had to fix his eyes on the plate and then keep them in that position. After a while he learnt that movement of the direction of sight had to be accomplished through moving the head. Wearing one of these devices, and without it, he noted what he could see on the fluoroscopic screen. The screen had been prepared in different ways. One arrangement was to stretch black threads on a rectangular metallic frame and then attach this frame to the screen. Another method was to cut out letters from the Monoyer or Snellen chart (the historical models for the still familiar ophthalmological test for determining visual acuity), and fix them to a glass plate which was placed in front of the screen. Which letters was he able to read? In some tests he placed a skeleton hand behind the screen, in others he observed the movements of the living hand, the lungs, and stomach of patients. With these arrangements he set out to investigate fluoroscopic viewing under different conditions of illumination.

First of all he tried to delineate the specificity of the radiologist's scotopic vision. As an explanation for the decrease of visual acuity, Béclère had stated that primarily the rods were at work in fluoroscopic examination. Forssell offered another interpretation. Vision in fluoroscopic examination is not *Dämmerungssehen* in the proper sense: 'It is true that we work in a dark room and with dark adapted eyes under fluoroscopic examination. But since we look at a surface which shines with an intensity of light, direct vision is possible. In this respect we use the eye's light apparatus as well as its dark apparatus.'[39] Forssell compares the situation with looking at stars, which cannot be seen in daylight, but only at night when the eyes are adapted to the dark.

Thus, both rods and cones are responsible for seeing during fluoroscopic examination. Visual acuity in the retina centre is just as high as in the periphery. To Forssell's surprise, however, there is a significant difference in the ability to perceive and judge details in images. Even when the periphery has the same visual acuity, the retina centre appears

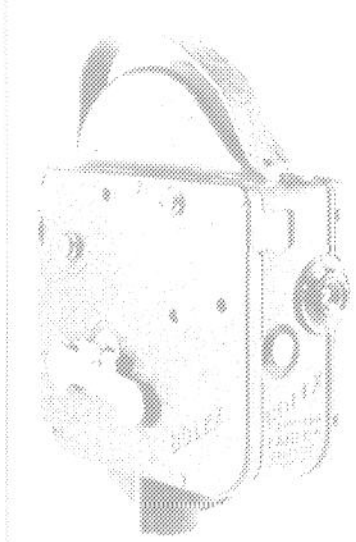

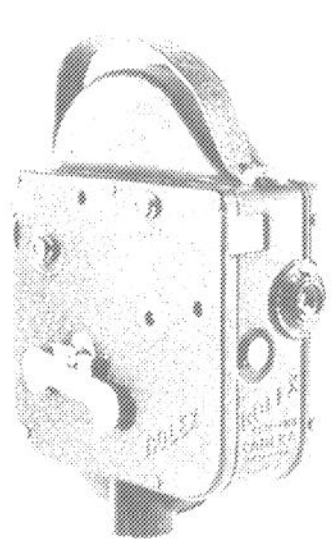

to enable more differentiated and accurate observation. Forssell was not sure how to explain this result. Not finding any description of the phenomenon in the literature, Forssell argued that the cause cannot be found solely in the eye's anatomical condition. Instead, Forssell proposed a psychological explanation wherein the retina periphery is characterised by its lack of endurance. Indirect vision, Forssell concluded, is 'psychically inferior'.

The real problem with fluoroscopic examination in the dark, then, is that peripheric vision creates a sense of fatigue that blurs the impression of the image and makes it difficult to draw firm judgments. The eye's constant fixation on a small surface on the bright screen can be extremely tiring to the observer. Forssell reported that on several occasions he experienced a black spot in the field of vision after a long stay in the laboratory, and once even 'total blindness' for a couple of seconds. Afterwards his eyes and head ached. There are two sides to this issue. On the one hand, the observer is in an extremely vulnerable position since how can we be sure that the radiologist makes accurate observations when his biological make-up is not adapted to the laboratory environment? On the other hand, something can be done since the observer, according to Forssell, can learn how to see in the dark. We are used to our attention being absorbed by whatever direct vision brings to our attention. But since this is partly a question of habit, it is possible to educate the eye, to direct its attention to peripheric impressions during fluoroscopic examination.

Forssell tries to characterise the trained eye of the fluoroscopist through comparison with other visual practices. The trained radiologist does not – as in the case of looking through a microscope with a fixed and steady gaze – obtain a grasp of details. Rather, he must constantly move his eye on the screen at the same time as the vacuum tube (and sometimes also the patient) is displaced. Forssell compares this kind of seeing with looking at a painting. The observer 'adjusts the eye to the screen, almost as when one adjusts the eye to survey a painting, without looking at a particular detail, and learns in this way how to get a general view of the picture and observe larger structures and movements'.[40] But when it comes to making a more differentiated observation, attention is directed towards a detail in the image. Forssell, who had a passion for hunting, compares the ability to make a quick apprehension to 'shooting a bird on the wing'.

I will now consider another series of experiments on dark adaption conducted during the same period but in a completely different setting: Statens biografbyrå, the Swedish Board of Film Censors.

Cinema and the spell of darkness

The legislation introducing censorship stated that no films should be allowed which were contrary to 'general law or good morals' or otherwise unfit for public exhibition. Films which were likely to excite the imagination of children under 15 years of age, or otherwise damage their spiritual development or health, should be forbidden. To avoid the harsh connotations of the term 'censor', the Board's officials, all of whom had campaigned against the corrupting influence of film, were described as controllers or examiners. One of the examiners, the physician Jakob Billström, contributed psychiatric expertise in difficult cases.[41]

In a letter addressed to the ministry for civil service affairs in 1911, the three censors requested the means to purchase 'an adaptometer, an apparatus for testing the condition of vision in the seeing of cinematographic images'. According to the censors, several reasons motivated this kind of investigation. First of all there was a general complaint about eye problems arising from film projection. Secondly, it would supply the Board with greater authority in the public eye if it could be shown to be seeking a more 'objective ground on which to stand'. The request was granted, and preparations for the experiments could begin.[42]

Billström himself conducted the experiments. The practical arrangements seem to have caused more trouble than he had anticipated. First of all, he had to find the adaptometer. An apparatus was constructed, but it took a long time before all the small parts were delivered, and the work was delayed. Another problem was that the screening room at the Board was not built to house experiments. It had to be transformed into a laboratory. An aperture was made in the door between the office and the screening room where Billström installed the adaptometer. Different technical solutions for darkening the room were tried, but dismissed because of expense. Finally, he had to give up the idea, and the experiments were made in the evening during the dark season.

Before an experiment started, all the lights in the room were lit and the test subject focused his or her attention on a chandelier for two minutes. Then the subject sat down on a chair and placed his or her head on a headrest in front of the adaptometer so that the eye was aligned for the apparatus. For an hour, readings were taken on the adaptometer at ten minute intervals and converted into graphs which showed the relative value of the lowest level of illumination that was required to create a perception of light.

The results were perhaps not what Billström had expected. The graphs for the test subjects were very different. Variation in individual response was simply too great to permit any general conclusions. However, Billström still thought his investigations had contributed something essential to the understanding of cinematographic vision. During the experiments he had noted a 'subjective sensation of fatigue'. Though he had not measured this phenomenon with quantitative methods as with dark adaption, he believed it had physiological or psychical causes. He described the experience as 'a sensation of shadiness and dazzlement, and sometimes twitches in the eye lids'. These symptoms had been most acute after seeing films portraying train or car journeys or rocking ships, but also more 'common' images. It seems that intense attention created fatigue.

All the people tested had experienced a 'subjective sensation of fatigue'. Interestingly, Billström adds a reflection on his own role as observer. As a censor, apart from the fatigue associated with viewing cinematographic films, he also suffered from the 'psychologically higher and more strenuous task of judging film'.[43] Once again, questions concerning the grounds upon which images were interpreted and judged are raised. How were the censors themselves influenced? Thus, what started as the censors' moral concern primarily for the younger members of the film audience was later expressed as an anxiety about their own role as observers of cinematographic images.

Conclusion: censuring the personal vs. the art of judgment

Why discuss darkness and dark adaption? As I have already indicated, the darkened rooms required for cinema and radiology appeared as deeply worrying places for individuals engaged in public culture at the turn of the century. Darkness, with its metaphorical connections, was associated with irrationalism, dogmatism, insanitary environments, sexuality, but above all with the powers of suggestion and illusion.[44] Anxiety for dangers lurking in the dark provided a motivation for prevention and intervention.

Although less evident than the fear of the audience spellbound in darkness, there was also, I think, a concern about the public image of scientific expertise. What dark powers were conjured up in the laboratory of the radiologist? How did the minds of the censors look (Fig. 5) after processing all those images of violence, decadence and perversity? For both Forssell and Billström, it was crucial that public respectability and authority were secured in order to establish their new institutions.

Forssell and Billström also shared a concern for securing the grounds for scientific objectivity. This was an important motive for their experiments: to establish parameters for the subjectivity of human vision so as to make experimental observation more reliable and authoritative. As Daston and Galison have shown, the history of objectivity can be told as 'how, why, and when various forms of subjectivity came to be seen as *dangerously* subjective'. They argue that a significantly new conception of objectivity was introduced in the middle of the nineteenth century which linked objective representation with a capacity for discipline and self-restraint on the observer's behalf. Fundamental to this idea

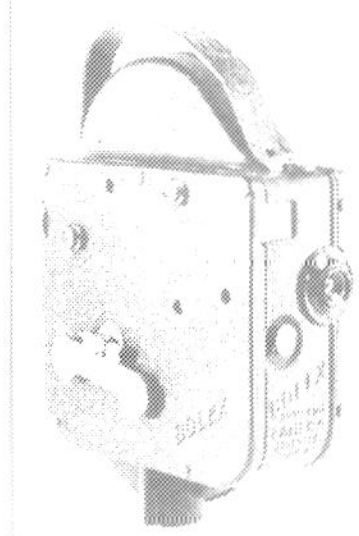

Fig. 5. A caricature by Gustaf Ljunggren portraying the three censors Jacob Billström, Marie Louise Gagner and Walter Fevrell before, and after five years of viewing and judging morally disputable films. [Source: Puck 12 *(December 1912).]*

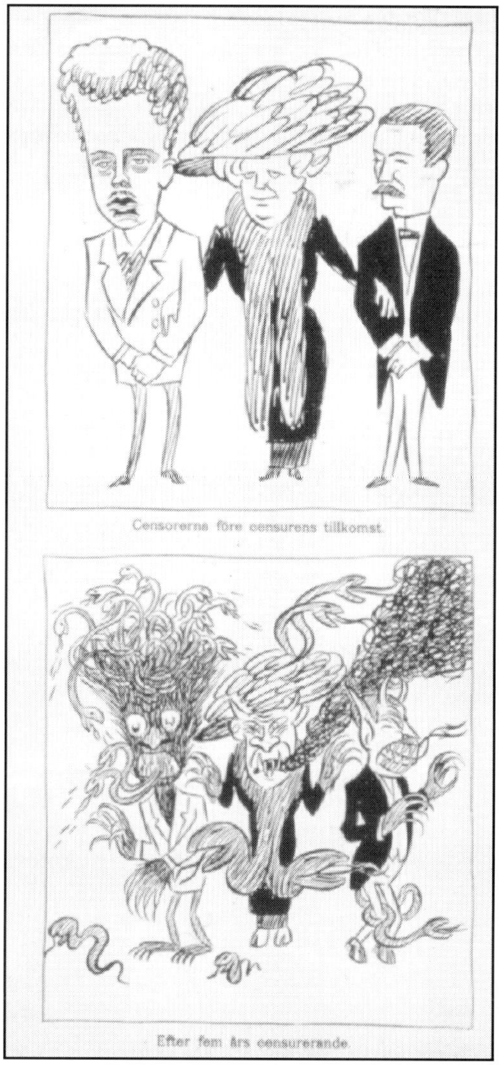

was the machine: the machine as a neutral and transparent mediator of natural phenomena, and as an ideal for the moral discipline of the scientists themselves. The observer was expected to practice a heroic asceticism in order to resist the temptations of judgment, interpretation and artistry.[45]

The ideal of mechanical objectivity can be related to the practices of the two physicians. Both disclose a fear for the tricks played by subjective vision. Illusion and error are seen as not so much a fault of the machine in itself, but arise from the limitations of the human eye. In the case of Forssell, this is manifested in his interest in determining the personal equation of the radiologist workers. He used himself and the other doctors in the laboratory as test subjects to find out individual variation in attention and fatigue during sessions in the dark. Interestingly, Billström too involves his own body in experiments on cinematographic vision. The adaptometer was used to measure the precise course of dark adaption. Similarly, Forssell checked and calibrated the observer's performance against the Snellen chart for visual acuity. Objectivity could only be guaranteed through self-surveillance.

As I have tried to indicate, however, there was more at stake. While Billström never really lost faith in the practices of mechanical objectivity, Forssell's view was more ambivalent. Despite his efforts to minimise and control the effects of subjectivity, he also saw the need for personal judgment and creative interpretation. Rather than give his whole-hearted support to the ideal of mechanical objectivity, he aligned himself with a new representational strategy that emerged in the first decades of the twentieth century. Galison, who recently has made a case for this turn in the history of objectivity, describes it as the event of 'the interpreted image'. The scientist now entered as expert, with the educated eye that could perceive patterns where others saw nothing.[46] To Forssell, as I have pointed out, the issue of education and instruction was crucial. Learning to recognise pathological changes from normal variation, to interpret the mysterious language of 'shadows', making meaning of movements seen in the dark, was all a matter of educating the eye.

In conclusion, Billström and Forssell draw our attention to the complexity of the history of scientific objectivity. While the observer of popular and artistic images stands as a watchdog guarding the values of mechanised science, the observer of scientific images celebrates the art of judgment.

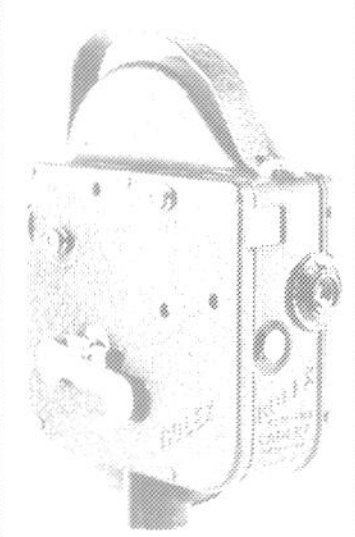

Notes

1. Lisa Cartwright, *Screening the Body: Tracing Medicine's Visual Culture* (Minneapolis: University of Minnesota Press, 1995). For discussion of radiology and the cinema, see chapter 5. For discussion of Macintyre, see especially 20–21.

2. Wolfgang Schivelbusch, *Disenchanted Night: The Industrialisation of Light in the Nineteenth Century* (Oxford: Berg, 1988), 206. A detailed account of the history of lighting techniques is given by Gösta M. Bergman in *Lighting in the Theatre* (Uppsala: Almqvist & Wiksell, 1977).
3. Barbara Maria Stafford, '"Fantastic" Images: From Unenlightening to Enlightening "Appearances" Meant to Be Seen in the Dark', in Frederick Burwick and Walter Pape (eds.), *Aesthetic Illusion: Theoretical and Historical Approaches* (Berlin: Walter de Gruyter, 1990). Also see her *Artful Science: Enlightenment Entertainment and the Eclipse of Visual Education* (Cambridge, Mass.: The MIT Press, 1994).
4. Schivelbusch, *Disenchanted Night*, 213–221.
5. Tom Gunning, 'The World as Object Lesson: Cinema Audiences, Visual Culture and the St. Louis World's Fair, 1904', *Film History* 6 (1994): 422–444, 423.
6. Anders Ekström, *Den utställda världen: Stockholmsutställningen 1897 och 1800-talets världsutställningar* (Stockholm: Nordiska museet, 1994).
7. Ekström, *Den utställda världen*, 190–191.
8. 'Experiment med röntgenstrålar', *Nordens expositionstidning* (20 August 1897).
9. Andreas Hasselgren, *Utställningen i Stockholm 1897* (Stockholm: Fröléen & Comp., 1897), 563–566.
10. Hasselgren, *Utställningen i Stockholm 1897*, 600. E.W. Wretlind, 'Stockholm-utställningen', *Hälsovännen* 12 (1897): 228.
11. The technical differences between X-ray photography (radiography) and fluoroscopy were outlined by Stenbeck in the first Swedish textbook on radiology, *Röntgenstrålarne i medicinens tjänst* (Stockholm: Wahlström & Widstrand, 1900).
12. Hasselgren, *Utställningen i Stockholm 1897*, 566–568.
13. Rune Waldekranz, *Levande fotografier: Film och biograf i Sverige 1896–1906*. Lic. avhandling, (Institutionen för teater- och filmvetenskap, Stockholms universitet, 1969), 35–52. Leif Furhammar, *Filmen i Sverige: En historia i tio kapitel* (Höganäs: Wiken/Filminstitutet, 1991), 12.
14. Ole Olsen, *Filmens eventyr og mit eget* (Copenhagen: Jespersen & Pios, 1940). See especially the chapters 'Karavanen' and 'Røntgen'.
15. 'Nytt från utställningen: Röntgens X-strålar', *Dagens Nyheter* (southern Sweden edition, 13 August 1896).
16. Hasselgren, *Utställningen i Stockholm 1897*, 566–568. 'Röntgenstrålar i Gamla Stockholm', *Göteborgs Handels- och Sjöfartstidning* (24 July 1897).
17. Ekström, *Den utställda världen*, chapter 5.
18. See, for instance, the description of the event in 'Röntgenstrålar i Gamla Stockholm', *Göteborgs Handels- och Sjöfartstidning* (24 July 1897).
19. Stenbeck's demonstration was reviewed in large circulation newspapers, see especially 'Röntgenstrålar i Vetenskapsakademien', *Dagens Nyheter* (20 November 1896).
20. Tom Gunning, 'The Cinema of Attractions: Early Film, Its Spectator and the Avant-Garde', in Thomas Elsaesser (ed.), *Early Cinema: Space, Frame, Narrative* (London: British Film Institute, 1990).
21. Tom Gunning, 'An Aesthetic of Astonishment: Early Film and the (In)Credulous Spectator', *Art & Text* 34 (Spring 1989): 31–45.
22. The standard description of the experience of the first X-ray pictures has been provided by Röntgen's biographer, Otto Glasser. According to him, the wife of Röntgen expressed horror and unease when she saw the first X-ray image of the human body: the skeletal image of her own hand, see Otto Glasser, *Dr. W.C. Röntgen* (Springfield, Illinois: Charles C. Thomas, 1945), 38. When commenting upon the reception of X-rays by the general public, Glasser views testimonies of fear as an expression of 'ignorance and pessimism'. See his article, 'The Discovery of the Roentgen Rays', in Otto Glasser (ed.), *The Science of Radiology* (Springfield, Illinois: Charles C. Thomas, 1933), 8–9. However, as I give a more detailed account in my doctoral thesis, the audience was well-prepared for this new kind of visual experience. In particular, I dwell on the traditions of the 'anatomy lesson', fantasmagorical shows, spiritist photographs, and séances. Placed in this historical context, the audience's experience reveals not ignorance but an awareness of the medium's capabilities.

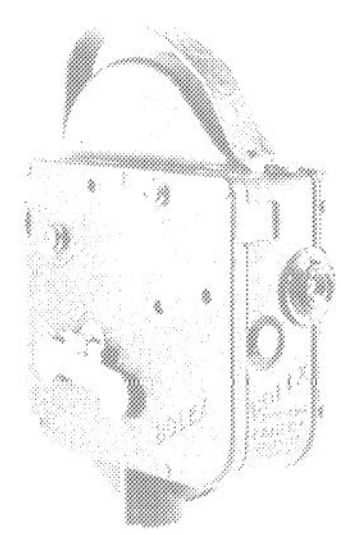

23. '"Undren" på utställningen: Experimenten med röntgenljus', *Stockholmstidningen* (10 August 1897).
24. See, for instance, 'Röntgenstrålar i Gamla Stockholm', *Göteborgs Handels- och Sjöfartstidning* (24 July 1897), and 'Experiment med röntgenstrålar', *Nordens expositionstidning* (20 August 1897).

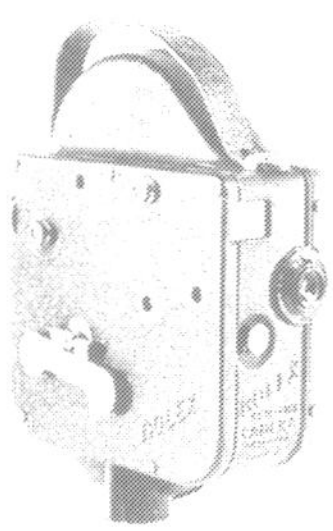

25. I have profited, in particular, from Simon Schaffer, 'Natural Philosophy and Public Spectacle in the Eighteenth Century', *History of Science* 21 (1983): 1–43, and Barbara Maria Stafford, *Artful Science: Enlightenment Entertainment and the Eclipse of Visual Education* (Cambridge, Mass.: The MIT Press, 1994).
26. These experiments are described in an unpublished manuscript by Gösta Forssell, 'Iakttagelser öfver seendet vid röntgengenomlysningar' (1912). Ms 818, Library of Karolinska Institute, Stockholm.
27. For a brief discussion about responses to the film screenings at the Stockholm Exhibition, see Ekström, *Den utställda världen*, 252. Accounts for the background to the establishment of the Swedish Board of Film Censors can be found in Erik Skoglund, *Filmcensuren* (Stockholm: Norstedts, 1971), Elisabeth Liljedahl, *Stumfilmen i Sverige: Kritik och debatt* (Stockholm: Proprius/Filminstitutet, 1975), and more recently in Jan Olsson, 'Svart på vitt: Film, makt och censur', *Aura* 1, 1 (1995): 14–46.
28. Jonathan Crary, *Techniques of the Observer: On Vision and Modernity in the Nineteenth Century* (Cambridge, Mass.: The MIT Press, 1992), 17.
29. Simon Schaffer, 'Astronomers Mark Time: Discipline and the Personal Equation', *Science in Context* 2, 1 (1988): 115–145.
30. Lorraine Daston and Peter Galison, 'The Image of Objectivity', *Representations*, no. 40 (1992): 81–128.
31. For an elaboration of the concept of 'art of judgment', see Peter Galison, 'Judgment Against Objectivity', in Caroline A. Jones and Peter Galison (eds.), *Picturing Science, Producing Art* (New York: Routledge, 1998).
32. Edwin G. Boring, *Sensation and Perception in the History of Experimental Psychology* (New York: D. Appleton-Century, 1942), 160–162, 191.
33. Boring, *Sensation and Perception in the History of Experimental Psychology*, 182, 203–206. The classical paper on the subject is Selig Hecht, 'The Dark Adaption of the Human Eye', in André J. Bruwer (ed.), *Classic Descriptions in Diagnostic Roentgenology*, I, (Springfield, Illinois: Charles C. Thomas, 1964), first published in *Journal of General Physiology* 2 (1920): 499–517.
34. R. Steven Turner, *In the Eye's Mind: Vision and the Helmholtz-Hering Controversy* (Princeton: Princeton University Press, 1994), 214.
35. 'Om seendet i ljus och mörker', *Industritidningen Norden* 32 (1904): 59.
36. William J. Tuddenham, 'Dark Adaption', in André J. Bruwer (ed.), *Classic Descriptions in Diagnostic Roentgenology*, I, (Springfield, Illinois: Charles C. Thomas, 1964), 741–742.
37. Antoine Béclère, 'A Physiological Study of Vision in Fluoroscopic Examinations', in André J. Bruwer (ed.), *Classic Descriptions in Diagnostic Roentgenology*, I, (Springfield, Illinois: Charles C. Thomas, 1964), first published in *Archives d'Electricité Medicale* 7 (1899): 469–489.
38. A biographical account of Forssell is given by Åke Åkerlund, 'Gösta Forssell 1876-1950: To the Memory of His Life and Work', *Acta Radiologica*, Suppl. 131 (1956): 1–50. In the following paragraphs, I refer to Forssell's unpublished manuscript, 'Iakttagelser öfver seendet vid röntgengenomlysningar' (1912).
39. Forssell, 'Iakttagelser öfver seendet vid röntgengenomlysningar' (1912), 39.
40. Ibid., 111.
41. Olsson, 'Svart på vitt'.
42. Riksarkivet (RA), Civildepartementet, huvudarkivet, konseljakter, 26 January 1912, no. 6. The results of the experiments were reported in a letter one-and-a-half years later, see RA, Statens biografbyrå, korrespondens i chefsärenden, brev från Jakob Billström till Civildepartementet, 21 June 1913. Dn 773/13.
43. Ibid.
44. For a discussion of darkness as metaphor in turn-of-the-century culture, see Elaine Showalter, *Sexual Anarchy: Gender and Culture at the Fin-de-Siècle* (London: Bloomsbury, 1992), 5–6.
45. Lorraine Daston and Peter Galison, 'The Image of Objectivity', *Representations*, no. 40 (1992): 82.
46. Peter Galison, 'Judgment Against Objectivity', in Caroline A. Jones and Peter Galison (eds.), *Picturing Science, Producing Art* (New York: Routledge, 1998).

The Bolex Motion Picture Camera

Carlos Bustamante
Institute for Timebased Media, Hochschule der Künste Berlin, Germany

The Paillard-Bolex H16 camera played a central role in the work of many avant-garde filmmakers from the 1940s through to the 1970s because of its precision and light weight, robustness and range of facilities,[1] and the high quality of its optics – especially the zoom lenses – and its simple operation, which made possible an infinite combination of creative cinematographic choices. The Bolex H16 is probably *the* camera which most influenced a generation of experimental and documentary/ethnographic filmmakers. Leafing through old issues of *Film Culture*, I note that the camera was also an important prop, a kind of status symbol, included in the portraits of Maya Deren (Fig. 1), Stan Brakhage (Fig. 2), Bruce Baillie, Jonas Mekas (Fig. 3), Marie Menken, Kenneth Anger, Gregory Markopoulos, James Broughton, Mike and George Kuchar, Robert Breer, and Paul Sharits: some of my *underground-film* heroes. The Bolex H16 (Fig. 4) seems, even nowadays, omnipresent in filmmakers' cupboards, and they still run perfectly, even after forty years of use and thousands of metres of film. I want to trace some of the factors which made the Bolex such a successful film camera.

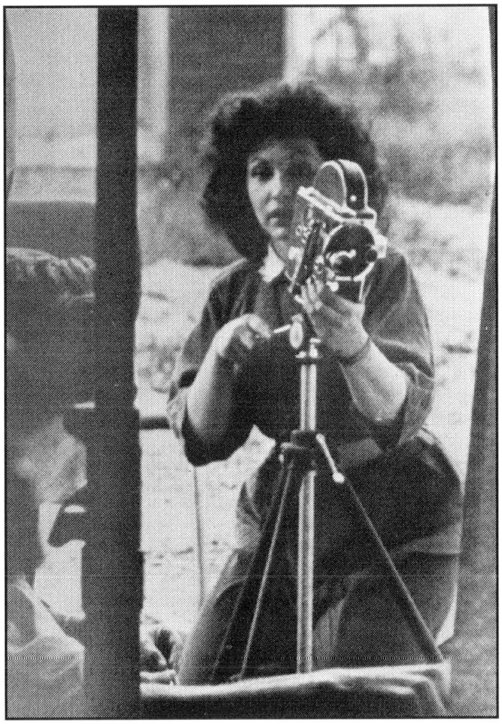

Fig. 1. Maya Deren with Bolex.

> In the year 1814 Ste. Croix was a small idyllic village with fewer than 2000 inhabitants, poor soil, a rough climate, and was far away from all the main thoroughfares. But these same hard conditions forced these mountain residents to fend for themselves, and educated them physically and morally for commercial activity. They were not satisfied with the country life of peasants, which would have only supplied them with meagre earnings, but instead they cultivated their in-born talents for manual dexterity. Back then, in Ste. Croix, the first clocks were manufactured in the residents' homes [*Heimarbeit*], and the brothers Paillard organised their distribution. And so it came to pass that in 1814 the foundation was laid down for the well-known factory which nowadays provides not only the residents of Ste. Croix with work and bread, but draws specialist workers from faraway parts of Switzerland and the French environs.[2]

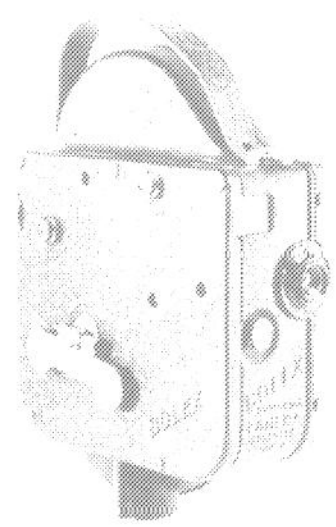

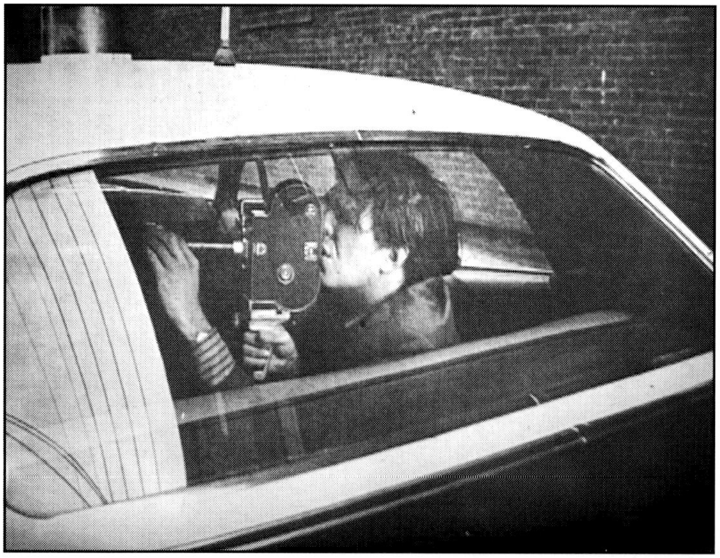

Fig. 2 (above). Stan Brakhage with Bolex.

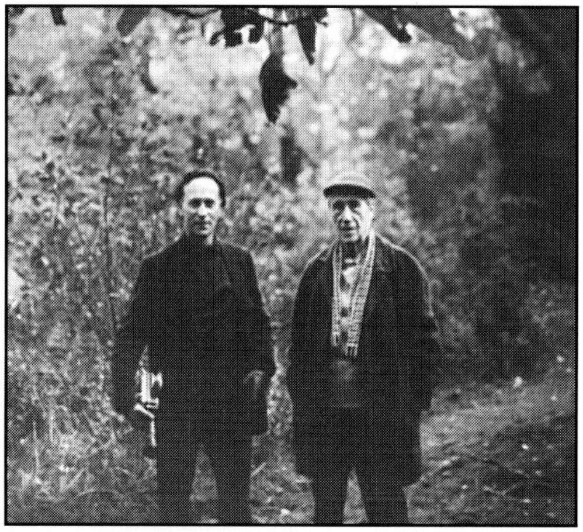

Fig. 3 (upper right). Jonas Mekas with Bolex and Hans Richter.

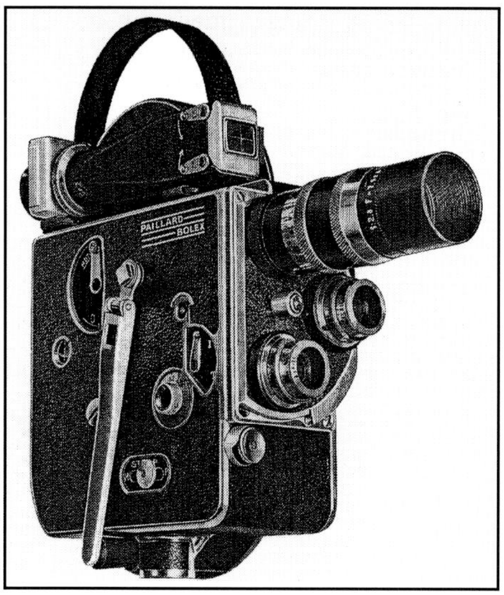

Fig. 4 (lower right). Bolex H16, 1935.

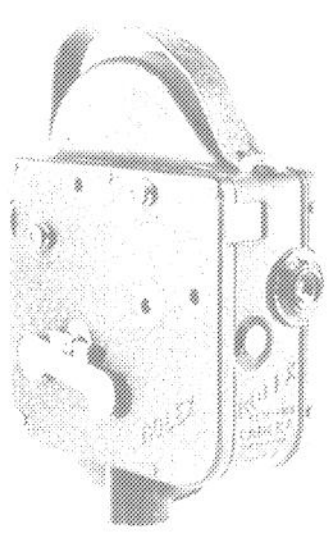

The article 'From the Story of the House of Paillard', which appeared in 1955 in the *Bolex Reporter*, a quarterly magazine published in Munich, does not go into many details about the company, but it does provide a sketch of the growth of the international reputation of their products: clockwork motors, music boxes, music cabinets, etc. The year 1895 was an extremely bad one for Paillard: the company's very existence was threatened by the phonograph. That same year, Ernest Paillard, his son, Albert, and his son-in-law, Eugene Thorens, took over the business and managed to save the company's ground capital. In 1898, they began the manufacture of the 'Echophon', a small cylinder phonograph, and in 1904, the Paillard-Gramophone. In 1920, they sold stocks in order to build a factory for the manufacture of typewriters in Yverdon. In 1935, they developed the 'Hermes Baby', a small portable typewriter, which was sold world-wide. The article concludes that 'because of this experience, the company took up more than twenty-five years ago the manufacture of small motion picture cameras'. Another article from the previous year, entitled 'Die Geburtsstätte der Bolex' ('The Birthplace of the Bolex'), was composed mostly of photos of some of the workers and the instruments they used to test the Bolex during the manufacturing process. These 'exacting tests offer the customer the absolute guarantee that not only is he buying an apparatus which he will use for years, but that he is acquiring an object of value (*Wertobjekt*) upon which he can rely – like a good watch – for decades to come'.[3]

One thing that puzzles me about these two articles as well as some short comments which accompany two pages of photos celebrating '150 Jahre Paillard', [4] or for that matter all of the articles in the *Bolex Reporter* and *Bolex Brief* from 1953 to 1966 available at the library of the Stiftung deutscher Kinematek in Berlin, is that no mention is ever made of the inventor of the Bolex. An article in the January 1973 issue of *film + foto* contains a short biographical sketch of Jacques Bolsey, engineer, inventor, and founder of the world's first amateur film club in 1923. Bolsey was born, apparently, in Kiev, Russia in 1895; his name was Jacques Bogopolsky which he later changed to Boolsky. He studied medicine and fine art in Geneva, Switzerland. In 1919, he invented the 'Cinegraphe Bol' for which he obtained a German patent on 24 March 1924. That same year he formed the company, 'Companie Bol, SA' to manufacture and distribute the 'Cinegraphe Bol'.[5] In England the camera was advertised as 'The Bol Cinegraph' (Fig. 5), and, as a 'cine-camera, projector and printer combined', was capable of taking and projecting both still and moving pictures, enlarging negatives, and printing motion picture positives on standard 35mm film stock.[6] The multipurpose design was not unique; there were other 35mm camera/projector/printer combinations available, for example, the Debrie Sept. 'The Bol Cinegraph' was very compact and, built like a clock out of brass plate, weighed 1.7 kg and could hold 10 metres of film. (The Debrie Sept weighed 1.8 kg and had a capacity of 5 metres.) Jacques Bogopolsky continued to improve his cameras, but the amateur film camera market was limited until two significant developments in motion picture technology became generally available: acetate film, and the 'Ciné Kodak Process'.

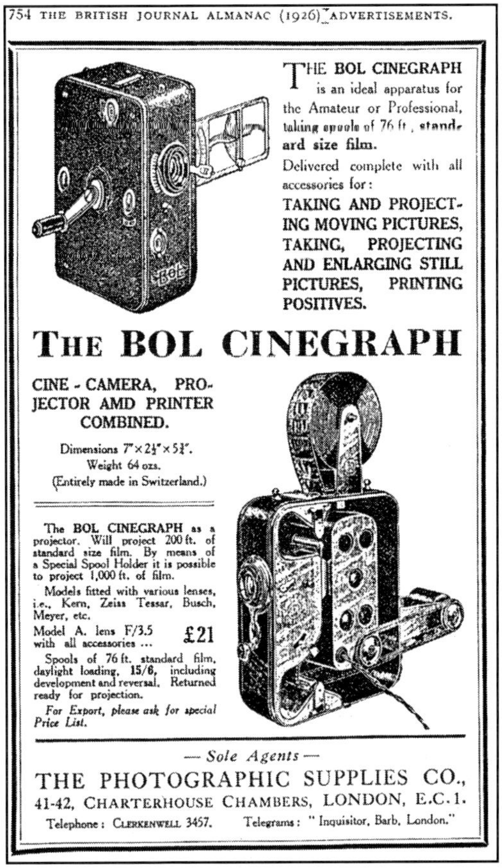
754 THE BRITISH JOURNAL ALMANAC (1926) ADVERTISEMENTS.

THE BOL CINEGRAPH is an ideal apparatus for the Amateur or Professional, taking spools of 76 ft., standard size film.

Delivered complete with all accessories for:

TAKING AND PROJECTING MOVING PICTURES, TAKING, PROJECTING AND ENLARGING STILL PICTURES, PRINTING POSITIVES.

THE BOL CINEGRAPH

CINE - CAMERA, PROJECTOR AMD PRINTER COMBINED.

Dimensions 7"×2½"×5¾".
Weight 64 ozs.
(Entirely made in Switzerland.)

The **BOL CINEGRAPH** as a projector. Will project 200 ft. of standard size film. By means of a Special Spool Holder it is possible to project 1,000 ft. of film.

Models fitted with various lenses, i.e., Kern, Zeiss Tessar, Busch, Meyer, etc.

Model A. lens F/3.5 with all accessories ... £21

Spools of 76 ft. standard film, daylight loading, **15/6**, including development and reversal. Returned ready for projection.

For Export, please ask for special Price List.

— *Sole Agents* —

THE PHOTOGRAPHIC SUPPLIES CO.,
41-42, CHARTERHOUSE CHAMBERS, LONDON, E.C.1.
Telephone: CLERKENWELL 3457. Telegrams: "Inquisitor, Barb. London."

Fig. 5. The Bol Cinegraph, 1926.

The Actien-Gesellschaft für Anilin-Fabrikation (AGFA) in Germany and the Eastman Kodak Company in the USA had been experimenting with acetate filmbases as early as 1909. AGFA, as a matter of fact, had begun manufacture of an acetate film, non-inflammable safety film in November 1909. However, the company encountered strong resistance in Europe due to higher production costs (the price per metre of the safety film was between 3 to 5 pfennings higher than normal nitrate film) and a more serious obstacle: exhibitors could not splice nitrate and acetate film materials together.[7] Both AGFA's and Eastman's sales strategy focused on political lobbying to persuade chiefs of fire departments and local governments to pass laws restricting the use of nitrate film, and to promote the use of acetate filmbase for motion pictures.[8] The First World War delayed these measures.

The growth of the amateur film market had been hindered not only by the dangers of nitrate, but also by the high cost of developing and printing negatives. Probably the lack of a 'standard' gauge for amateurs did not help matters either. J.G. Capstaff of the Research Laboratory at Eastman Kodak began experimenting with reversal processing in 1916. He had noticed that images made by reversal processing to obtain a direct positive image had a finer grain than those developed from a negative. By the latter part of 1922, work on reversal processing had progressed sufficiently for the Kodak management to place the 'Ciné Kodak Process' on the market the following year. The 'Ciné Kodak Process' consisted essentially of the following four steps: development of the film to form a negative; bleaching of the image with acid bichromate or acid permanganate to change the negative image to a soluble salt of silver that can be dissolved out; a full re-exposure of the film to light, and a final development of all of the remaining silver halide to form a positive image. Special equipment and skill were required to develop the film, but it was

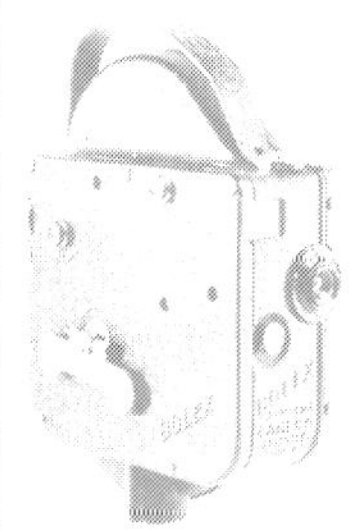

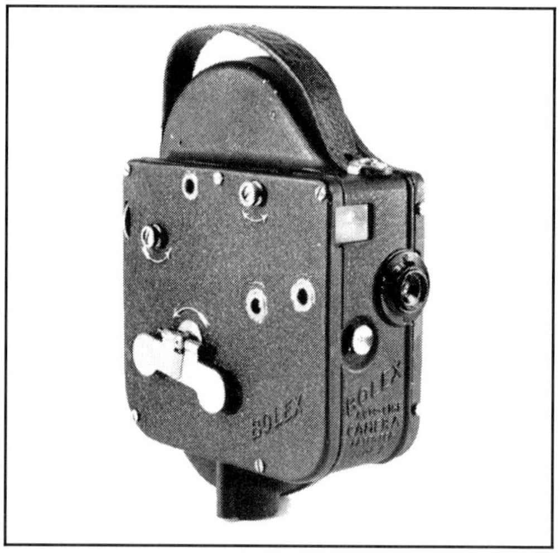

Fig. 6. Bol Auto Cine Camera, 1929.

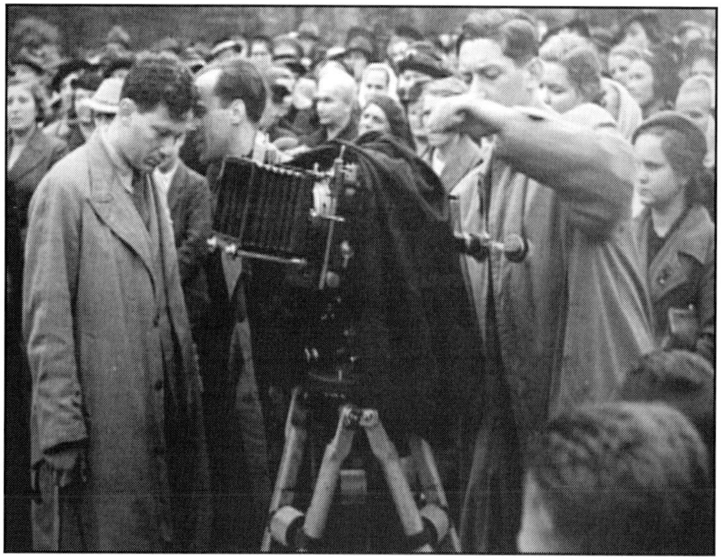

Fig. 7. Alexander Hammid with Debrie Parvo.

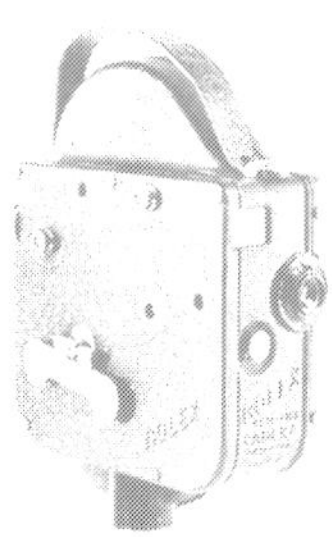

estimated that the new process cut the cost of motion pictures to one sixth that of the negative-positive method.[9] An advertisement for the 'Ciné Kodak Process' which appeared in the *New York Times* on 5 July 1923 announced:

> The Ciné-Kodak for motion pictures. You can now make motion pictures the Kodak way. Ciné-Kodak offers a new simplicity, a new economy, a new compactness. It opens wide the door to personal motion pictures.

Jacques Bogopolsky designed and built the 'Bol Auto Cine Camera', the Bolex, a 16mm clockwork-driven cine camera in 1928 in response to this new market (Fig. 6). However, the stockmarket crash of 1929 affected both the Paillard business as well as Bogopolsky's. He approached the Paillard company in mid-1930, and convinced Paillard to take over the Bolex; an agreement was signed on 29 September 1930.[10] He supplied the designs, the brandname and stayed on as adviser for several years. Paillard manufactured the 'Bol Auto Cine Camera B' (Models I, II, III) until 1935, at which time the Bolex H16 was introduced.

The Bolex H16 retained the basic shape of the 'Bol Auto Cine Camera B' but it was a much more sophisticated motion picture camera. As Andrew Alden observes:

> The body of the H16 was made from cold pressed aluminium alloy with a pressed aluminium side panel covering the film chamber which contained the 30 m (100 ft) daylight loading film spools, the twin sprocket film transport and the automatic film threading assembly. The flat panels of the body were covered in black leathercloth and the exposed edges of the camera were highly polished.
>
> Filming speed of the camera was variable between 8 and 64 f.p.s. plus single frame, the shutter also had an 'I' (instantaneous) and 'T' (time) setting for use during single frame operation. ... A button release was fitted to the front of the camera for continuous running and a slide release was fitted to the side of the camera allowing continuous ('M' = *Marche*) or single frame ('P' = *Pose*) running, an accessory adapter for cable release was available for the side and later the front release.
>
> The most striking feature of the new camera was the D-shaped three lens turret on the front, the lens mount now standardised to take 'c' type screw mount lenses.[11]

In 1938 a variety of lenses was made available, including 'Hektor' from Leitz, 'Cinor' from Som Berthiot, and 'Tessar' and 'Sonnar' from Zeiss, an assortment ranging from 15 mm to 150 mm with f-stops from 1:4 to 1:1.4.[12]

The Bolex H16 Camera was intended for the amateur film market, but this market was not limited to the makers of *home movies*; the marketing strategy for the Bolex H16 emphasised the wide range of uses for the camera, especially under extremely adverse conditions: from the snow-covered Swiss Alps to remote safaris in tropical African villages; from scientific laboratories and industrial complexes to studios with glamorous models. The *Bolex Reporter* however makes no mention of the *avant-garde* or 'poetic' cinema.

Maya Deren and Alexander Hammid bought a Bolex for $400.00 in 1943 in order to shoot *Meshes of the Afternoon* (1943).[13] Hammid, an experienced 'professional' cinematographer, had used the Debrie Parvo extensively in Czechoslovakia (Fig. 7). Dziga Vertov's *Chelovek s kinoapparatom* (The Man with a Movie Camera, 1929) pays hommage to the Debrie Parvo and demonstrates this camera's infinite capabilities. Hammid's choice of the Bolex seems to confirm the camera's reputation, reliability and flexibility, as well as the importance of the 16mm format for personal and experimental films. The quality of the cinematography of *Meshes of the Afternoon* bears witness to Hammid's talent, experience, and understanding of the Bolex. Perhaps one of the most effective (if obvious) uses is that the camera can be hand-held with one hand. There are many different ways in which the hand-held camera is used in *Meshes of the Afternoon*. Amongst the most striking is the point-of-view shot which shows a right hand moving quickly towards the phonograph player, lifting the needle off the record, and stopping the machine; or, the magnificent shots of Maya Deren in which she seems to float/struggle against the ceiling of the staircase.

The precision of the Bolex can be witnessed in the matte shots which show *two* Maya Derens. However, Hammid revealed in an interview with Miriam Arsham that one of the matte shots (shot 59, Fig. 8)[14] was not perfect, 'In one shot, if you look very carefully, you see a little jiggle. But only in one of them, the first one.'[15] The three other matte shots (shots 102 (Fig. 9), 106 (Fig. 10), 112 (Fig. 11)) in *Meshes of the Afternoon* are perfect, which is amazing for a camera without a registration pin. The hand crank and frame counter made

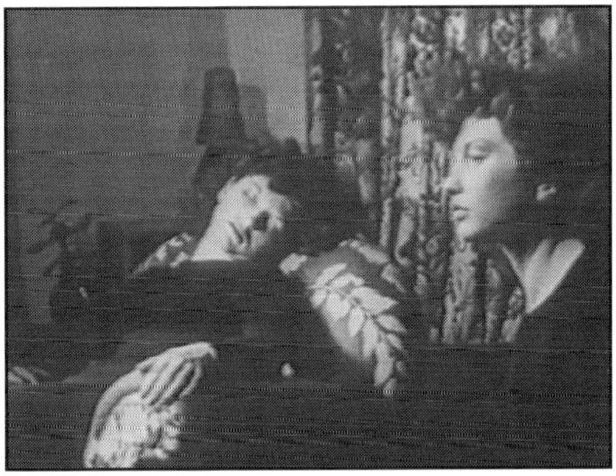

Fig. 8. Meshes of the Afternoon, *shot 59.*

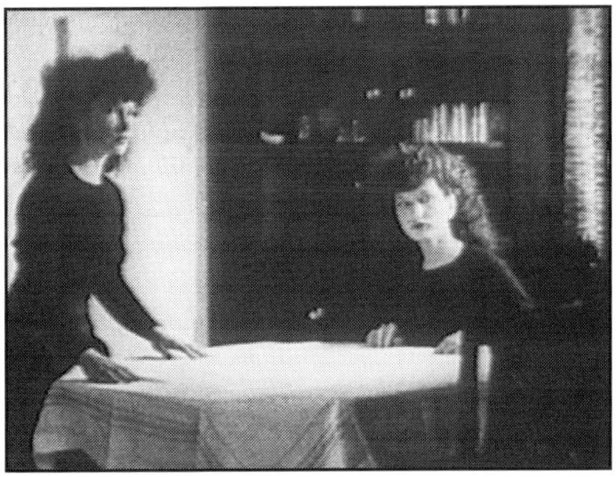

Fig. 9. Meshes of the Afternoon, *shot 102.*

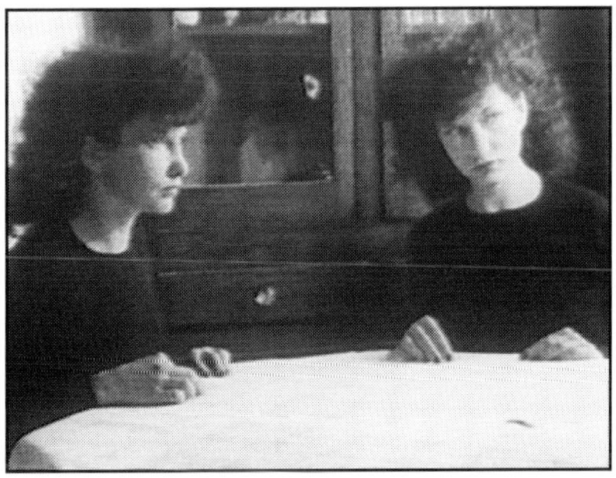

Fig. 10. Meshes of the Afternoon, *shot 106.*

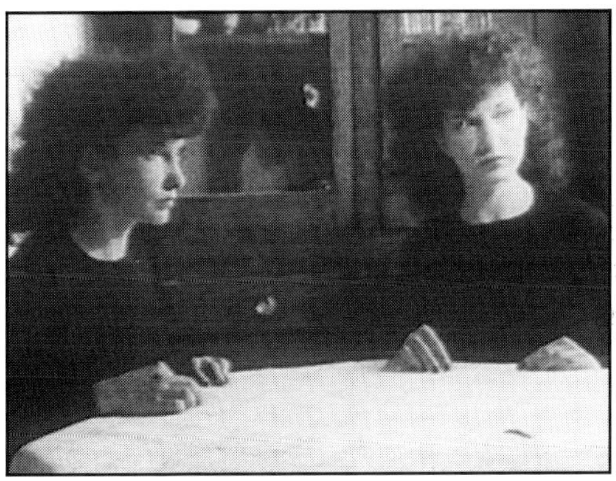

Fig. 11. Meshes of the Afternoon, *shot 112.*

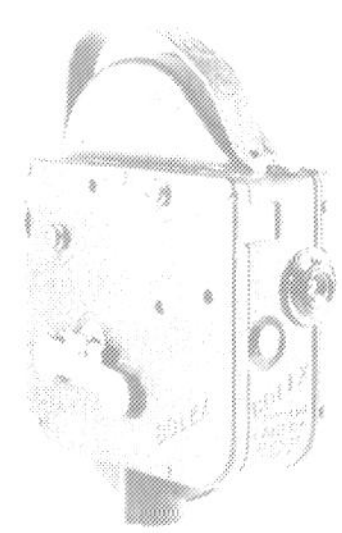

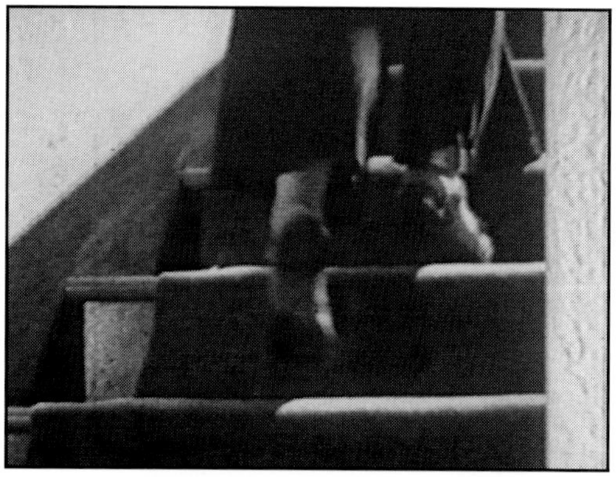

Fig. 12. Meshes of the Afternoon, *shot 40.*

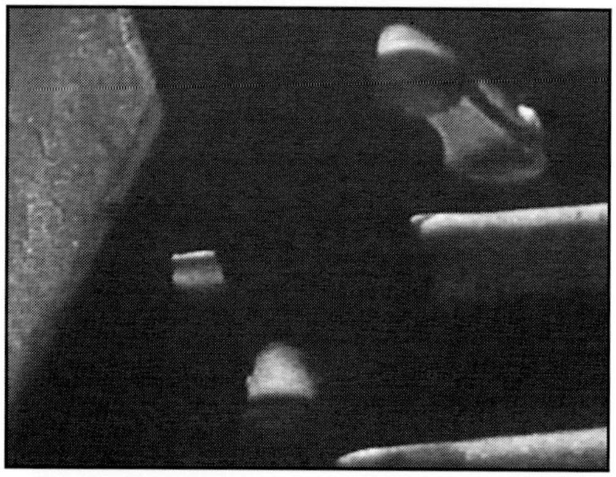

Fig. 13. Meshes of the Afternoon, *shot 41.*

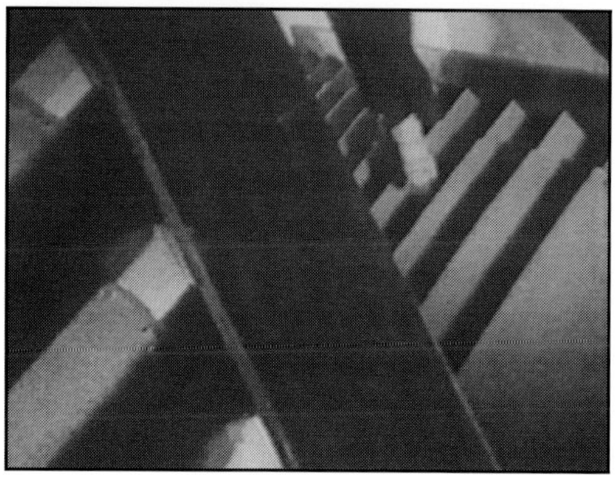

Fig. 14. Meshes of the Afternoon, *shot 42.*

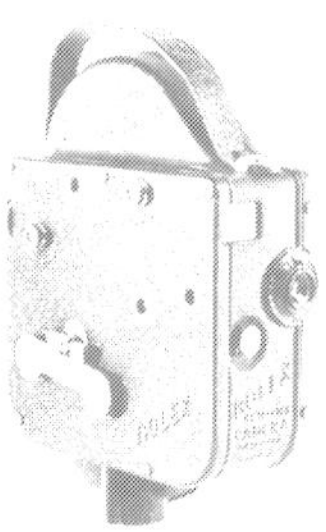

it possible to rewind the film, switch mattes, and shoot the second exposure precisely.

The Bolex's variable speed facility was used most effectively in the shots which show Maya Deren's sandaled feet climbing up the stairs in CU and in slow motion (shots 40 and 41). These were taken at 64 frames per second. The first shot (40, Fig. 12) was probably made with a 25 mm lens, and the second shot (41, Fig. 13) was 'taken with a three inch lens',[16] i.e. 75 mm. A third shot (42, Fig. 14), a reverse wide angle shot looking down the stairs, was probably also taken at 64 f.p.s. 'When these three shots are edited into a sequence, the effect on the beholder is that of an endless climb.'[17] Maya Deren revealed that the continuous running capability of the Bolex was used in 'the scenes where Sasha and I appear together [since] we did it by starting the camera and jumping in front of it.'[18]

Maya Deren and Alexander Hammid did not exploit all of the Bolex's innummerable 'trick' possiblities; they used restraint. Only one sequence (Shots 83–92) seems flawed, 'And those jump cuts, on the stairs, where she gets smaller and smaller, and then bigger and bigger – that bothers me', Hammid told Miriam Arsham in 1976. 'It doesn't make much sense. ... It seems like a gimmick for the sake of a gimmick. It doesn't seem to have any meaning.'[19] Alexander Hammid edited *Meshes of the Afternoon*, and a close analysis reveals that many shots for this sequence were in fact edited in the camera; some were later re-spliced, perhaps trying to improve it.

In the course of preparing this essay, I had the opportunity of watching some 'poetic' films which were shot with the Bolex. Marie Menken's film *Notebook* (1962) still conveys a fresh approach to single-frame exposure: for the section 'Moon Play' she used the 'I' exposure and for 'Nightwriting' the 'T', creating delightful and playful transformations. In his film *Sorrows* (1969), Gregory Markopoulos achieves a sombre overtone through his use of the fader, the hand crank for rewinding the film, and the frame counter to keep track of complex patterns for multiple exposures. By comparison, Jonas Mekas' films *Reminiscences of a Journey to Lithuania* (1971–1972) and *Scenes from the Life of Andy Warhol* (1990) seem simpler, more spontaneous in structure, but nonetheless exciting and revealing of moods, relationships and situations. The films of Rose Lowder, such as *Bouquettes 1–10* (1994–1995), are exhilarating through her careful choice of colours and the creation of complex structures, all 'edited' in the camera. The Paillard-Bolex H16 is indelibly linked to the films of the *avant-garde* from the 1940s through to the 1970s. Only the pressures for synch-sound *cinéma vérité* and for more complex optical effects seem to have shifted filmmakers' preference to other cameras.

Notes

1. Spring-wound motor, reflex viewer, variable shutter, fader, frame counter, variable speed control, single frame instant and time exposure, continuous running, hand crank, matte box, and turret head for three lenses.
2. Unsigned article, 'Aus der Geschichte des Hauses Paillard', *Bolex Reporter*, 1/1955: 4.
3. Unsigned article, 'Die Geburtsstätte der Bolex', *Bolex Reporter*, 4/1954: 5.
4. Unsigned article, '150 Jahre Paillard', *Bolex Reporter*, 2/1964: 3, 24.
5. Unsigned article (Max Abegg), 'Wie es zur Weltmarke Bolex kam', *film + foto*, 1/January 1973.
6. Advertisement, *British Journal Almanac*, 1926, 754.
7. For a more detailed description of AGFA's strategy, see Carlos Bustamante, 'AGFA, Kullmann, Singer & Cie. et le Cinéma Français des Premiers Temps', in Jean A. Gill, Michèle Lagny, Michel Marie, and Vincent Pinel (eds.), *Les vingt premières années du cinéma français* (Paris: Presses de la Sorbonne Nouvelle, 1995), 53–66.
8. For a discussion of Eastman's conviction that only safety film should be used for the amateur film market, see Glenn E. Mathews and Raife G. Tarkington, 'Early History of Amateur Motion-Picture Film', in Raymond Fielding (ed.), *A Technological History of Motion Pictures and Television* (Berkeley: University of California Press, 1974), 131.
9. In their 'Early History of Amateur Motion-Picture Film', Glenn E. Mathews and Raife G. Tarkington explain that with the film emulsions then available, it was decided that a picture

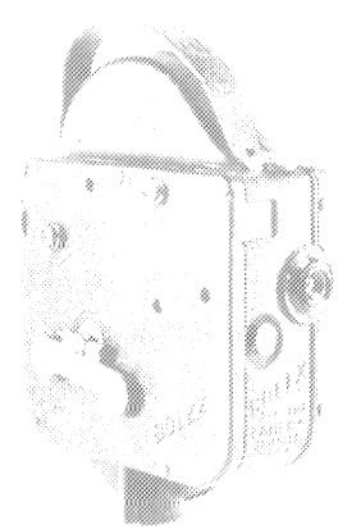

2
The Observer

Sore Society: The Dissolution of the Image and the Assimilation of the Trauma

Bent Fausing

Department of Aesthetics and Visual Communications, University of Copenhagen, Denmark

Why digital images? In the following discussion, the new media are regarded as a turning point in representation. The symbols and self-esteem in the public sphere show us a house divided in itself, and this plays a role in creating media and in constructing new images that are detached from referential and representational bonds. The *public* has difficulties in approaching the *private* with explanations and images. There is a tendency that expression goes in the reverse direction, from the private sphere to the public media. The point at which you live clearly expresses that you are made of flesh and blood, and is found in affect and trauma. Images coming from these private spaces now fill the public media. This is a tendency that can also be more generally seen in the aesthetics and theory of (moving) images. The characteristics of digital images and the use Orlan makes of them, as well as the crisis in representation that she and the media mark, are linked with the notion of 'sore society'. Why is there a displacement from representation towards the presentation of the real and the trauma? Why is the place for physical and psychic affect and their present dissolution sought in visual aesthetics? In answering this question, my essay attaches importance to affect and the wound as essential critical features.

The image is true (if you want it to be true)

We readily accept digital recordings of sounds and digital writing on a computer screen, but we hesitate to accept computerised images. Why? Because in our culture we are afraid of letting go of our attachment to the referent. Most people connect digital images with manipulation, and manipulation with bodies. The digital image has been called (alluding to Walter Benjamin's celebrated essay that first appeared in 1936), *the photograph in the age of digital manipulation*.[1] It is more appropriate to speak of *electronic montage*, where pixels from one image are invisibly fused with another, as in the case of the ice skaters, Tonya Harding and Nancy Kerrigan, who appeared together on the ice rink on the front page of *New York Newsday* (16 February 1994), even though they had not skated together for a long time. The desire to see them together was apparently quite strong, and through digital manipulation, it was possible to show them together on the same ice rink. Similarly, *Time* (vol. 143, no. 21, 1993) could compose a digital portrait of a non-existent young American woman comprising 15 per cent Anglo-American, 17.5 per cent Orient-American, 17.5 per cent Afro-American, 7.5 per cent Asian-American, 35 per cent South-

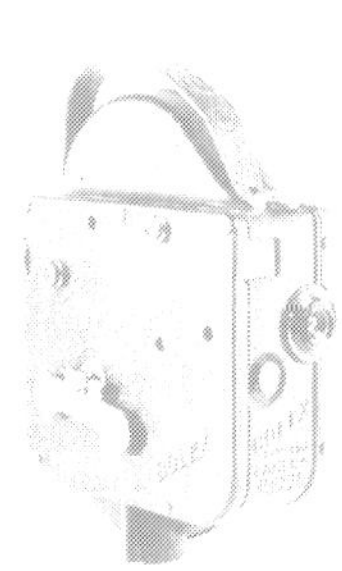

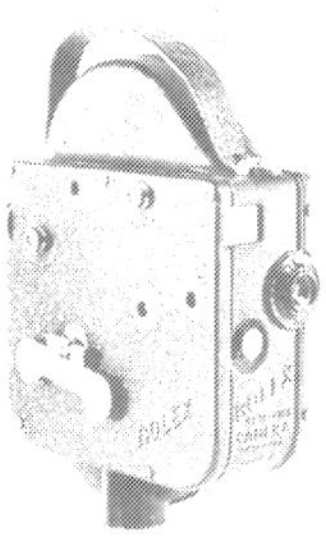

ern European-American, and 7.5 per cent Hispanic-American.

In 1997 the Swedish King and Queen reported the weekly magazine *Se & Hör* to the police for digitally manipulating a photographic image. The magazine had shown their daughter with the body of a model posing in different bathing suits. William J. Mitchell demonstrated a similar process when he showed TV-hostess Oprah Winfrey in the body of actress Ann-Margret.[2] The problem of digital manipulation entered the intrigues of movie fiction at an early date. In Philip Kaufman's film *The Sun Rises* (1993), a digital image, which at a closer inspection is clearly seen as manipulated, is used as decisive evidence.[3] The Japanese singer Kyoko Date is absolutely virtual. She was conceived by Japan's largest multimedia company, HonPro, in 1995, and makes music videos, CDs and gives interviews, though she only exists as a digital image. Sherry Turkle sees a parallel between the virtual bodies of digital images and the process of 'on-line self-representation'[4] in which users can manipulate their sex and visual characteristics on the internet. Those who live such parallel lives through their computer screens are still tied, however, to the wishes, pains and morals which belong to their physical presence.

It is significant that in all these cases we are dealing with arranging or (ex)changing bodies. Clearly there is nothing new about arranging photographs. The pioneer in portrait photography, Félix Tournachon Nadar, for example, employed six people to retouch photographic negatives and three to retouch positives. Photographs have been arranged via so-called combination copying (like Oscar Gustave Rejlander) and photo-montage (like John Heartfield), just as it was common – more simply with scissors, glue and retouching – to remove parts of the motif. In this way the Gang of Four could be made more invisible on a photograph taken in Beijing in 1976.

Instead of emphasising the ways in which a photographic image may be manipulated, I would like to point out that all media can lie. Another and more constructive way of saying the same thing is to emphasise that the image is true, if one wants it to be true. The problem thereby changes from asking how the photographic image was made to asking, how has it been used and with what effect? New digital technologies do not just open up more invisible tricks: they create, to put it in a more sympathetic way, possibilities for continuous creation, flexibility, mobility and accentuation. Post-photography is not just a new tool to create virtual images: it is also an indication of a deep conflict within representation itself for representing something altogether *different*. Digital technologies and the ways in which they are used make it clear that *all* images, first and foremost, are *productions* and *creations*.

The boundary between internal and external

The body can be a terrifying reminder that we are mortal and are therefore prone to mutability. Our skin forms a thin membrane between a mobile exterior and a fluctuating interior. The disintegrated body and the fragmented ego have become symbols in the art and popular culture of our time, and an ever greater challenge to the individual. But they are also expressions of the longing for a completely *different* body. This emerges in the search for the armoured body (or 'hard body' as it was termed in relation to the *Terminator* films)[5] and in images of the disintegrated body. Expressed this way, the aesthetics of discontinuity correspond to our anxiety that the body may be broken, or promote the fear that its contents may overflow.

The body is the point from which our world starts, and the point at which the external world makes contact. The body marks the boundary between exterior and interior, and their interconnection. In Anthony Aziz's and Sammy Cucher's computer-based portraits, the sensory orifices of faces are sealed so that it is impossible for the observer to enter these persons. Aziz and Cucher point out the senseless or, as they put it in their series, *The Dystopia* (1994), 'blind' belief in the future.[6]

Digital technology and biotechnology will make the boundaries between the natural and the artificial change, since the smallest element of the image, the pixels, need only be worked into the smallest discrete part of the body, the genes of the cells.

The digital body

Perception and recognition is related to our own body size, to the sensations experienced on the surface of the body (cleanliness, softness and warmth, and their opposites), and emotions which we recognise from our physical experience of bodily movement, weight, balance and rhythm. We perceive the world through the senses and experiences of the body. We cannot be in two places at once, as we say, since we are tied to our bodies. It is from here that we remember the past, look forward in time, and perceive things in relation to a here and a there in space. Even when our thoughts are elsewhere, our bodies are still present. We may be 'beside ourselves', as indicated by the word 'ecstasy', and forget our bodies, but the body will still be bound to stasis, i.e. a state related to a point in time and a locality which it can momentarily transcend, thereby confirming its attachment to that state. No matter whether it seeks to be itself or to transcend being, a body has a space and a time, otherwise it does not exist. According to English usage, where there is no body there is no person, 'nobody'.

Digital images, however, also shake our certainties on this point. Nancy Burson's persons do not exist since they are only present as composite bodies. What emerges forcefully from her work is that our concept of the body's unity also promotes an apprehension that we can no longer control it. The ultimate referent is no longer reality but an *image*. Nancy Burson's images are actually present before our eyes, as fabulous bodies and chimeras from the age of computing, but at the same time, they do not exist. Nevertheless, we react *bodily* towards them as if they were real just as we do when confronted with Anthony Aziz's and Sammy Cucher's 'senseless' digital portraits without eyes, mouths and ears, or Inez van Lamsweerde's sharp and smooth digital commodity bodies. The modern way of asking 'who am I' hazards the body in one way or the other; the last referent is put to a hard test, both *physically* and *visually*.

Photo-realism

The passage of light through the lens to a light-sensitive emulsion, where the image is *recreated* in the form of light, shade and colour, is the basis of photography. This foundation is being replaced by digital recording and storage on CD-ROM or hard disk, where different images can be mixed to form hitherto unseen tableaux by working with pixels from various image sources. In a commercial for a digital camera the new possibilities are explained in a way which also draws our attention to a related issue, the abstract place and time where the imaginary body can be created and moved:

> No film. No negatives. No development. No enlargement./Minolta RD-175 is an ordinary advanced reflex camera in the professional end of the scale. But with this little addition that it does not use film. Normally you can shoot 36 times before you have to reload. With a Minolta RD-175 you can go on until the counter stands at 114. And the pictures are stored digitally on a so-called PC-card … in a resolution up to 152 x 1146 pixels. When you have finished your shots, you are able to send the pictures for viewing or further treatment on your screen via an SCSI-cable, you can send the pictures to your customer on the other side of the planet with a modem.[7]

The fascination of conventional photography resides in its apparent simplicity and banality since it resembles reality. The basic idea of photography, its noem, for Roland Barthes and others, is its undeniable *it has been*.[8] That is why, using new digital technologies, and despite the fact that they dissolve the temporal and spatial associations of images, people make such efforts to imitate the temperament of the photographic image. The digital image simulates photography to give the appearance of something that exists or has existed in time and space. Direct reference to reality is displaced, as in the mental images from dreams, imagination and creativity which freely associate and are processed into new expressions that comment on reality. The classic photograph is characterised by the fact that *that which was* (in reality) *is* (as an image on the film); and at the same time that *that which is* (as an image) *was* (once upon a time, in reality). Reality's referent, which adheres to the photographic image, is replaced in the

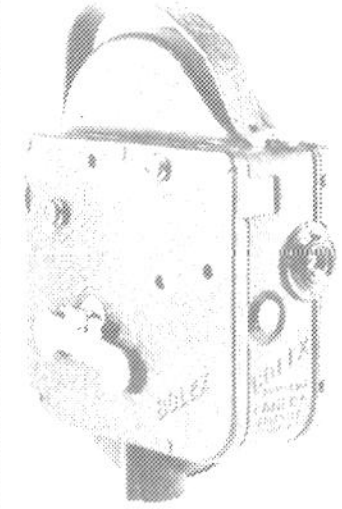

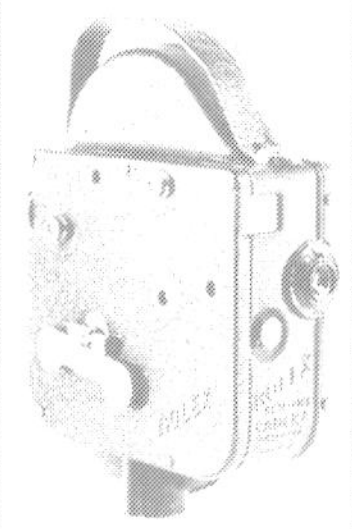

digital versions of post-photography by a *mental referent*. This mental referent lacks, needless to say, no link to reality, neither in its inspiration from reality nor in its comment on it. The realism which belongs to the referent is brought from the external towards the internal, the spectator's *mental* and *bodily* stock of images.

Computer images have, apparently, put aside the optical trace of reality and its realism. Their goal is not realism, but *photo-realism*, i.e. the recreation of a type of realism that is associated with the photographic image in contrast to that of reality. Digital images eliminate photographs and at the same time, celebrate and immortalise the specifically photographic by imitating photographic realism. The digital image *simulates* reality instead of representing it. The drawback to the computer image is that it is *too* real, *too* perfect, *too* sharp. That is why such images are digitally 'shaken' to make them approximate the kind of reproduction attained by conventional photography. Keith Cottingham and Inez van Lamsweerde are two artists who exploit this hyper-sharpness to obtain a surreal play between *motif* and *materia*. Victor Burgin shakes – like Wim Wenders in his *electronic paintings* – his digital images, so that they look more 'abstract' and 'picturesque'.[9]

Nancy Burson gives her black-and-white portraits such a shake that they appear to be blurred, leaving the observer to attempt to fine-tune his or her senses and imagination when decoding them. Burson's images often have – like Valie Export's works – the title *Without title*, and not without reason since the persons, non-existing, anonymous, without sociality and nationality, are assembled through conventional and unconventional elements. Their human aspects can be drawn from celebrities or from unknown individuals, to be crossed in the next moment with animal or lifeless objects. Her digital images are often reproduced by using Polaroid film which increases the black, blue or green depths of the computer screen, and accentuates the dry, inhuman material and pale appearance of her portraits.

The body and a mass of media

Orlan raises further fundamental questions concerning reference, and touches many raw nerves in the body and soul. She uses digital images to design her new appearance. Who is Orlan? Which is the correct referent? Is the professor of philosophy who plays Orlan, the person who changes her appearance (and the new appearance that changes yet again), or is she the media image created from all these persons? Serious confusion has been created between referent and representation.

Orlan is not Orlan's real name. The face is not her face. Neither, at some point in time, will her body be her body any longer. To date, Orlan has carried out/undergone nine performance operations. Orlan makes herself into a variable image (or, to use Marcel Duchamp's term, a ready-made) in her living installations. Duchamp's presentation of himself as the woman Rrose Sélavy is emphasised by Orlan as inspiration for her drag show. Orlan's masquerade involves more than merely clothes and make-up since her entire body participates as a living sculpture which is being formed and which can change at any time. Her performance takes place in operating theatres, and involves plastic surgery and a plethora of media. One of her performances has the title *Omnipresence*, and a number of performances are titled *The Rebirth of St. Orlan*. Like Burson, the body and modern electronics fuse in the creation of a new image, an image that is her own body. Orlan differs, however, from Burson by being the image *herself*, an image which is modified by calculations carried out by computer, and executed with the help of the latest advances in plastic surgery. As in a performance by Yves Klein or in Christo's packages, art is not just encapsulated in the works themselves. Art also resides in the peripheral event: in filming and video recording, in photographing or collaborating with actors or with other participants, with doctors and nurses, as well as in the choreography of the scenery and the dressing of space. The artistic act pervades the entire process of self-production.

The name 'Orlan' was not chosen at random. It contains allusions to the artificial material, *orlon*, to the Maid of *Orléans* (Joan of Arc) and to *Orlando* (Virginia Woolf's androgynous character in the novel of the same name). The artificial, the religious, and a sense of new identity are thereby evoked and united. The process of self-production is also filled with obvious religious connotations. In the first place, this is achieved through Orlan creating herself under the influence of the mythological female figures of a large number of Renaissance and post-Renaissance artists – the Diana of the Fontainebleau school, Boucher's Europa, Botticelli's Venus, Gérôme's Psyche, and Leonardo's Mona Lisa. In the second place, through the creation of the image, to play God at the computer, and then to compose the ideal image using the techniques of plastic surgery.[10] Zeuxis, the Greek artist, created the ideal woman by combining elements taken from a variety of female figures in much the same way as Orlan achieves this in the operating theatre, and as Nancy Burson and Inez van Lamsweerde now create their tragic-comic ideal figures through the use of computers.

Orlan chooses her feminine prototypes not only for their beauty, but also for their mythological-symbolical values. Diana, because she was a powerful adventuress who would not submit to men; Psyche, because of her need for love and spiritual beauty; Europa, because she looked towards another continent and allowed herself to be led into an unknown future; Venus, because of her association with fertility and creativity; Mona Lisa, for her androgynous quality. As on other occasions, Orlan is extremely conscious of her choices, so conscious that the choices, according to Parveen Adams, seem to be 'no more than rationalisations'.[11]

Orlan calls her exhibitions *carnal art*, thus alluding to flesh and carnival (*carne*) in order to distinguish them from *body art*. In her provocative approach she follows the same path as others in recent times – Andy Warhol, Joseph Beuys, Jeff Koons and Cindy Sherman – where the artist is just as relevant as the art itself. Warhol's insistence that surface is the vital element is decisive for Orlan. Still more decisive, however, is Joseph Beuys' inclusion of the artist as a shaman whose *sores* represent the sickness of the whole of society. Orlan carries on Barbara Kruger's and Cindy Sherman's work on the scopic drive and the subject-object roles of women. Barbara Rose describes Orlan's sculpturing of the body through operations as a mix of agonising martyrdom (Joan of Arc) and the suffering caused by plastic surgery for the sake of beauty.[12] Where Kruger and Sherman use the individual image, Orlan uses the *course* of the operation, film and video recordings, as well as video transmission. The sequence is peculiarly capable of revealing *metamorphosis*, that an image becomes another image, that a body becomes another body. Furthermore, Orlan contributes something new to the current use and perception of video as a medium. It is not pleasant to watch Orlan's operations; on the contrary, it hurts. Thus she undermines the comfortable role of the observer which is associated with the medium of video and electronic media in general.

In the history of art, the scene of the operation is, in itself, a location for sexual tension, social power, irony and transformation – as may be seen in the work of Andrea Mantegna, Rembrandt van Rijn, P.-A. Brouillet, Jean-Martin Charcot/Albert Londe, and in the present period, Hannah Wilke, David Lynch, Krzysztof Kieslowski and Lars von Trier. In Orlan's version, however, much more happens. The patient, doctors and nurses are dressed in creations by Paco Rabanne or other famous designers. There are plastic fruits and flowers, crucifixes, large-scale enlargements of previous operations on Orlan. Telefaxes are sent to the whole world, dances are performed, poems and theoretical works are read aloud by Orlan during the operation (which for this reason, always takes place under local anaesthetic), and the entire show is then transmitted as video to galleries and television stations around the world: 'The operating room becomes my studio from which I am conscious of producing images, making a film, a video, photos, and objects that will later be exhibited. These works attempt, by varying degrees, to be autonomous.'[13]

In the operating theatre, Orlan has an interpreter present to translate the readings and the progress of the operations into sign language for the deaf. The interpreter also has another function: to make the performance yet more *mimetic*, and to remind us, as Orlan explains, that at certain instants we are all

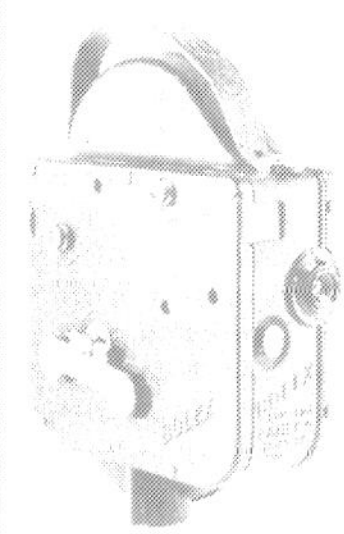

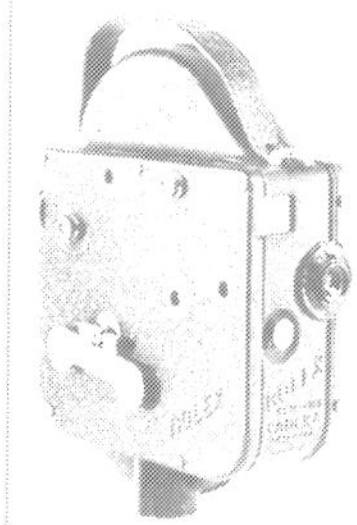

unable to use our senses, the condition that Aziz and Cucher present as a permanent situation in their dystopian images.

Many cultural echoes can be heard in Orlan's *tableaux vivants*. They do not merely approach a blasphemous scene, they also refer us to Mikhail Bakhtin's theories of the carnivalesque and grotesque transgression in which the body's orifices and mutability are emphasised, which process, with its sense of exaggeration, points towards mannerism and the circus. In addition, Orlan, instead of permitting herself to be controlled by the modern electronic media, controls them herself by arranging the setting and the process of recording.

The essential image

One of Orlan's favourite books for the readings she gives during operations is Julia Kristeva's *Pouvoirs de l'horreur*. For Kristeva, horror and its power are linked to the body's orifices and secretions, physical oneness and separation. Anna Price compares Orlan's performances to Antonin Artaud's *Théâtre de la Cruauté* of 1935; they gamble with life and death in Orlan's operating theatre.

We are homeless in Orlan's universe, the last thing we have to hold on to, the body, is in dissolution. In the movement between formlessness and new form, her performances open up a gap and point concretely at the piece of meat with which we are constituted just as her performances also point towards the flesh that senses, feels, desires, aches and dies. Orlan's work both accentuates and attacks representation and images. What is at stake is the distance between fantasy and reality which is manifested in two ways. First, through the fantasy about the perfect body which she enters and at the same time denies; second, through what we see, the scenario itself, which is not just imagination, but reality, a *real performance*. This composition opens up many fantasies and (aggressive) projections.[14] Our vision, however, also causes us to close our eyes or look away: death, a body being opened, a hypodermic needle stuck into an arm and – for some people – birth. On such occasions, it is as if there were an immediate connection between the body which is being observed, and the body which observes since pain, observed in one body, enters the other directly as physical pain.

Pain seems to be an essential and provocative part of Orlan's 'cosmimesis'[15] and passion plays. Pain is also an essential concept of the Western world, as in Christ on the cross and catharsis, 'com-passion' (Lessing). Orlan's voice, which reads constantly during the operations, not only has the function of uttering a passage and expressing an idea. She uses a droning voice as an indication that there is (still) life, and that the Word is on the way to becoming flesh. At the same time, the various scenes are brought together to the droning of this monotonous voice, which has the effect of an aesthetic anaesthetic and a consciousness expander, and of a draining of the object (Orlan) which is being drained in a different way as she becomes another. At the same time, Orlan – as shameless as a small child – gazes into one of the cameras. While the observer may be preparing to turn away in distaste, Orlan continues to gaze at him or her. It is provocative, and Orlan's form is the aesthetics of confrontation rather than the Theatre of Cruelty. Her constantly watching eyes have the same function as the sealed eyes of Aziz and Cucher, to provoke the observer's sensory perception.

Western belief considers the body to be divine and, therefore, sacrosanct. If the body is touched and arranged, it loses its divine status and becomes earthly, profane. The interior can be changed to adapt it to the exterior. Orlan blasphemously strikes out on the opposite path: that the exterior can be adapted to the interior, as a kind of inverted psychoanalysis. The body is not to be formed from within, but from without. Orlan's operations set aside the boundaries between interior and exterior. She is not like a Russian doll that has yet another doll inside, or like part of a collection of anatomical specimens where the organs lie passively and wait for searching eyes. Orlan's interior becomes her exterior, and the exterior becomes the interior in her metamorphosis. She is active and defiantly asserts that identity, and therefore the body is a choice, and is not preordained by Nature and God. She makes herself into God, into an autonomous authority that forms and emphasises the body anew.

Underlying this *blasphemy* is also a desire to reactivate classical, religious concepts of the body, and the transformation of bread and wine in Holy Communion, rites of materialisation and incarnation. Secular forms are used to delineate a religious metamorphosis: that the Word can become flesh and the spirit incarnate. Orlan often relates that through a process of psychoanalytic reversal (wherein the body is adjusted to the internal, and not the other way round), she tries to come back to an image that is one with her persona. In her own secular-sacral way we find in Orlan's credo traces of the *essential image* of the body as defined by Henri Bergson.[16] According to Bergson, identity between image and the world is created through *extension* towards objects. For Orlan, identity is created in the opposite movement by *cutting into* the body and thereby adjusting it to the internal and essential image, which process has very strong religious undertones.

When boundaries become fluid

The authorities and the Church especially are, as Orlan also proclaims, no longer able to provide satisfactory interpretations that create understanding for the discontinuities in society and the individual. People are trying, almost in desperation, to keep the last referent together, the body, which remains when all of the other, more abstract, referents are gone – God, common sense, morality and progress. The elimination of referents is, of course, not new: its roots go back to the secularisation that took place during the last century. What is new are the technological possibilities that exist for dissolving – and also for creating – the body, and for showing these processes and movements.

Images of the ruined, fragmented body correspond to discontinuities in society. The cracks in society spread into the physical feeling of integrity itself, and assail the last referent by fragmenting our image and our sense of body. That is why the disintegrated and destroyed body has become the focal point in pictorial art, not least in film in recent years. This is not solely the case in splatter films, but also in such feature films as *Silence of the Lambs*, *Cape Fear*, *Terminator*, *Dead Again*, *Strange Days*, *Total Recall*, *Alien*, and (significantly titled) *Body Parts*.

The experience of discontinuity has overtaken the experience of identity. The aesthetics of discontinuity have replaced the aesthetics of identity, in both high and popular culture. Discontinuities are emerging everywhere – in the family, in society, in the psyche, and in the body. The disintegrated body (and the urgent search for its images) can be considered as a longing to emerge on the other side of the experience of vulnerability and alarming emptiness, and to emerge into a more complete and satisfying life as in Orlan's blasphemous religiosity. At the same time – as expressed particularly clearly in Burson's digital images – identity resides less and less in the static body. Forming the body, which once more becomes identical to creating and being an image, creates identity.

This game with the virtual is not a new theme; it is familiar from many films throughout the history of cinema. The most durable of these, such as *Frankenstein* and *Dr. Jekyll and Mr. Hyde*, have been filmed over and over again. The viewing experience presents an opportunity to enter *another* body and identity, virtually, while the *observer's* body remains in reality. Such films thematise the virtual dream in which a new appearance attempts to erase an old identity (for instance, *Seconds* (which in Denmark was titled: 'The Man Who Changed Face'), John Frankenheimer, 1966, and *Zelig*, Woody Allen, 1984), or that the implantation of microchips and computer programmes (as in *Total Recall*, Paul Verhoeven, 1990 and, most recently, *Strange Days*, Kathryn Bigelow, 1996) can create another life. It is the very boundary between interior and exterior which, as observed above, is challenged. Thus, it is not only the interior and exterior that vary but the *boundary itself*, since it is the boundaries between the two states that define the body as such. These boundaries declare that something is outside and something within. Orlan exhibits this variable transition by displaying herself in the throes of becoming an image, a fiction that can become a new image, a new fiction, a new identity. Burson creates her computer-based

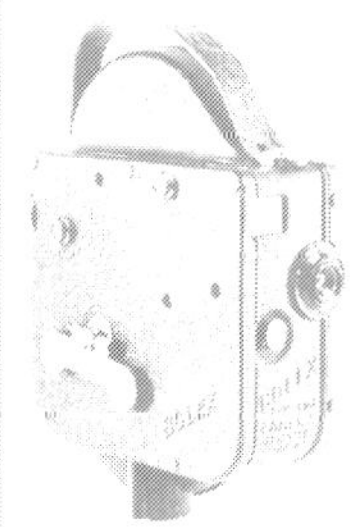

digital images of people who do not exist in the same way that she merges the pixels of five or six people into a non-existent archetype: power, man and beauty; plastic surgery carried out with computer technology.[17] *Mankind* (1983–84) consists of reproductions of races combined to form one face. *Evolution* (1984) combines animal and man. Burson seems to point to the body as a common tie across race, sexuality and gender, and also connects with our animal origins. The novelty of these examples is that they show not only the disintegration – as has so often been the case in recent years in the works of Andres Serrano, Cindy Sherman, Kiki Smith and, in Denmark, Christian Lemmerz – but at the same time the recreation of the body, its new form. The body becomes an image that is created, dissolved, and created anew, over and over again.

Mental presence

Digital images mark the aesthetics of changeability, where documentary photography fuses together with art photography. One can repeatedly rearrange pixels as in the classical arts where one could go on adding lines, figures, and colours to a painting, or take away meaning in a sentence. Modern re-photographed photography, so-called appropriation aesthetics, and digital images put the referential relationship to reality aside. Developing Walter Benjamin's aura thesis with an *aura of absence*,[18] Douglas Crimp's discussion of staged photography (like Cindy Sherman's) or photographic appropriation aesthetics (like those of Sherrie Levine and Richard Prince) also indirectly points towards a far more fluid understanding of the relation between photography and post-photography. The aura, Crimp points out, is an aimed at essence of absence. The image has indeed been evacuated of its authenticity in relation to classical photography. Nevertheless, traces remain in the form of a presence or 'ghost', which points to the fact that the lie (i.e. the fiction) is a part of truth and reality. Benjamin's indication of the duplicity of closeness and distance in the concept of aura has become just *absence* of referents.

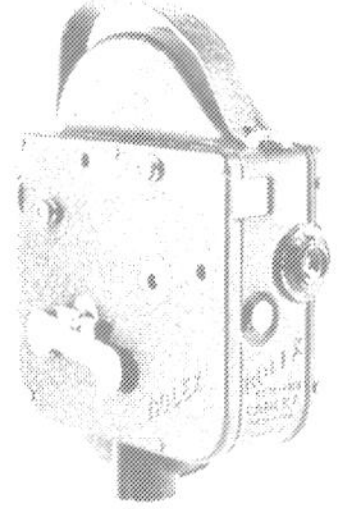

On the referential level, Crimp's observation is correct. If the problem, however, is seen as one involving a mental referent, then it is not only the issue of absence that is under discussion, but also one of increased *presence* which takes place in the observer. Lamsweerde's deformed bodies, for instance, are absent, non-existent. Nevertheless, they will always be compared with an existing and observing body's presence. In that way, the image will be present through a 'ghost' or, to express it differently, through the mental referent.

Representation and trauma

My examples have been moving between an *aesthetics of representation* and an *aesthetics of confrontation*. Representation is not sought out in the old-fashioned way in order to gain knowledge, but the *scene* of trauma is sought out in order to find gaps and dissolutions in the symbolic order, the state which I loosely referred to as an 'aesthetics of discontinuity'. Hal Foster sees this as a common occurrence in modern visual aesthetics and theory. A general shift 'in conception – from reality as an affect of representation to the real as a thing of trauma'.[19] The reality of trauma is not only – as one might first think – bound to weakness, but arises from the strength and authority which is implied in being able to witness trauma and its conditions.

Orlan is not the only artist who represents this shift in modern visual aesthetics, but she goes a step further since not only does she show the dissolution of the body, but she emphasises its re-creation. Cindy Sherman clearly demonstrates a similar development from the staged photographs of the 1970s to the traumatic-abjective photographs of the late 1980s and 1990s. Sherman's latest photographs show monstrous doll-figures, fragmented bodies, and disgusting body liquids. In these pictures she is not concerned with the play of illusion; she is concerned with the seamy sides of disguise and the (im)materiality of the body. These are pictures which *exist* more than they *represent*. The weighting is in the bodies' *presence* as bodies rather than as *roles*.

The other examples that I have mentioned all show signs of the movement from *representation* towards *presentation* consequent upon another definition of the real. As a reason for the change, Foster points out that there is a deep disbelief in the textual model of the real. Belief in reality understood as a text or a simulacrum has ceased, and a repressed reality has returned as *trauma*, i.e. as a *bodily* shock arising from vigorous external experience. The repressed returns as a traumatic physical event in pictures and in visual media. Trauma and its body are proof that this reality exists, that it is not just a phantom. Also a number of other external and unavoidable realities have revealed the real in a traumatic way: the continuing threat of aids, the fact that formerly life-threatening diseases (which had supposedly been eradicated) have reappeared, an almost permanent state of pollution accompanied by increasing poverty, an ever more violent society, and a faltering welfare system. The dissolved body remains the real martyr. Therefore we are not only faced with a yearning for trauma, but also for a sense of bodily dissolution in the form of an aesthetics of discontinuity.

This process is mirrored theoretically in the concept of abjecton and in a number of theories that have been concerned with the relation between perversion and creativity.[20] A *lingua trauma* (Foster) characterises a more general tendency in culture: talk-shows; 'live soap operas';[21] the 'shock of reality' in advertisements and fashion articles;[22] the ways in which the 'true crime' is privileged;[23] a concentration on biography and autobiography; dissolved bodies and narratives in literature and film. As regards the latter, Foster mentions Paul Auster, Dennis Cooper, Atom Egoyan and Terry Gilliam where the real is reconstructed analytically; reversed, discordant or sporadic. In any event, one is uncertain as to whether the trauma has already happened or is about to be created.

Sore society

Around the last turn of the century shock was warded off through body armouring and shock-alertness. Now there is no mitigation of trauma. On the contrary, an assimilation is taking place. This shift has many roots.[24] One of the principal reasons is that the distinction between private and public has dissolved. Second, society has changed from using relatively simple media (such as the pen, typewriter or telephone that relate the private sphere to the public) to large public media which, figuring the private sphere in supposedly private moments, broadcast to a vast international audience events such as the funeral of Princess Diana. Third, the relation between man and machine is no longer characterised by extension and prosthesis, but by incorporations which change the relation between occurrence and representation. The relation between biosphere and mechanosphere has, for a number of reasons, changed radically. The first alludes to the fact that the private and public spheres are no longer well-defined unities. Mark Seltzer has argued that there is a connection between the modern open body and the new openness (Habermas' *Öffentlichkeit*) in the public sphere. This sphere is not so much characterised by administration and power, but by the task of passing on a 'wound culture' which shows itself in 'the excitations of the torn and opened bodies, the torn and exposed individual, as public spectacle.'[25] In public we witness social and mental assaults which challenge the traditional interpretation of trauma as simple representation. The confusion concerning the sore and the cause of the trauma indicate a confusion of the public and the private: is it physical or mental, private or public, a question of representation (fantasy) or a question of presentation (incident)?

The second reason marks a continuation of the problems concerning representation. Trauma and shock are most often linked to direct presentation. In order to address these issues you have to speak about the private in public, and represent them. Thereby a high degree of personification characterises public life. The mimetic relationship between subject, trauma and mimesis also exists in psychoanalysis, but only as internal matters which now become public through the visual media. In this pathological public sphere, where internal and external have collapsed, the photographic imprint is sought as the presentation of violation itself. As with memory, trauma and shock do not exist purely and simply, but are constantly interpreted, adapted, assimilated and symbolised.

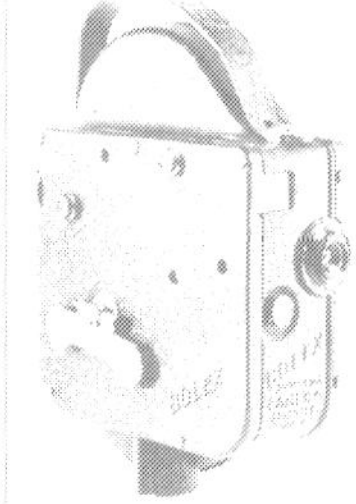

The third reason points to complex media-

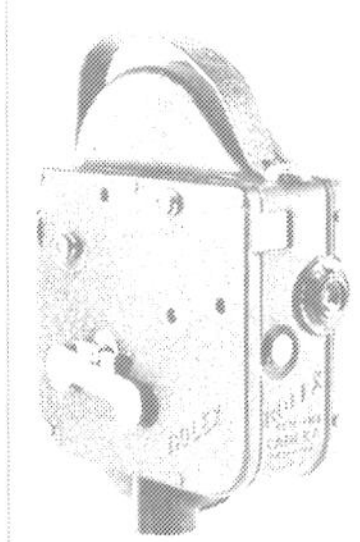

technological conditions. The first industrial revolution at the end of the last century regarded and used *machines* as an alternative to human energy. The second industrial revolution – or electronic era – is, on the contrary, characterised by the fact that man and machine have an influence on each other as two forms of communication in a continuous process. The exchange between body and information, neither permanent nor inviolable, is marked by the movement from *preventing* to *assimilating* the sore.

There is an attendant problem here because if the sore is connected to mimesis and media (which were associated with presentation), what does it mean in the era of digital media to *take* something literally? While creating a desire to make visible, to materialise oneself and one's story visually, this process is not only a problem for the traumatised, but one implicit to digital culture itself.

The place for total bodily and psychic affect

Out of all this *dissolution* also comes *solution* because, on the one hand, the subject is emptied, an abject; on the other hand, it is also elevated. Since you cannot challenge somebody else's trauma, you have to believe it, identify with it, or not. In the biography and autobiography, in the performance, in therapy culture, in the modernity of secretion, in talk-shows, memory-lane programmes, and in movies, the trauma is treated as an event that guarantees the *subject*, that the subject is present, no matter how ruined or injured it is, because it returns as a survivor and an (eye)witness. With Freud and Benjamin in mind, one could say that the shock-experience is encountered without any defence mechanism in order to occupy a place of total bodily and psychical affect while at the same time being drained of affect. In this process, the focal point transfers from the object to the subject, and not just the subject's perception and affect, but also its whole bodily constitution.

What we see is a return to the big affect prior to language against which, as Julia Kristeva has argued, language and representation protect. Susanne K. Langer has also considered the issue to the extent that the great existential themes concerning life, death and bodily affect are particularly suitable for symbolisation.[26] Trauma and loss are, according to Hanna Segal, an essential part of what one could call *the creative position*,[27] which can be paralleled with Christopher Bollas' notion of 'the aesthetic moment'.[28] In connection with the concept of aura, Walter Benjamin refers to the insignificant point of chance where 'reality so to speak glows' because the photograph is not filled with consciousness but 'filled up with unconsciousness'.[29] So we are indeed dealing with an essential and unavoidable existential and aesthetic point.

These traits should also be seen as an attempt to solve contradictory demands about defusion and fusion. On the one hand, the imperative from deconstructive analysis; on the other hand, the imperative from the multicultural stories and connections in the media. On the one hand, the imperative that is created through a shattered subjectivity arising from a disorganised society; on the other hand, the imperative to confirm subjectivity, identity and authority at any price.

Critical strategy

The notion of the real was introduced by Lacan in order to catch a psychic register which remains inaccessible to the subject. Seen in the context of the body and visualisation, it means that the entry into language and the symbolic order is identified with the loss of the body only to return as *image*; but the image is, of course, not a body, it is an image, meaning. The real is the invisible register which the loss of the body creates, and which a sequence of images tries to visualise and embody.

Where Foster evaluates the real and the culture of the trauma as *negative*, which characterises the culture of the 1990s, I am nowadays less certain that it is *only* a symptom. Here I will briefly point to investigations on the cultural history of shock, fatigue and hysteria by Walter Benjamin, Anson Rabinbach and Elaine Showalter.[30] Furthermore, and contrary to Foster (and Lacan), I regard the real respectively the trauma as a *critical strategy* passed on by a surplus of *signifiées* which have been sent to this place because they could not be in the symbolic. Where Foster – together with Lacan[31] – only sees the tragic, loss, and trauma in the real, I consistently seek out the energy, the surplus, and abjection. Some of Lacan's own formulations, however, aim at the fact that what has been refused by the symbolic order returns in the real. In this context, Judith Butler regards the real as a place where a social abjection takes place, and connects Kristeva's term together with a critical interpretation of the real.[32] The different works mentioned here attempt, in various ways and with varying degrees of success, to investigate the critique that is embedded in the real.

The pure reality

At the same time, the question arises whether or not such a real space without symbols actually exists, a place where reality speaks without symbolisation. This is symbol understood solely as veil and power that is the symbolic order (Lacan). Nevertheless there exists (as Langer and others have pointed out) another use of symbols, a revealing, expansive, and far-seeing use, which is present in both the real and in traumatic scenarios.[33] The real is not totally an un-communicated and asymbolic affect-reality; it is also communicated in image-bodily symbols.

Elsewhere I have written about the visionary place where the image all at once opens up and invites us in, and I have written about the different names this place has received, including *troumatique* (a pun upon the French word for 'hole' (*trou*)) and trauma.[34] The hole is not to be regarded as empty, but as a place where something can be poured continuously. What we have, in my opinion, is rather an ideology or 'open public sphere' (*Öffentlichkeit*) which postulates that we are speaking our mind openly, without representation or symbol. At the same time, as Parveen Adams points out in relation to Orlan, ideology uses symbols as a form of rationalisation to conceal pictorial expression or, as in Orlan's case, religious symbolisation. Orlan creates a confrontation with the real, and she does this by using many symbols, a process which should not be mistaken for her conscious visualisation. Parveen Adams is troubled by an emerging tendency to deny the separation between reality and representation. She fears the unification of expression wherein art becomes reality, and reality becomes art. In this way society lacks the exchange that should occur from one area to the other. Discontinuity ought to be the real distinctive feature of democracy.[35] If the same thing is uttered in all of the different means of expression available to a given society, then society will lack the internal dynamic and critique that ought to exist between its various means of expression.

At an external level it is correct to say that there is now a tendency to remove all symbolisation from reality and try to pass it off as pure reality. However, Adams does not seem to notice that symbols are not erased in this process but pour out of the open hole. Rather, one could say that we are living in a period of high symbolisation which gives the impression of being both a direct and genuine *presentation*.

As a general definition Hal Foster's thoughts seem to be correct, yet they also show signs of ambivalence marked by a strong sense of distance and intense sympathy with the trauma which blurs motives and goals.[36] His thesis can be both absolute yet narrow, which is especially evident when we look at concrete theoretical and analytical examples. Theoretically, trauma is absolved by the whole: abject, aura, the real. Thus there should be nothing wrong with trauma playing such an important role, which Foster acknowledges (if only by negation) and thus denies.

I will try to explain myself further by considering a film by Paul Auster and Wayne Wang. In *Smoke* (1995), the main character, Auggie Wren, has been taking photographs on the same street corner in Brooklyn outside his shop every morning at 8 o'clock for the past twelve years. For another character in the movie, the writer Paul, Auggie's photos are all exactly alike as he hastily and absent-mindedly looks through them one evening in the shop. For Auggie, each day is distinct, has a different quality of light, special people, changeable weather, changing seasons. Paul is suddenly confronted with one of the pictures in the numerous albums that have been

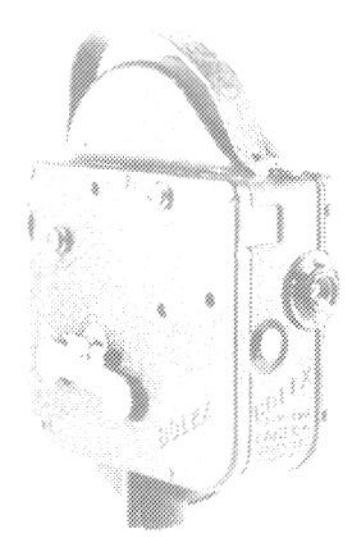

made over the years. Paul's late wife is frozen in a photograph which both changes his understanding of Auggie's photographic project and at once confronts him with *chance* and *trauma* since, by chance, she had passed by at the very moment that Auggie took his photo one morning four years previously. We are not witnessing that type of external assault so typical of a hold-up or an armed robbery in American film. We see an *internal* attack as Paul breaks down as a result of his casual glance at the photograph since an image, which he has carried unconsciously inside, has suddenly found its external representation. It is not just the internal attack, his trauma, that is of interest in this scene; it is the collision between his indifference to Auggie's photographs, and the sudden sharpened sensation evoked by involuntary memory that has touched his sore. That which cannot be *seen* in the symbolic order *shows* itself in the real. This sensation – recalling Henri Bergson – accords with the essential picture which the body contains, and which enables it to adapt to early experience. It is also an attempt to shorten the distance between these fundamental images which are at play in the displacement from representation to traumatic presentation.

The analogy reveals further theoretical implications. Trauma and its related theoretical terms are central to Hanna Segal's theory of *re-creation* and to the work of writers as diverse as Melanie Klein, Marcel Proust and Ella Sharpe. It is present in the theory of Jacques Lacan, and in the work of Julia Kristeva and Susanne K. Langer. More indirectly, it recurs in Benjamin's notion of *aura*, Barthes' *punctum*, and Bergson's concept of *durée*, all of which point to the fact that loss is a gain. Trauma is an essential and unavoidable part of the creative act, which the scene from *Smoke* silently and urgently depicts. Foster seems so caught up in the abjective and traumatic nature of expression that he does not see that a number of contemporary images show re-creation not as a traditional representation, but as a symbol-arranged *presentation* from the real, thereby elucidating the continuity between the condition of lack, creativity and transformation. Behind all this, however, there is also a growing realisation that we are always personal; that the essential image and the essential body in each and everyone – sooner or later – will seek its manifestation, both aesthetically and theoretically.

Notes

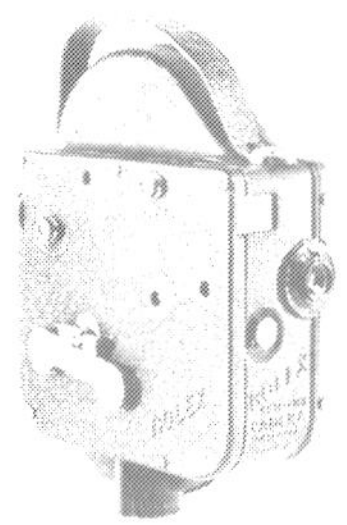

1. Wolfgang Coy, 'Mit fotografischen Gedächtnis' in Alexis Cassel *et al.* (eds.), *Fotografie nach der Fotografie* (Dresden, Basel: Verlag der Kunst, 1995), 71. For further consideration of the relation of cultural anxiety to the loss of the referent, see Thomas Elsaesser, 'Digital Cinema: Delivery, Event, Time' in Thomas Elsaesser and Kay Hoffman (eds.), *Cinema Futures: Cain, Abel or Cable?: The Screen Arts in the Digital Age* (Amsterdam: Amsterdam University Press, 1998) and Thomas Elsaesser, 'Truth or Dare: Reality Checks and Indexicality, or The Future of an Illusion' in Anu Koivunen and Astrid Söderbergh Widding (eds.), *Cinema Studies into Visual Theory?* (Turku: D-Vision, 1998).
2. William J. Mitchell, *The Reconfigured Eye: Visual Truth in the Post-Photographic Era* (Cambridge, Mass.: The MIT Press, 1994 (1992)), 209.
3. This movie – like Ridley Scott's *Blade Runner* (1982) – casts doubt on the referential premise. In earlier films concerned with representation, the media play an important role in the narrative, not least in Michelangelo Antonioni's *Blow-Up* (1967). In *Blow-Up*, where the conditions of referentiality are not questioned, the main character increasingly finds, through a series of enlargements of a photograph taken in a park, a (hidden) referent (which of course would end in nothing if the process of enlarging the photograph continued ad infinitum). Such a process is impossible with a digital image where reality is increasingly revealed by enlargement. *Blade Runner* is the first movie to introduce the digital problem and to acknowledge the dissolution of referentiality, cf. Bent Fausing, *Emotion in Motion: On Affect and Images* (forthcoming).
4. Sherry Turkle, *Life on the Screen: Identity in the Age of the Internet* (New York: Simon & Shuster, 1995), 267.
5. Susan Jeffords, *Hard Bodies: Hollywood Masculinity in the Reagan Era* (New Brunswick: Rutgers University Press, 1993).

6. Anthony Aziz and Sammy Cucher, 'Nachrichten aus Dystopia' in Alexis Cassel et al (eds.), op. cit., 126.
7. *Markedsføring* no. 8, 1996: 7.
8. Cf. Roland Barthes, *La chambre claire: note sur la photographie* (Paris: Cahiers du Cinéma Gallimard Seuil, 1980), 148 and also Roland Barthes, 'Rhétorique de l'image', *Communications* no. 4, 1964: 47. This point of view has its clearest and deepest elaboration in Roland Barthes' *La chambre claire* because of the autobiographical element in his discussion of photography. A similar view may also be observed in other theories of photography. Thus John Berger writes in his excellent article, 'Understanding a photograph', that 'Every photograph is in fact a means of testing, confirming and constructing a total view of reality' (John Berger, 'Understanding a Photograph' (1968) in Nicos Stangos (ed.), *Selected Essays and Articles: The Look at Things* (Harmondsworth: Penguin Books, 1970), 182), where the photograph, as he also points out, comprises 'records of things seen' (Berger, op. cit., 179). This argument can also be found in the work of Susan Sontag, who declares that 'a photograph is not only an image (as a painting is an image), an interpretation of the real, it is also a trace, something directly stencilled off the real, like a footprint or a death mask' (Susan Sontag, *On Photography* (New York: Dell Publishing, 1977), 154). The death-mask metaphor goes back, at least, to the 1945 essay by André Bazin, 'The ontology of the photographic image', where Bazin compares photographs with mummies and relics, objects which display 'transference of reality from the thing to its reproduction' (André Bazin, 'The Ontology of the Photograph Image' in A. Trachtenberg (ed.), *Classic Essays on Photography* (New Haven: Leete's Island Books, 1980), 241). Photography 'affects us' (André Bazin, op. cit., 241) just like things in nature. Finally, the viewpoint that photography is the trace of reality emerges in the writings of Vicki Goldberg who observes that 'bearing witness is what photographs do best; the fact that what is represented on paper undeniably existed, if only for a moment, is the ultimate source of the medium's extraordinary powers of persuasion.' (Vicki Goldberg, *The Power of Photography: How Photographs Changed our Lives* (New York: Abbeville Press, 1991), 19). The connection between photography and death – and not just the death-mask – is, according to Martin Jay, (almost) as old as the photographic medium itself, and was noted in 1841 by Ralph Waldo Emerson (Martin Jay, *Downcast Eyes: The Denigration of Vision in Twentieth-Century French Thought* (Berkeley, Los Angeles, London: University of California Press, 1994), 134).
9. Victor Burgin's own explanation in a conversation with Thomas Dreher, see Thomas Dreher, 'Victor Burgin: Angelus Novus' in Alexis Cassel et al (eds.), op. cit., 144.
10. Orlan has, however, been joined by Cindy Jackson, who in the last ten years has gone through a series of plastic operations in order to look like her ideal, the Barbie Doll, yet with lips like Julia Roberts', cheeks like Claudia Schiffer's etc. Where Orlan makes use of and comments on the representations, Cindy Jackson only uses ideal representations, cf. the TV-programme, 'Yderzoner', TV-Danmark, 9 September 1997.
11. Parveen Adams, 'Opération Orlan' in Duncan McCorquodale (ed.), *Orlan: ceci est mon corps … ceci est mon logiciel/Orlan: this is my body … this is my software* (London: Black Dog Publishing Limited), 1996, 88.
12. Barara Rose, 'Is it art? Orlan and the transgressive act', *Art in America*, February 1993: 84.
13. Orlan, 'Conférence' in Duncan McCorquodale (ed.) op. cit., 90.
14. Michel Onfrey has collected reactions from those who, as he puts it, prefer to judge instead of understand: 'Megalomania, delusions of grandeur, histrionics, mystical delirium, self-mutilation, depersonalisation, fetishism, activism, narcissism, masochism, hysteria, schizophrenia, paranoia, perversion, you name it, it's there.' Oh God, oh Sigmund Freud, what a litany of sins!', Michael Onfrey, 'Esthétique de la chirurgie' in Duncan McCorquodale (ed.), op. cit., 35.
15. The term is coined by Parveen Adams in Parveen Adams, 'Opération Orlan' in Duncan McCorquodale (ed.), op. cit., 58.
16. See Bent Fausing, *Emotion in Motion: On Affect and Images* (forthcoming).
17. Another aspect of Nancy Burson's work is to adjust old photographs to the contemporary state. This work is done in collaboration with the National Center for Missing Children and the FBI in connection with children who, for many years, have been missing. Burson also makes so-called reversals, i.e. Elvis Presley or Marilyn Monroe as babies created from their adult appearance. Burson's computer programmes have also been used in connection with plastic surgery where doctors operate

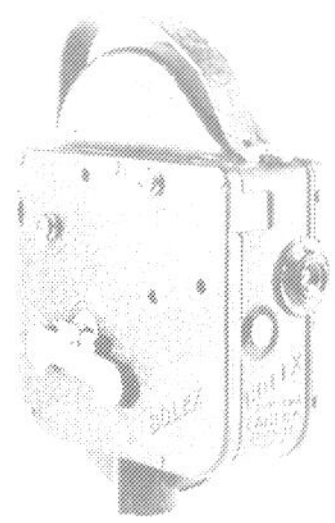

experimentally on the artificial, digital bodies and faces, see Rebecca Buselle, 'A Defining Reality: The Photographs of Nancy Burson' in *Metamorphoses: Photography in the Electronic Age* (New York: Aperture, 1994), 73ff, and Jacques Clayssen, 'Digitale (R-)Evolution' in Alexis Cassel *et al.* (eds.) op. cit., 79.

18. Douglas Crimp, *On the Museum's Ruins* (Cambridge, Mass.: The MIT Press, 1993), 127.
19. Hal Foster, *The Return of the Real* (Cambridge, Mass.: The MIT Press, 1996), 146.
20. In *La chambre claire*, Roland Barthes uses the term 'punctum' (which is the Latin word used to represent, in Greek, 'trauma') to denote 'wound' or 'sore'.
21. I am referring to the series of so-called reality television programmes and psycho-shows which are nowadays on most TV channels. For more detailed explanation, see Bent Fausing, *Emotion in Motion: Affect and Images* (forthcoming).
22. As with Benetton's series of advertisements, *Shock of Reality*, 1991–1995, showing Albanian boat refugees, a dying aids patient, a new-born baby, etc. or *Damernes Verden* (Women's World) which staged fashion photographs with 'Ulrike Mainhof' as a model from the real world, see *Damernes Verden*, no. 3, March 1997. A common strategy in many of these advertisements and fashion photographs is that they seek to induce trauma by motivating the consumer, first and foremost, through shock and affect, not (as previously) solely through glamorous commodity aesthetics.
23. Text on a sign at the Coliseum Bookstore, New York, May 1997. The genre also exists on TV where real life disasters are dramatised or depicted by means of amateur video-recordings (camcorders) or tapes from emergency calls.
24. In this section I am primarily employing concepts from Ewa Lajer-Burchardt, and Mark Seltzer (see Ewa Lajer-Burchardt, 'Real bodies: Video in the 1990s', *Art History* 20 (1997) and Mark Seltzer, *Serial Killers: Death and Life in America's Wound Culture* (New York, London: Routledge, 1998), and my forthcoming study, *Emotion in Motion: On Affect and Images*.
25. Mark Seltzer, op. cit., 253.
26. Susanne K. Langer, *Philosophy in a New Key: A Study in the Symbolism of Reason, Rite, and Art* (Cambridge, Mass.: Harvard University Press, 1942), 144ff.
27. My term, see Bent Fausing, *Emotion in Motion: On Affect and Images* (forthcoming).
28. Christopher Bollas, 'The Aesthetic Moment and the Search for Transformation' in *The Annual of Psychoanalysis*, 6, 1978, 386.
29. Walter Benjamin, 'Kleine Geshichte der Photographie' (1931) in R. Tiedemann and H. Schweppenhäuser (eds.), *Walter Benjamin: Gesammelte Schriften*, vol II, part 1 (Frankfurt: Suhrkamp Verlag, 1977), 371.
30. See Walter Benjamin, 'Das Kunstwerk im Zeitalter seiner technischen Reproduzierbarkeit' (1935) in R. Tiedermann and H. Schweppenhäuser (eds.), *Walter Benjamin: Gesammelte Schriften*, vol 1, part 2 (Frankfurt: Suhrkamp Verlag, 1974), Anson Rabinbach, *The Human Motor: Energy, Fatigue, and the Origins of Modernity* (San Francisco: Basic Books, 1990), and Elaine Showalter, *Hystories: Hysterical Epidemics and Modern Culture* (London: Picador, 1997).
31. Jacques Lacan, *The Four Fundamental Concepts of Psycho-Analysis* (New York, London: W.W. Norton, 1981), 53ff.
32. See also Julia Kristeva, 'The True Real' (1979) in T. Moi (ed.), *The Kristeva Reader* (New York: Blackwell, 1986). In this article Kristeva describes how the true-real, which is the real in Lacan, despite all the defences which have been put up, nevertheless manages to get through, and that is why its appearance is a constant challenge since unwanted truth slips through the social agreements of the symbolic order.
33. Compare my analysis of 'Pigen med rotten' (1988) in Bent Fausing, *Drømmebilleder*, second revised edition (Copenhagen: Tiderne Skifter, 1993), 94ff.
34. Bent Fausing, *Emotion in Motion: On Affect and Images* (forthcoming).
35. Parveen Adams, *Emptiness of the Image: Psychoanalysis and Sexual Differences* (New York, London: Routledge, 1996), 57.
36. Hal Foster does, however, seem more open and relativistic when, in an interview with Staffan Schmidt, he states that 'I wonder what it is really all about ... but perhaps we have to go through all this – whatever it is', 'Konsten behöver publiken', *Paletten* nos. 2–3, 1996: 29.

Closing In: Telescopes, Early Cinema, and the Technological Conditions of De-distancing

Jan Holmberg

Department of Cinema Studies, Stockholm University, Sweden

But what happens when what you see, even though from a distance, seems to touch you with a grasping contact, when the matter of seeing is a sort of touch, when seeing is a contact *at a distance?*

Maurice Blanchot[1]

Figure 1 shows an illustration from Jules Verne's short story 'Un Drame dans les airs' (1874). Although this example is rather obscure, air balloons like this one are a common means of transportation throughout Verne's œuvre, from his first novel, *Cinq Semaines en ballon* (1863) to *Maître du monde* (1904). This is hardly a coincidence. Although perhaps not often acknowledged as such, the air balloon too is a quintessentially *modern* vehicle, if not as emblematic as, say, the train. Nevertheless, balloons promised to change the perception of the world, and hence fit nicely into Verne's project whose ambition, in the words of his publisher, was to 'summarize all geographical, geological, physical and astronomical knowledge gathered by modern science and to retell, in a delightful and picturesque form ... the history of the universe.'[2]

A common motif in postcards or as props in fairground photo studios, balloons were indeed a great part of popular culture during the nineteenth century. Among other literary examples besides Verne, we find balloons in some of the lesser-known Edgar Allan Poe stories such as 'The Balloon-Hoax' and 'Hans Phaall'.[3] That Etienne-Gaspard Robertson, inventor of another modern form of entertainment – the phantasmagory – was also a balloon enthusiast (holding the world altitude record of 7,170 metres in 1803), or that Nadar dedicated his later career to aerial photography from his balloon 'Le géant', also indicate how balloons were perceived as important scientific instruments as well as a new kind of popular entertainment. It is not hard to see the reasons for this: never before had man reached this kind of altitude; never before had such a substantial *distance* been established between the human body and terra firma (as constantly emphasised in Verne's and Poe's stories).[4]

If we now return to the illustration from 'Un Drame dans les airs', there seems to me to be something very ambiguous, almost disturbing, about the picture. What is going on here? Two men have distanced themselves from the face of the earth, only to *compensate* for this distance, this loss of immediate, tactile contact, by the aid of an optical device: the telescope. It seems, however, that telescopes or binoculars were standard equipment in balloons, as there are often elaborate descriptions in the stories to which I referred of what the heroes see through their telescopes from above.[5] In simultaneously using two modern

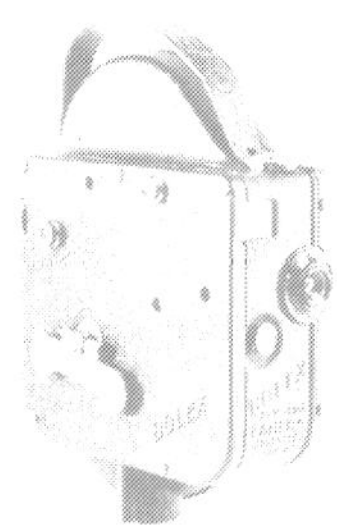

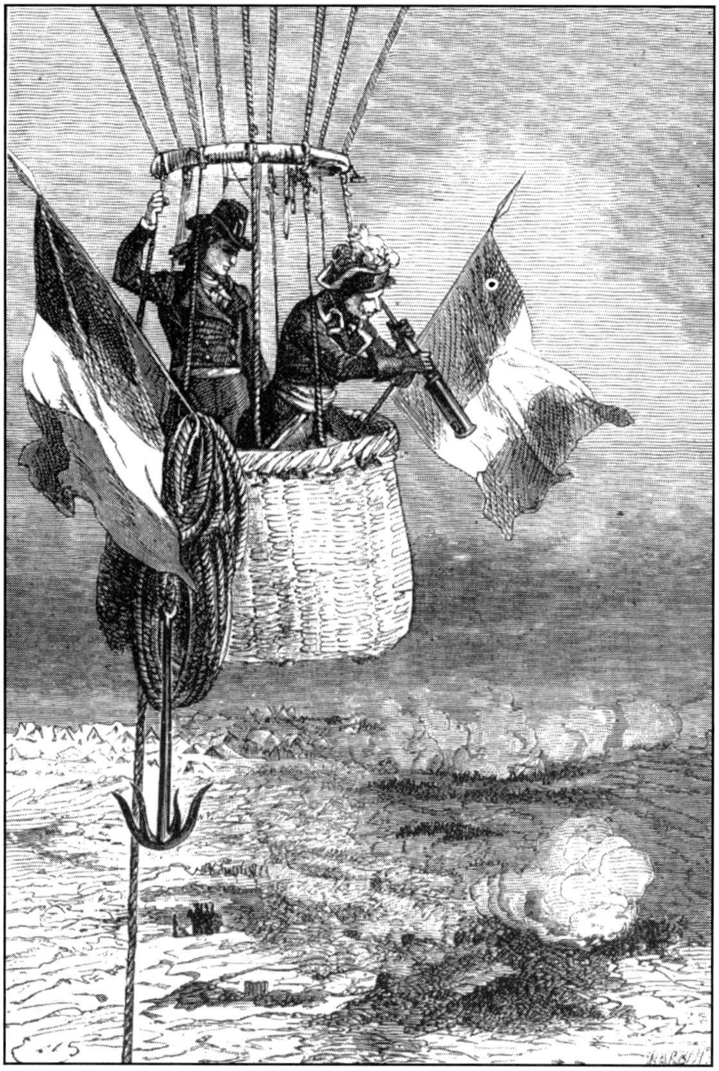

Fig. 1. Jules Verne, Un Drame dans les airs, *1874.*

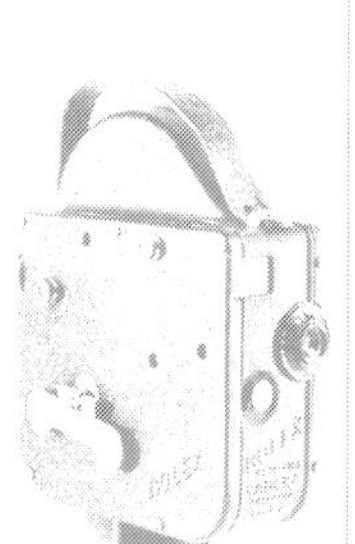

technologies where one is the prerequisite of the other (the one obtaining distance and providing a general view, the other simulating proximity and providing a narrow view), the characters in this image seem to illustrate Martin Heidegger's concept of *Ent-fernung*, or 'de-distancing':

> De-distancing means making distance disappear, making the being at a distance of something disappear, bringing it near. Da-sein is essentially de-distancing. As the being that it is, it lets beings be encountered in nearness. De-distancing discovers remoteness.[6]

In order to make 'sense' of the surrounding world, existence brings the world closer according to Heidegger not by simply approaching it, but *negatively*, by making the distance disappear, by distancing distance. It is also important to note that the distance that is being annihilated is not measured in 'geometrical' terms. Rather, our relation to things and other beings is determined by a 'circumspect approaching, a bringing near as supplying, preparing, having at hand.'[7] To 'have at hand' hence does not necessarily mean to have something at arm's length (a yard); it is an epistemic rather than a spatial concept. This is, in essence, what differentiates Heidegger's conception of space from the traditional Cartesian one. Rather than a mathematical, objective co-ordinate system with given variables, space is, for Heidegger, a matter of distance between the subject and the world around. As 'the distantiation of distance', the ambiguity of de-distancing is summarised by Jacques Derrida as a 'remote proximity'.[8] This seemingly paradoxical description captures the ambiguity of the actions of Verne's heroes in the illustration, as well as it describes an important aspect of modernity. In order to see (in both senses of the word), we have to be close *and* distant. De-distancing promises a scrutinising close-up view of minute details without letting go of 'the big picture'. This is what we see in Fig. 1. The two men are both looking in the same direction, but their views complement each other. On the one hand, optical magnification, on the other hand, a general view.

Many of the phenomena we most often associate with modernity have to do with a narrowing of space, or coming closer: the tactility of the crowded city street; the idea of 'the world within reach' in the ideology of the world exposition; the discovery of new, formerly unseen parts of reality through new technology, on a micro- as well as a macro-level (the microscope, the X-ray, the telescope); and of course, other technological innovations such as the railway, the telegraph, the telephone, and so on and so forth. (The concentration camp – the 'apotheosis of modernity' as Jean-François Lyotard puts it – is in a sense the horrible culmination of this tendency.[9]) But as numerous studies have pointed out, all of these phenomena are important aspects of a particular *modern* experience, and 'tele-' seems to be the privileged prefix of the era.[10] Thus it might be argued that a certain 'dialectic' of distance and proximity plays a crucial part in modernity, in the specifically modern experience.

***

In 1935, Walter Benjamin wrote a letter to his friend Werner Kraft, discussing his on-going project. In describing his work, Benjamin chose a certain metaphor which should come as no surprise:

> As for me, I am busy pointing my telescope through the bloody mist at a mirage of the nineteenth century that I am attempting to reproduce based on the characteristics it will manifest in a future state of the world, liberated from magic. I must naturally first build this telescope myself and, in making this effort, I am the first to have discovered some fundamental principles of materialistic art theory. I am currently in the process of explicating them in a short programmatic essay.[11]

The result of this telescopic vision is, of course, his famous essay, 'The Work of Art in the Age of Mechanical Reproduction'. What interests me here, however, is why Benjamin is using an optical technology as a trope for his project. Why is it only through the aid of a *telescope* that the distance between his own time and the nineteenth century can be overcome?

Heidegger gives only a few examples of how de-distancing takes place, and we can note that all of these examples[12] are technological:

> All kinds of increasing speed which we are more or less compelled to go along with today push for *overcoming distance*. With the 'radio', for example, Da-sein is bringing about today de-distancing of the 'world' which is unforeseeable in its meaning for Da-sein, by way of expanding and destroying the everyday surrounding world.[13]

Compare this to Benjamin:

> [T]he desire of contemporary masses to bring things 'closer' spatially and humanly ... is just as ardent as their bent toward overcoming the uniqueness of everyday reality by accepting its reproduction. Every day the urge grows stronger to get hold of an object *at very close range* by way of its likeness, its reproduction.[14]

But again, the 'distance' that is to be overcome is not necessarily literal. As with Heidegger's notion of de-distancing, Benjamin's concept of the aura is defined by a kind of double-bind, as indicated by an often cited footnote:

> The definition of the aura as a 'unique phenomenon of a distance however close it may be' represents nothing but the formulation of the cult value of the work of art in categories of space and time perception. Distance is the opposite of closeness. The essentially distant object is the unapproachable one.[15]

Various technologies (the radio for Heidegger, or for Benjamin, cinematic devices such as close-ups and slow-motion) are pre-requisites of de-distancing, since they are *epistemic* prostheses as much as optical ones since, as Benjamin writes, 'the enlargement of a snapshot does not simply render more precise what in any case was visible, though unclear: it reveals entirely new structural formations of the subject'.[16] In order to understand, we need to come closer, although the proximity may not (*or must not*) be literal, but technologically mediated or optically simulated. Unfortunately, however, modern technologies of approaching do not ensure a truthful perception of reality since closing 'in' necessarily leaves something 'out'. This sense of compromise is at the core of 'modern perception'.[17]

In what follows, I will try to demonstrate this ambiguous tension between distance and proximity – so aptly captured by the seemingly paradoxical concept of de-distancing – by giving a few examples of what I will tentatively call the 'figure of closing in' in cinema. By this, I mean a concrete figure of approaching, but with epistemological ambitions: the desire to know, comprehend, or understand by approaching. Obvious examples of this would be classically dissected scenes that conform to the pattern of: (a) establishing long shot; (b) medium shots; and (c) close-ups

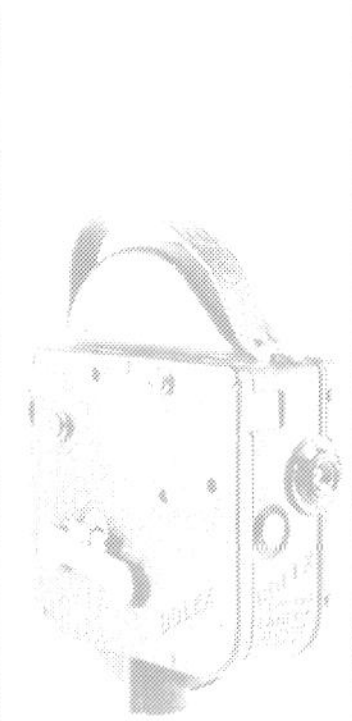

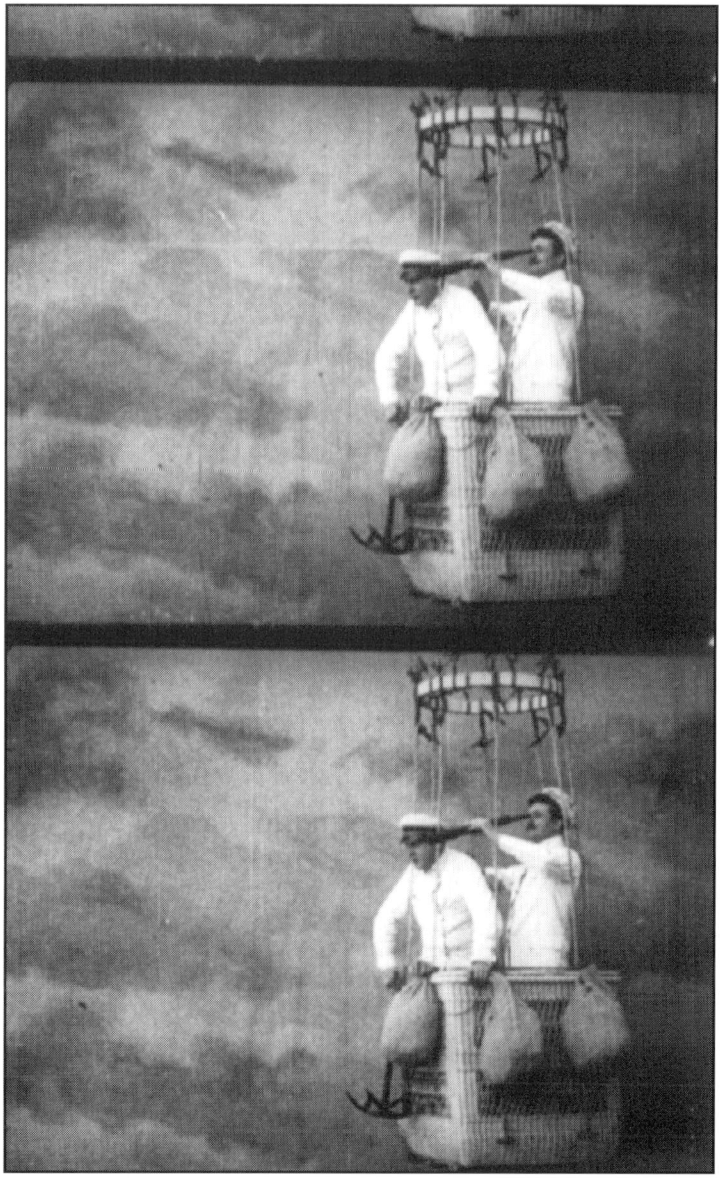

Fig. 2. Un Drame dans les airs, *Pathé, 1904.*

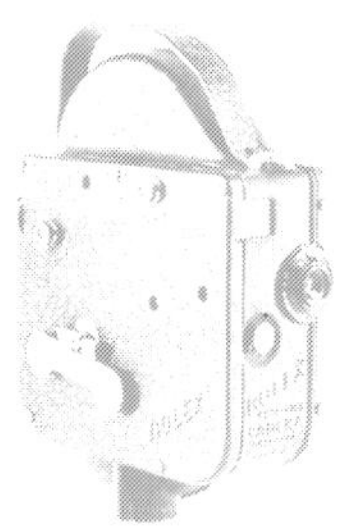

'revealing' character psychology. Here, gaining knowledge is correlated with closing in. A more spectacular example is the famous scene in Alfred Hitchcock's *Young and Innocent* (1937), where a single tracking shot from long shot to extreme close-up reveals the killer. These are examples of the figure of closing in *par excellence*, with the uncomplicated equation *learning by approaching*.

The 1904 Pathé film, *Un Drame dans les airs*, is rather loosely related to Verne's story. But as its possible literary origin, the narrative of the film is essentially a pretext for the insertion of spectacular actuality shots of impressive cruisers and sublime sea storms. (Verne's story, on the other hand, is a pretext for telling a history of ballooning.) As such, this film is an example of what Richard Abel calls the 'bricolage model' of early cinema, negotiating between a sketchy narrative and the attractions of early cinema.[18] (In this context, we should not fail to note that Tom Gunning's concept of 'attraction' also has a spatial dimension of coming closer, something which I think even further adds to the precision of this term.) The alternation between actuality footage and fictional scenes of the bricolage model requires a diegetic technological relay, in this case a telescope. As many scholars have pointed out, telescopes and binoculars often served this purpose, as in the Swedish Pathé film *Katarinahissen* (The Katarina Lift, 1908) or the unidentified film known as *Dr. Ams Tram Grams kikkert* (The Telescope of Dr. Ams Tram Gram, c. 1908).[19] These films also illustrate the ambiguity of closing in. On the one hand, telescopic vision enlarges the view of the world, visually and epistemically. On the other hand, the relation between the character's point of view through the optical device and the reverse actuality shots are weak, 'a mis-match' as later cinematic practices would call it. As opposed to classical point-of-view shots, these 'sight links', to use Gunning's term,[20] do not merely give us a closer or better view, but a *different* one, from another angle, another perspective (Figs. 2-4).

In the films of Georges Méliès, telescopes abound. Typically, they motivate closer framings as in *Un voyage dans la lune* (1902) or *L'Eclipse* (1907), but sometimes, as in *Les quatre cents farces du diable* (1906), telescopes are seemingly excessive parts of the *mise-en-scène*, promises of simulated proximity never carried out. But whether functioning as a relay between shots in the bricolage model, or as diegetic excuses for closer framings, or even as a merely decorative part of the backdrop, the frequent use of telescopes seems to underscore a dimension of near and far in early cinema.

As we have seen, however, the telescope does not necessarily ensure a more perfect vision.

The attitude displayed in these films towards the diegetic technologies seems to be ambiguous, not least since they often end in disasters (balloons crashing, telescopes running amok, astronomers falling out of windows, etc.). Moreover, a point is reached when the mediated closeness offered by different tele-technologies seems to *enhance* the frustration over distance, leading to the discovery that optical or auditive 'presence' was nothing but an illusion; the simulated proximity amplifies the anxiety of distance. The sense of delusion that attends technology surfaces in a number of films, exemplified by the many early adaptations of André de Lorde's Grand Guignol play, *Au téléphone*, where a husband overhears the murder of his family over the telephone. However, this very emblem of modern technology fails its purpose since, as Gunning has observed, '[r]ather than allowing [the husband] to overcome space and time, the telephone torments him with distance and impotence.'[21]

Later examples of this kind of implicit or allegorical critique of the approaching technologies redolent of modernity include, of course, Hitchcock's *Rear Window* (1954), and its paraphrase, Brian De Palma's *Body Double* (1984) as well as Wes Craven's *Scream* (1996). In the latter, and in other contemporary films where cellular phone technology offers the possibility for the interlocutor to move about freely, the simulated proximity of the telephone call is often eventually upgraded to an *actual* encounter in space. There are numerous examples in contemporary cinema and television of this interplay between distance and proximity, to which any regular viewer of *The X-Files* will testify.

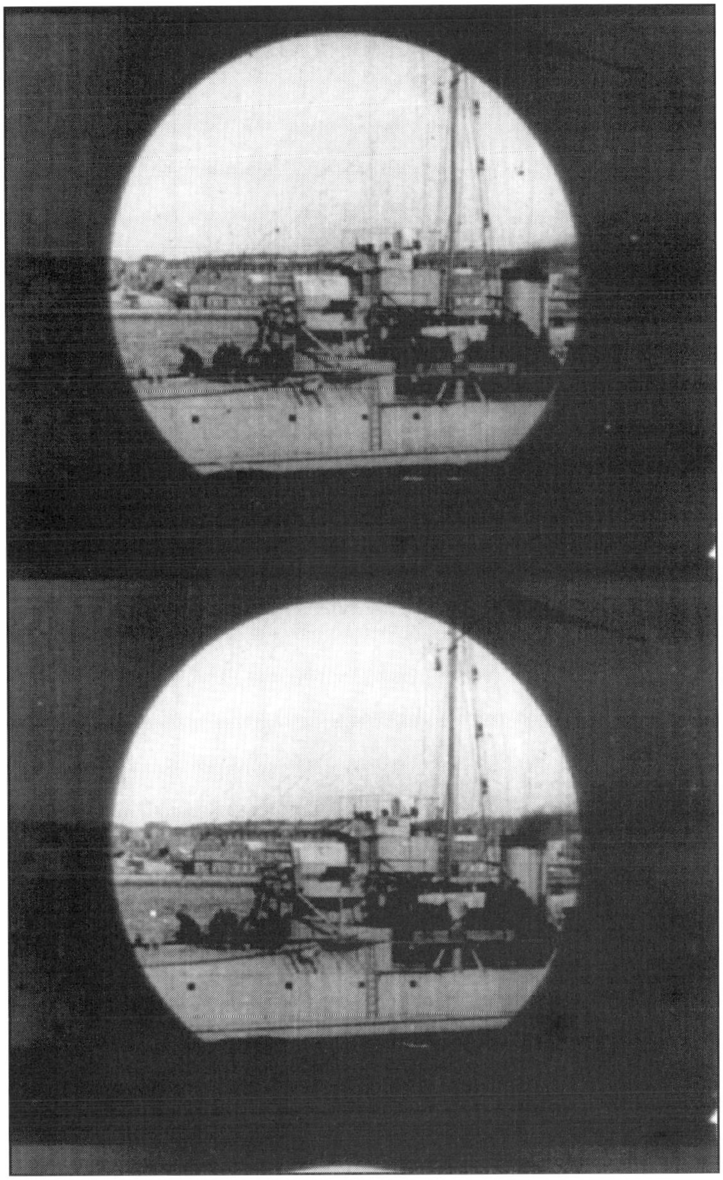

Fig. 3. Un Drame dans les airs, *Pathé, 1904.*

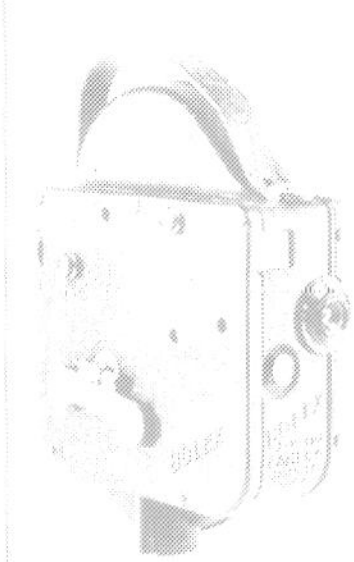

Michelangelo Antonioni's *Blow-Up* (1966) illustrates another problem with the figure of closing in. When the protagonist gradually enlarges the photograph he has taken, something hidden to the naked eye seems to be unveiled. But the 'closer' he gets, the more uncertain he becomes of the alleged murder in the park. In getting *too* close, we cannot distinguish any features at all: everything gets blurred, dissolved. (This seems to be the other side of the coin of Benjamin's idea of the enlargement revealing 'entirely new structural formations of the subject'.) The ambiguity of distance and proximity which I have tried to sketch might, in fact, be the aporia defining that which we give the name 'modernity'.[22]

***

In his story 'Le Horla' (1887), Guy de Maupassant formulated the dream *par excellence* of his time – to see the unseen:

> How profound that mystery of the Invisible is! We cannot fathom it with our

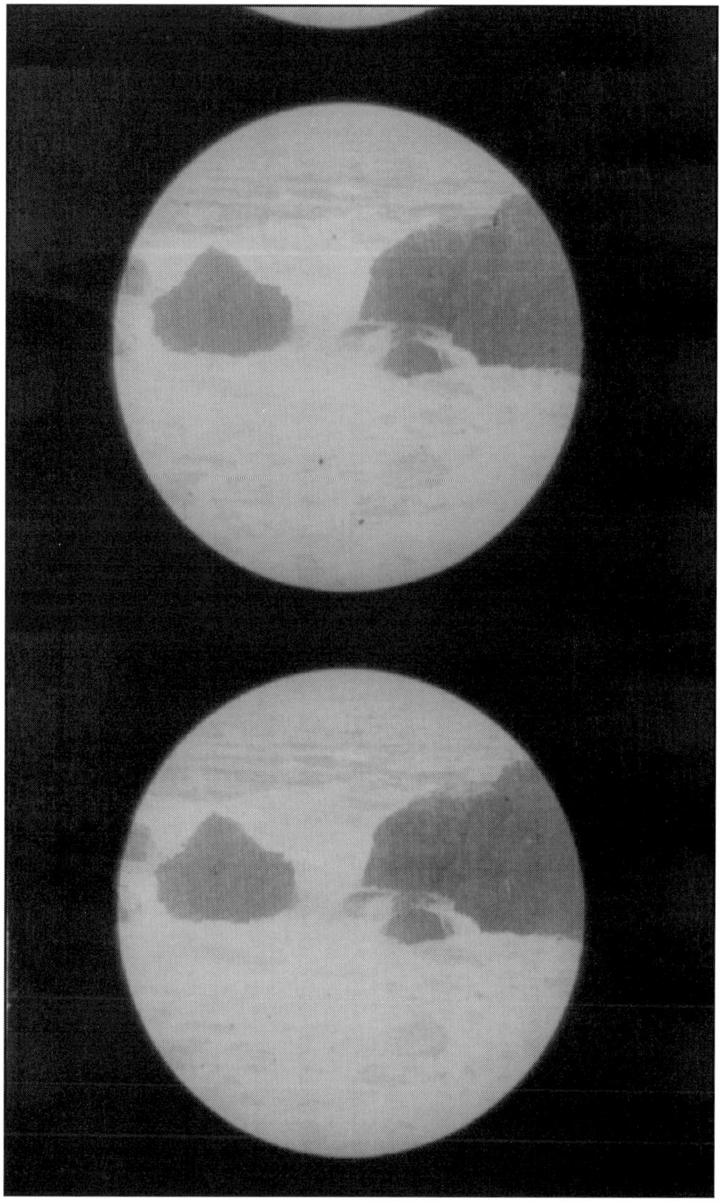

Fig. 4. Un Drame dans les airs, *Pathé, 1904.*

paltry senses, with our eyes that can distinguish neither what is too tiny nor too large, too near nor too far, neither the inhabitants of a star nor those of a waterdroplet. ...

Ah! If we only had other organs to perform for us, how much more we could discover in the world around us![23]

Modernity's various technologies of approaching – the telescope, the microscope, the close-up – promise to be such supplementary organs of which Maupassant dreams, disclosing 'hidden aspects of the world about us' as Siegfried Kracauer puts it.[24] But the dilemma is still there: in magnifying the small or bringing the distant close, we still do not see it, since it is now 'too large'. The mystery of the Invisible remains profound indeed.

Acknowledgement: I would like to acknowledge my debt to Tom Gunning and Jan Olsson for reading and generously commenting upon earlier versions of this essay. For their helpful observations, I am most grateful. Thanks are for several reasons also due to the Swedish Television Archive, John Fullerton, Bart van der Gaag, and my friends and colleagues Patrik Sjöberg and Martin Thomasson. Finally, I wish to record my gratitude to my wife, Sara Albrecht, who came up with the original idea.

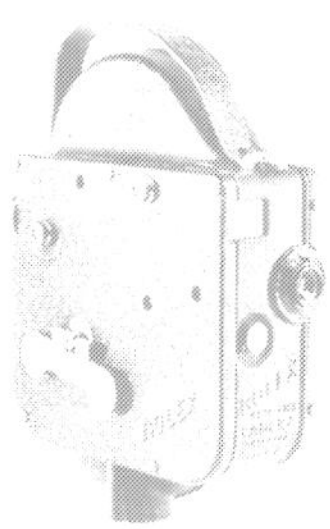

Notes

1. Maurice Blanchot, *The Gaze of Orpheus*, ed. P. Adams Sitney, trans. Lydia Davis (Barrytown: Station Hill, 1981), 75.
2. Pierre-Jules Hetzel, 'Avertissement de l'Editeur', *Voyages et aventures du capitaine Hatteras* (Paris: Ed. Hetzel, 1866), ii, quoted in Arthur B. Evans, 'The Extraordinary Libraries of Jules Verne', *L'Esprit créateur* 28, 1 (Spring 1988): 75-86. My translation.
3. The latter was much appreciated by Verne, as indicated in his essay 'Edgard Poë et ses œuvres', *Musée des Familles* 31, 7 (April 1864): 193–208.
4. One example will have to suffice: 'In about ten minutes after starting, the barometer indicated an altitude of 15,000 ft. The weather was remarkably fine, and the view of the subjacent country – a

most romantic one when seen from any point – was now especially sublime.' Edgar Allan Poe, 'The Balloon-Hoax', in *Tales of Mystery and Imagination* (London: J. M. Dent & Sons Ltd., 1963), 373.

5. It is also significant that Nadar, in the famous photograph of himself in a balloon gondola, is posing with a pair of binoculars in his hand.
6. Martin Heidegger, *Being and Time*, trans. Joan Stambaugh (Albany: State University of New York Press, 1996), 97.
7. Ibid., 98.
8. Jacques Derrida, *Spurs: Nietzsche's Styles*, trans. Barbara Harlow (Chicago, London: University of Chicago Press, 1979), 50–51.
9. I am grateful to Michael Renov for bringing Lyotard's notion to my attention.
10. To my knowledge, one of the first formulations of this aspect of modernity may be found in Walter Benjamin's revised Baudelaire essay, where he describes modern experience in terms of an extreme, involuntary proximity, or 'shock', see 'On Some Motifs in Baudelaire', in *Illuminations* ed. Hannah Arendt, trans. Harry Zohn (London: Fontana Press, 1973), 152–196. More thorough studies on the subject range from Wolfgang Schivelbusch, *The Railway Journey: Trains and Travel in the 19th Century*, trans. Anselm Hollo (New York: Urizen Books, 1979) to Stephen Kern's *The Culture of Time and Space 1880–1918* (Cambridge, Mass.: Harvard University Press, 1983). In the field of cinema studies William Uricchio has skilfully demonstrated this 'tele-discourse' in different contexts, see, for example, his 'Television, Film, and the Struggle for Media Identity', *Film History* 10, 2 (1998): 118–127.
11. Walter Benjamin, 'To Werner Kraft', *The Correspondence of Walter Benjamin 1910–1940*, ed. Gershom Scholem and Theodor W. Adorno, trans. Manfred R. Jacobson and Evelyn M. Jacobson (Chicago, London: The University of Chicago Press, 1994), 516.
12. The examples are: the radio; a pair of spectacles; a telephone receiver, and, simply, the city street, see *Being and Time*, 98. For a discussion of the primordial technological or prosthetic nature of de-distancing, see Bernard Stiegler, *Technics and Time 1: The Fault of Epimetheus*, trans. Richard Beardsworth and George Collins (Stanford: Stanford University Press, 1998), 250–252.
13. Martin Heidegger, *Being and Time*, 98 (emphasis added).
14. Walter Benjamin, 'The Work of Art in the Age of Mechanical Reproduction', in *Illuminations*, 216–217 (emphasis added). I owe much of my comparison between Benjamin and Heidegger to Rebecca Comay's essay 'Framing Redemption: Aura, Origin, Technology in Benjamin and Heidegger', in Arleen B. Dallery and Charles E. Scott with P. Holley Roberts (eds.), *Ethics and Danger: Essays on Heidegger and Continental Thought* (Albany: State University of New York Press, 1992), 139–167.
15. Walter Benjamin, 'The Work of Art in the Age of Mechanical Reproduction', 236.
16. Ibid., 229–230.
17. I use the term 'modern perception' aware of its oversimplification. By this formulation, I want to make clear that I do not think that perception *per se* is subject to substantial change.
18. Richard Abel, *The Ciné Goes to Town: French Cinema 1896–1914* (Berkeley, Los Angeles, London: University of California Press, 1994), 105f.
19. For a discussion of the ocular peculiarities of *The Katarina Lift*, see Jan Olsson, 'Förstorade attraktioner, klassiska närbilder: Anteckningar kring ett gränssnitt', *Aura, filmvetenskaplig tidskrift* 2, 1–2 (1996): 34–79, in particular 42; and John Fullerton, 'Seeing the World with Different Eyes: Cinematographic Vision and Turn-of-the-Century Popular Entertainment', in John Fullerton and Jan Olsson (eds.), *Nordic Explorations: Film Before 1930* (Sydney, London: John Libbey, 1999).

    The only known print of *Dr. Ams Tram Grams kikkert* is in the Swedish Television Archive sorted under this Danish title. Although unclear, the film might be of French origin. For an excellent discussion of the technological thematics of the film, see Trond Lundemo, *Bildets oppløsning: Filmens bevegelse i historisk og teoretisk perspektiv* (Oslo: Spartacus forlag, 1996), 217–218.

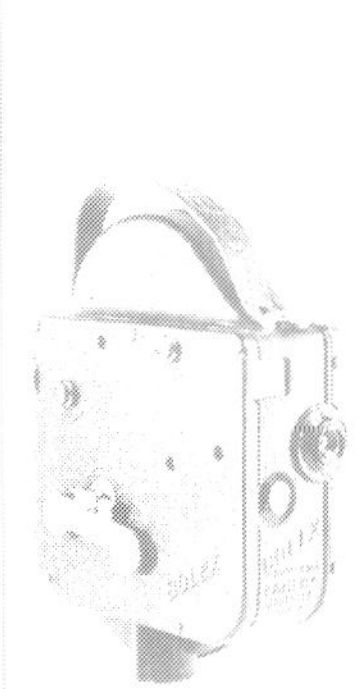

20. Tom Gunning, *D. W. Griffith and the Origins of American Narrative Film: The Early Years at Biograph* (Urbana, Chicago: University of Illinois Press, 1994), 169.
21. Tom Gunning, 'Heard Over the Phone: *The Lonely Villa* and the de Lorde Tradition of the Terrors of Technology', *Screen* 32, 2 (Summer 1991): 184–196, 192.
22. For an account of the aporia of 'bringing closer' in Benjamin's definition of the aura, see Samuel Weber, 'Massmediauras', in *Massmediauras: Form, Technics, Media* (Stanford: Stanford University Press, 1996), 88.

23. Guy de Maupassant, 'The Horla', trans. Joan C. Kessler, in Joan C. Kessler (ed.), *Demons of the Night: Tales of the Fantastic, Madness, and the Supernatural from Nineteenth-Century France* (University of Chicago Press, 1995), 285 (translation modified).
24. Siegfried Kracauer, *Theory of Film: The Redemption of Physical Reality* (Oxford: Oxford University Press, 1960), 49. The supposed ability of the close-up to reveal unknown territories analogous to those discovered by the telescope, the microscope, or the X-ray, has been a common idea, witness Béla Balázs' observation that 'The search for the literature-free "pure film style" led directors not only to travel with the camera into unknown distances, but also to penetrate into a yet undiscovered and unknown nearness', *Theory of the Film: Character and Growth of a New Art*, trans. Edith Bone (London: Dennis Dobson, 1952), 164.

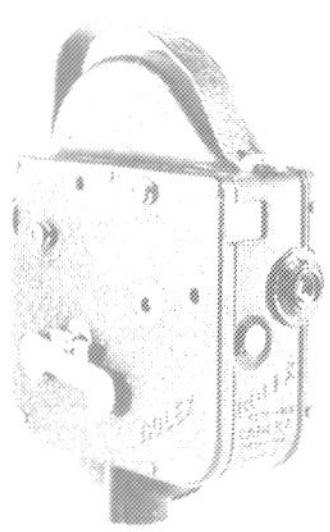

'We Partake, as it Were, of His Life': The Status of the Visual in Early Ethnographic Film

Alison Griffiths

Department of Speech, Baruch College, City University of New York, USA

Introduction

In her speech as outgoing president of the American Anthropological Association in 1960, pioneering visual anthropologist Margaret Mead urged her colleagues to make greater use of the technologies of the still camera, audio tape recorder and motion picture camera. According to some observers, the reaction among Mead's professional audience was decidedly mixed. Her appeals were greeted with 'restless stirrings and angry murmurs ... as these notebook-oriented scholars expressed their irritation at this revolutionary suggestion.'[1] In some respects, it is hardly surprising that Mead's colleagues balked at the idea of using tape recorders and motion picture cameras in the field; beyond Mead's and Gregory Bateson's own pathbreaking forays into ethnographic film and the isolated projects undertaken in France under Jean Rouch and at the Institute für den Wissenschaftlichen Film in Germany, there had been little intellectual or professional support for ethnographic filmmaking before 1960, and visual anthropology as a whole occupied a precarious position within the larger discipline.[2] The prohibitive cost of film equipment and stock within the modest budgets of most fieldwork expeditions, together with the lack of training in 16mm filmmaking and editing, put off all but the most determined of would-be ethnographic filmmakers. But even if anthropologists had been able to resolve the challenges of financial expense and technical inexperience, they were still left with the question of what to do with film footage once they returned from the field. Would the footage function primarily as a visual supplement to written fieldnotes, or would it impinge in more direct ways upon anthropological theory by being used not only as an extension of the eye, to paraphrase David MacDougall, but as an extension of the mind, a way of furthering conceptual understanding of indigenous cultures?[3] Should the unedited footage be made available to a small group of specialised researchers? Should the material be edited to produce a work of interest to commercial distributors for theatrical audiences, or should it be distributed non-theatrically for the exclusive use of anthropological teaching to non-specialists?

Anthropologists have been interested in (and perplexed by) the legibility of the photographic and cinematic sign as ethnographic evidence for most of this century. However, paradigmatic shifts in the history of the discipline, including the waning of social evolutionism after 1900 and the rise of the Malinowskian concept of extended fieldwork in the late 1910s and 1920s, caused anthropologists to turn their attention away from forms of data that could be represented visually toward the production of written texts.

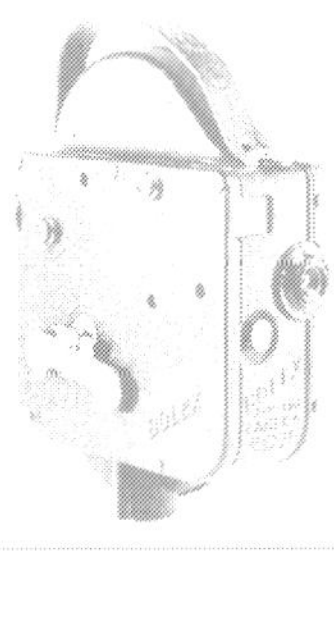

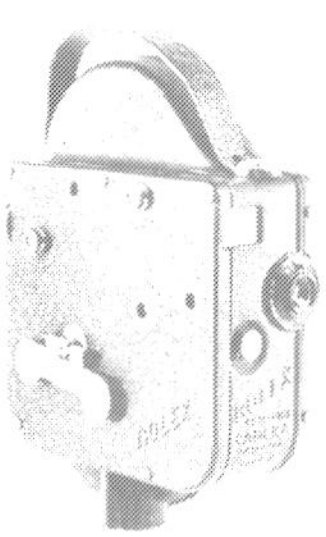

Anthropology's changing methodological, theoretical, and institutional discourses created scepticism about the uses of visual media. The increased emphasis upon the production of written text in the form of the anthropological monograph, which aimed to convey information not available through the examination of visible evidence, marginalised photography and film within ethnography, further discouraging anthropologists from adding mechanical recordings to their arsenal of fieldwork techniques. For most anthropologists, the declining interest in visual recording was particularly significant for cinema's place in the discipline, since its emergence and institu- tionalisation coincided with this reconsideration of the anthropological value of visual evidence.

In order to appreciate the kind of intellectual challenges motion picture technology presented to the nascent discipline of anthropology, we should first consider the effect photography had upon what Jonathan Crary calls the 'intelligibility of visuality';[4] in other words, before jumping to any conclusions about anthropologists' perceptions of cinema, we should take a closer look at the discursive construction of photography as a scientific recording device within the professional anthropological press of the late nineteenth century. In particular, it would be beneficial to consider how corporeal evidence collected by mechanical means was compared with traditional modes of ethnographic inscription, such as engravings, sketches, diagrams and written accounts. Before cinema's invention, photography was able to meet the intellectual and methodological demands of the nascent discipline in the latter part of the nineteenth century and was hailed in both British and American anthropological circles as the *sine qua non* of advanced scientific practice. But while enduring for almost half a century, this love affair with photography was destined not to last; by the time motion pictures were fully institutionalised in the mid-1910s, photography was no longer viewed with the same epistemological reverence as it had been in the latter half of the nineteenth century.

With these reservations in mind, I begin this essay with a brief overview of how photography was rhetorically constructed by late nineteenth-century anthropologists before looking at two early examples of ethnographic filmmaking produced at very different moments in the history of ethnographic film: the filmmaking of British anthropologist Alfred Cort Haddon (1855–1940) from 1898, and films shot by Norwegian anthropologist Carl Lumholtz (1850–1922) as part of his four-year expedition to Borneo between 1913 and 1917. By looking at examples of ethnographic filmmaking from either end of the early cinema spectrum, my aim is to evaluate how a different kind of anthropological object might be produced by cinema, as opposed to photography and the written monograph. Here, in particular, I will consider the impact of the moving image upon the human sensorium and the kinds of challenges film presented to traditional models of sight and epistemology within anthropology. I end the essay with a brief discussion of the prospective impact of new digital technologies on anthropology, given the contentious history of visual anthropology over the past century.

The still camera as a 'truth machine'

Of all the human senses, sight assumed a position of unquestionable dominance in nineteenth century anthropology; it was tales of what intrepid travellers, colonial officials, and missionaries had *seen* on their trips to the distant corners of the globe in the form of drawings, sketches, engravings, water-colours, paintings, and cultural artifacts that thrilled the general public and scientific societies alike, and that provided the much sought-after evidence of racial inferiority and so-called barbaric practices. After the invention of photography, however, such forms of visual evidence were judged by some to have been limited by the eyes *and* imaginations of their producers, mere *interpretations* of what had been seen rather than scientific proof. According to such critics, a sketch or water-colour of an ethnographic scene, no matter how carefully rendered, lacked the immediacy and objectivity of an anthropological photograph. As the Curator of the British Museum, William Henry Flower, argued in 1882:

Photographs ... with their histories care-

> fully registered, of any of the so-called aborigine races, now rapidly undergoing extermination or degeneration, will be hereafter of inestimable value. Drawings, descriptions and measurements are also useful, though in a far less degree.[5]

While painterly renderings of indigenous life might continue to be 'useful', they could not compete with the 'inestimable value' of the photograph. Needless to say, to contemporary observers, nineteenth-century ethnographic photographs are no less tainted by the romantic imaginings of nineteenth-century scientists than their artistically rendered counterparts. Beyond our current awareness of the inherent manipulation of the photographic image, historians have pointed out that a great many of such photographs were subjected to extensive retouching and other techniques that altered the relationship of the image to its referent.

Flower was not alone, however, in endorsing the scientific utility of photography; innumerable anthropologists appeared wholly seduced by photography's mimetic capacities in the second half of the nineteenth century. C. Read, author of the preface to the ethnography section of the influential anthropological handbook *Notes and Queries*, endorsed Flower's view when he argued in 1899 that researchers should devote as much time as possible to taking photographs. 'By these means', Read advised, 'the traveller is dealing with facts about which there can be no question, and the record thus obtained may be elucidated by subsequent inquirers on the same spot'.[6] Photography was here seen as a means of validating the accuracy and impartiality of field data and as a comparative resource for future researchers. Despite being criticised by American anthropologists for its antiquarianism and evolutionary precepts, the handbook achieved widespread recognition within British anthropological circles, and was personally endorsed by Haddon, who wrote the photography section of the 1899 edition.

At a time when evolutionism dominated nineteenth-century anthropological theory, systems of racial classification and measurement occupied the efforts of many anthropologists, and photography's ability to render objective and verifiable evidence was frequently asserted by anthropologists, especially within the practice of physical anthropology. The near-obsessive measuring, classifying, charting and ranking of human physiognomy that preoccupied so much of physical anthropology (as well as pseudo-sciences such as phrenology and craniometry) were responses to the wider challenge of how to make sense of observable physical and cultural differences among the peoples of the world. J.H. Lamprey claimed that his system of classification, consisting of photographing native peoples in front of perpendicular lines suspended across a wooden frame,[7] 'greatly helped ... the study of all those peculiarities of contour which are so distinctly observable in each group' and served as 'good guides to their definition, which no verbal description could convey, and but few artists could delineate'.[8] As a mode of scientific inscription, anthropometric photography was premised upon the notion that a mechanically produced image of the body obtained via standardised photometric methods would permit the recovery of 'reliable and comparative morphometric data'.[9] Photography lent a veneer of scientific respectability to anthropometric studies because the indexicality of the image offered the appearance of legitimate evidence.

Many nineteenth-century anthropologists endorsed anthropometric photography for what they saw as its reliable, systematic, and scientific collection of uncontaminated 'facts'. One goal of anthropometric photography was to make the native body legible as an ethnographic sign, since the detection and measurement of individual anatomical features were seen as offering the perfect solution to the problem of how to guarantee objectivity and 'truth' in anthropological investigation. Cultural theorist David Spurr argues that within this cartographic gaze, 'the eye treats the body as a landscape ... proceed[ing] systematically from part to part, quantifying and spatialising, noting colour and texture.'[10] The promise of mathematical precision in the diverse measurements obtained in a single sitting convinced many scientists of anthropometry's utility in the project of racial classification and criminal identification.[11] Flower, for example, argued that 'physical

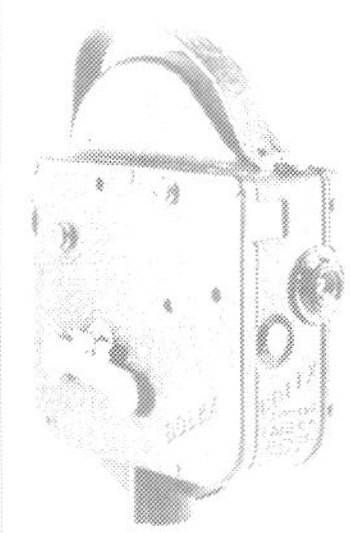

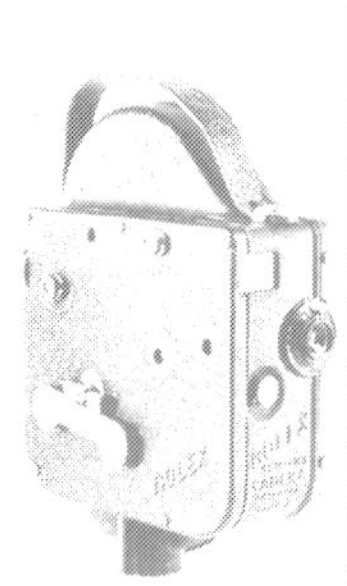

characters are the best, in fact the only true tests of race, that is of real affinity; language, customs etc., may help or give indications, but they are often misleading'.[12] Within nineteenth-century theories of racial classification, cultural difference was signified by physiology, an indexical sign free, in the minds of most anthropologists, of the threatening ambiguities of other markers of cultural difference. In an operation of optical empiricism, Flower reads from the mute and non-reciprocating bodies of the native peoples the signifiers of race, understood as indelible manifestations of an inner character, vital to the project of racial and social classification.[13]

However, if the scientific patina of photographs of naked bodies in standardised poses alongside measuring instruments in front of grid-like backdrops appealed to many anthropologists, the process of extracting anthropological data from photographs was fraught with technical difficulties. The problem of ensuring the verticality of the anthropometer, coupled with the difficulties of exact and uniform positioning of a subject's head and arms, led to disputes within anthropometry over what and how measurements should be consistently applied.[14] If the indexical properties of the photographic image promised an optical trace of what had passed before the camera lens, the photograph nevertheless remained silent about what these visual data might signify in broader ethnographic terms. Despite these difficulties, proponents of anthropometric photography invested a great deal of faith in the visual sign, what Christopher Pinney calls 'the truth of photographic scientism', as a means of bolstering prevailing social and evolutionary theories, which used photographs as evidence of the racial inferiority of native peoples.[15]

At the same time, nineteenth-century ethnographic photographs produced in more naturalistic settings relied upon captioning as a way to imbue the image with scientific value and disassociate the object from commercially produced photographs. Given the extensive interpenetration of scientific and popular ethnographic photographs in the nineteenth century – countless photographs created within the contexts of missionary work and anthropological expeditions found their way into commercial photography markets – no single textual feature distinguished non-anthropometric photographs from other types. Instead, the scientific authority of the 'legitimate' ethnographic photograph had to be established by means of the specific exhibition context or through captioning. Under these circumstances, it is not surprising that the ethnographic status of nineteenth-century photographs seems less a matter of textuality than a product of the specific ideological contexts and reception communities in which they circulated.

Perhaps in response to the permeability of boundaries between the scientific and commercial contexts of ethnographic photography, anthropologists often went to considerable lengths to distinguish their work from the 'fakery' and sensationalism of commercial photography, an effort which can also be detected in discourses surrounding early ethnographic film. The often highly-prescriptive interpretative meanings imposed upon the photograph via captioning or discursive context (popular press illustration, stereograph, *carte-de-visite*, or scientific plate) is reminiscent of the contextual determinants of the reception of early ethnographic film. The ethnographic status of filmic actualities was to a large extent shaped by the film lecturer and the exhibition context, and in the case of both photographs and films, the intended meanings of visual texts were rarely trusted to 'speak for themselves', but were insistently circumscribed by the exhibition context, be it a caption accompanying a photograph, a didactic intertitle, or a comment uttered by a lecturer.[16] The provenance of an ethnographic photograph may have mattered less to most anthropologists than the kinds of illustrative or pedagogic uses to which it could be put. When accompanied by a detailed caption in a scholarly text, a photograph produced for the tourist market would undergo a process of de-contamination and become acceptable as ethnographic evidence. In similar ways, commercially shot films of native peoples exhibited within a scientific lecture might acquire the status of scientific evidence by force of the scholarly prestige of the lecturer or the elite connotations of the venue. However, despite the complex and interconnected relationships surrounding film and photographs, these technologies differed in their perceived ability to represent indigenous lives: they placed very different demands, both practical and ideological, upon anthropologists, and for these reasons must be understood on their own terms. With this in mind, I will examine some early examples of anthropological filmmaking and consider what happens to the ethnographic image when it is animated by the motion picture camera.

Moving bodies in space: film and the vicissitudes of sight

Over three days in September 1898, Alfred Cort Haddon shot five brief films of Torres Strait Islanders and Australian Aborigines on Mer Island during a nine month expedition to the Torres Strait in 1898.[17] Haddon had very little confidence in the *cinématograph* and no experience using film, although he was, as I have pointed out, an accomplished photographer. Rather than discuss all of Haddon's films, I would instead like to focus exclusively on Haddon's film of the climax of the Malu-Bomai ceremony (the *zogo-le* dance)[18] and consider the challenges this film presented to traditional ways of seeing within anthropology.[19] At stake here is an enduring paradox in the history of visual anthropology, a tension between the apparent sufficiency of the ethnographic image, its excess of visual detail on the one hand, versus its discursive insufficiency, the fact that while on the surface it may appear to tell us a great deal about a particular social or cultural practice, it nevertheless remains 'annoyingly mute', in David MacDougall's words, about what these cultural forms and symbols might actually mean in broader anthropological terms.[20]

Haddon had been interested in the Malu ceremonies since his first trip to the Torres Strait in 1888, but it was not until the second trip ten years later that he persuaded two Mer Islanders, Wan and Enter, to reconstruct the Malu-Bomai ceremony for the motion picture camera.[21] Making models of the elaborate masks worn during the ceremony out of cardboard used for packing supplies was a symbolically charged act for the Islanders, since these re-enactments entailed rekindling ritual beliefs and practices that had been suppressed through missionary activities and the Islanders' conversion to Christianity some thirty years prior to Haddon's second visit to the Islands in 1898. The intercultural negotiations involved in preparing for the ceremony to be recorded on wax cylinders as well as film were complex. For example, Wan and Enter wanted to donate the 10 shillings Haddon paid them for making models of the masks to the annual missionary meeting and ended up concealing them from church officials when they brought the masks to show Haddon at the weekly prayer meeting.[22] Traditional Islander beliefs, Christian faith, and salvage ethnography all collided around the Malu-Bomai ceremony, a fact not lost on Haddon who, according to curator and anthropologist Anita Herle, was 'acutely aware of the contrast between traditional Island Custom and the realities of Islander life, which was strongly influenced by traders, colonialists, and missionaries'.[23]

But Haddon had another set of issues to contend with besides the logistical arrangements involved with the Malu ceremony reconstructions, namely the ambiguous status of the dances in relation to popular ethnographic spectacle and neutral scientific demonstration. As visual anthropologist Elizabeth Edwards argues, this tension is played out in the contrast between Haddon's suppression of the lyricism and exoticism of the Malu-Bomai film in his description of the ceremony in his Expedition volumes intended for a professional readership[24] (the six-volume *Reports of the Cambridge Anthropological Expedition to the Torres Straits*, published between 1901 and 1935),[25] and the evocative description of the same dance in his account of the expedition published for a general readership, *Head-Hunters: Black, White and Brown* (1901): 'The grotesque masks worn by ruddled men, girt with leafy kilts, had a strange effect as they emerged from the jungle and very weird was the dance in the mottled shade of the tropical foliage, a fantasy in red and green, lit up by spots of sunshine'.[26] Anthropological knowledge is thus evoked quite differently in the *Reports* version of the dance than in *Head-Hunters* (one cannot imagine for a minute that Haddon would have

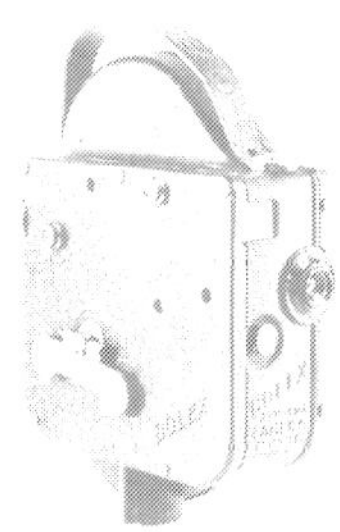

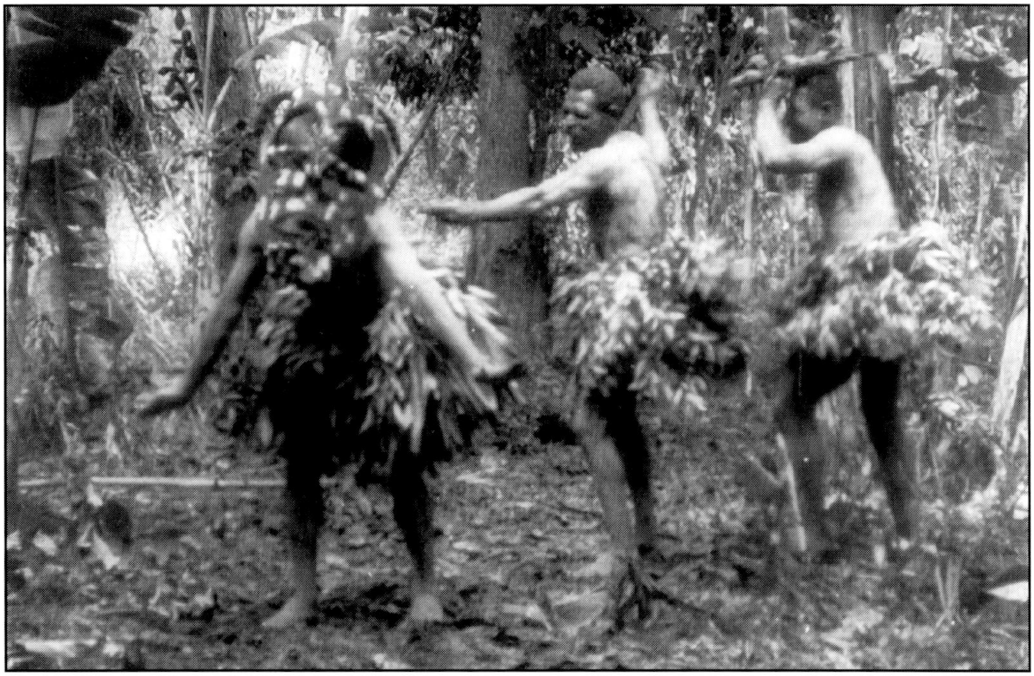

Fig. 1. Frame enlargement of Mer Islanders performing climax of Malu-Bomai ceremony. The film was one of six shot by Alfred Cort Haddon in the Torres Strait in September 1898. [Copyright Cambridge Museum of Archaeology and Anthropology.]

referred to the dance as a 'fantasy in red and green' for scientific readers of the *Reports*: this contrast, Herle notes, 'reveals a different way of looking and understanding [and] highlights the difficulty in translating revelatory experience into academic text as well as Haddon's attempt to engage different audiences at a time when field-based anthropology was attempting to position itself'.[27]

Haddon's staged fragments of Islander dances and industries exist as much as aestheticised versions of the salvage ethnography paradigm, with its 'vanishing race' ideology and melancholic 'structure of feeling', as they do scientific articulations of ethnographic information about native communities.[28] As staged re-enactments, the dances and ceremonies transport the intended viewer back into the mythic past of Islander ceremonies, or rather, as Edwards puts it, 'transports the mythic past to the present through performance'.[29] But these constructions of anthropological knowledge through visual means also suggest the phenomenological differences between photography and film as tools of ethnographic inscription: Haddon's *Head-Hunters* description of the Malu-Bomai ceremony attempts to capture the fantastical qualities of the live experience (qualities that are better conveyed by the moving as opposed to the still camera), whereas the dispassionate and descriptive *Reports* account more closely parallels Haddon's still photographs of the same event. Haddon therefore succeeded in evoking in the *Head-Hunters* description something of the visceral and tactile quality of the dance which could only be hinted at in the photographs. Moreover, Haddon's cinematic study of the Malu-Bomai ceremony provides a compelling sense of 'being there' among the Islanders, a feeling of being co-present at an event which is absent in photographs of the same scene (Fig. 1). One is also struck by the way the film negotiates several different and potentially contradictory modalities: the theatricality of the performance, intimations of the subjectivities of the Islanders, and the putative certitude of scientific knowledge. Edwards argues that the tension between the conditions of possibility for *scientific* knowledge and the conditions of possibility for the subjective experience of a native community is 'enhanced by the nature of [film] and photography, the realism of which heightens and makes theatrical, yet paradoxically allows the impression of authenticity of experience through mimetic scientific devices'.[30] But where Edwards does

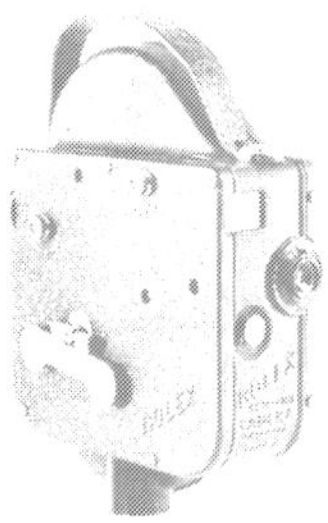

not differentiate between photography and film's mimetic potency as tools of ethnographic reconstruction, cinema, with its heightened verisimilitude and kineticism, raises pressing questions about the readability of cinematic and photographic inscription as anthropological knowledge.

As the disparity between the description of the Malu-Bomai ceremonies in *Head-Hunters* versus the *Reports* reveals, at the same time cinema offered a solution to the problem of ethnographic reconstruction, its mimeticism created a new set of issues that had less to do with methodology and scientific accuracy than how the visceral qualities of the cinematic image could be reconciled with principles of ethnography. It was cinema's ability to render the ethnographic body with such fidelity, ocular pleasure and tactility that perhaps presented one of the greatest challenges to anthropologists (and which partially explains why Haddon was only one of a handful of anthropologists to display an interest in filmmaking during the early cinema period). Film, with its inalienable life-like qualities, may have posed troubling questions to the privileging of sight as the paroxysm of anthropological knowledge. But what was it exactly about film that touched a nerve?

Looking at Haddon's film of the Malu-Bomai ceremony, one is struck by the tactile quality of the cinematic image, the way in which the flat spatial composition and surface textures of the image seem to affect the spectator not just visually, but through a simulated sense of touch. This sense of tactile closeness to the cinematic image has been identified with so-called haptic or haptical cinema, defined as 'relating to or based on the sense of touch'.[31] Among the first theorists to explore cinema's haptic qualities was Noël Burch, who saw in early cinema's stylised, presentational mode of address a tactility which would resurface in experimental cinema of the 1960s.[32] In different terms, Francis Hubbard Flaherty – wife, photographer and cinematic collaborator of documentary filmmaker Robert Flaherty – in 1960 characterised the viewing encounter between ethnographic subject and film spectator in a manner evocative of haptic cinema. Flaherty describes an ethnographic film of a potter as a quasi-spiritual conjoining of the cinematic spectator with the on-screen performer, a coming together of the virtual body of the performer with the real body of the viewer: 'The motion picture camera can follow ... movements closely, intimately, so intimately, that as with our eyes we follow, we come to feel those movements as a sensation in ourselves. Momentarily we touch and know the very heart and mind of the [performer]; we partake, as it were, of his life, we are one with him'.[33] Laura U. Marks has also argued that there is a more 'dynamic subjectivity between looker and image' in haptic visuality than in optical visuality, because in the former, the spectator's eyes are akin to organs of touch and seem to reach out and 'feel' the image through a form of 'embodied perception'.[34]

Following the movements of the Malu-Bomai dancers in the lush forest setting, the rhythm of the circular dance has a tactile and somewhat mesmerizing effect, as we expectedly await the completion of the dance pattern which returns the performers to their original positions. Moving in a circle before the camera, the dancers seem to oscillate between their ontological status as two dimensional figures merging with the background and bodies that come toward us through the circular movement of the dance, thus creating a tension between foreground and background. Could it be that early cinema's predilection for offering the spectator a sensorially rich ethnographic experience (richer, one could argue, than even the still photograph) threatened to turn ethnographic knowledge into something that defied scientific logic, since it was more closely affiliated with the senses than the intellect? In any event, early cinema, with its kineticism and tactility may have threatened the certainty of ethnographic knowledge at a time when theories of kinship and language, theories that were inherently resistant to visual representation, were displacing the pursuit of the visual correlates of racial hierarchies within early twentieth-century anthropology.

In this regard, Haddon's films display little of the obsessive care with which French physician Félix-Louis Regnault arranged the profilmic and filmic elements of his 1895 comparative physiological studies of the na-

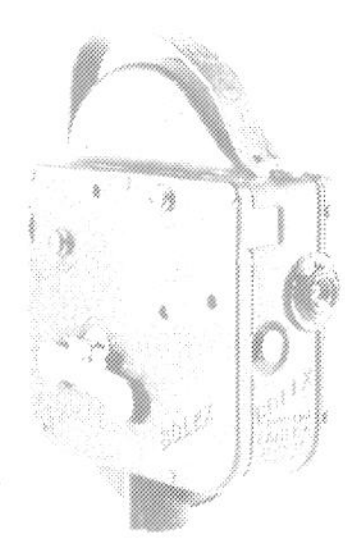

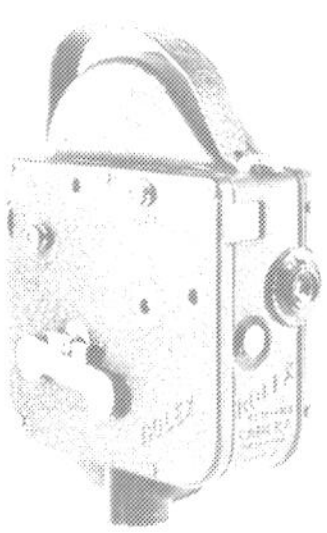

tive body, including background, blocking, and shot scale.[35] Distinct from Regnault's goal of the rigorous scientific recording of a taxonomy of human movement, Haddon's filmmaking seemed motivated more by a romantic longing to preserve Islander culture than a need to categorise people on the basis of human typology. However, if Regnault's and Haddon's respective interests in motion photography and cinematography were driven by their own specific research agendas (we mustn't forget that Regnault's background was as a physician), they both nevertheless saw in motion picture technology a *demonstrative* function which, Edwards argues, was akin to the visual thinking and planning taking place in the natural sciences. Like Regnault's chronophotographs, Haddon's films and photographs were undertaken primarily to verify the scientific evidence gathered first-hand on the Islands. As Edwards explains: 'The development of central photographic strategies was not intended as mere illustration, but as an integral part of the presentation, proof and transmission of evidence'.[36] But anthropological images are never produced in a social vacuum; in considering how a corpus of photographs or films materialises in ethnographic expeditions, we must also disentangle the intricate web of social relations which always enmesh image making. This is often easier said than done, although in cases where fieldwork diaries and other records are extant, it is possible to identify the kinds of social interactions that took place around visual technologies and their place in the fabric of indigenous lives. Surviving fieldwork diaries and other contemporaneous evidence suggest that photography was already familiar to Torres Strait Islanders and formed the basis for complex social interactions between Islanders and European scientists; as Edwards points out, 'the display, projection through magic lantern shows, exchange and gift of photographic images played a central role in the Expedition's social relations with the Torres Strait Islanders'.[37] Furthermore, as Edwards notes, the status of the photograph as a vehicle for social interaction was anticipated by Haddon in his 1898 trip to the Torres Strait, evidenced by his bringing along two magic lantern projectors and a collection of lantern slides of both general-interest subjects and of images taken of Islanders during the 1888 expedition.[38]

Other issues are more difficult to solve by recourse to the extant records of the expeditions, including the question of Haddon's own attitudes toward and use of his Torres Strait films. Writing from the field, Haddon admitted to the managers of his London film laboratory (Neuman and Guardia) that his inexperience and the formidable logistical difficulties of filmmaking would probably make his efforts futile and, indeed, Haddon did not attempt to use the film camera during the rest of his expedition. Additionally, Haddon never wrote about the ethnographic significance of his films, and exhibited them on no more than a handful of occasions, which may say something about their marginal status as ethnographic evidence within the corpus of materials brought back from the Torres Strait. Haddon, may, however, have had his own set of reasons for not promoting, exhibiting, or discussing the films upon his return, and it is difficult to say with certainty why the films were not afforded greater significance. Haddon's inability to feel fully in control of the cinematic apparatus, may, therefore, have been only partly to blame for his subsequent disinclination to exploit the films more fully. Lacking intertitles and entirely dependent on expert contextualisation in order to be made legible to both scientific and lay audiences, the films may have seemed less amenable to scientific explication than Haddon's photographs, which were used as note-taking devices and stylistically altered to enhance their ethnographic detail. This impediment suggests a new set of questions: What happens to ethnographic knowledge when film intertitles are introduced? Is a different kind of anthropological object produced when visual imagery is discursively mediated by the subjective first-person voice of the anthropologist-narrator in the form of intertitles? And what happens to ethnographic authority when the anthropologist appears in the frame? To consider such questions, let us turn to the ethnographic filmmaking of Norwegian anthropologist Carl Lumholtz.

In Borneo: The Land of the Head-Hunters: ethnography for the masses

An accomplished explorer and ethnographer,[39] Lumholtz set off for Central Borneo in 1913 when he was sixty-three years old, the first of a number of trips to the area he made between 1913 and 1917 (Fig. 2).[40] Although Lumholtz's training was grounded in nineteenth-century evolutionary theory and physical anthropology, he practiced a form of participant observation which was more the norm among twentieth-century professional anthropologists than among their nineteenth-century armchair-theorist predecessors. In 1903, Lumholtz's Mexican research received high praise from Haddon, who wrote that the Norwegian anthropologist had 'added a new chapter to the history of man'. According to Haddon, 'the native tribes in these parts of Mexico were hitherto unknown to anthropologists, and it is very fortunate that they had such an able and sympathetic observer as Dr. Lumholtz'.[41]

Lumholtz's expedition to Borneo was sponsored by Norwegian authorities, with supplementary funding from British and Dutch learned societies and American and British friends.[42] According to anthropologist Victor T. King, who wrote the introduction to the 1991 edition of Lumholtz's popular account of the expedition, *Through Central Borneo*, Lumholtz collected:

> ... anthropometric data from more than 200 Borneo natives; he assembled a large amount of cultural material on several different Dayak groups, including the Ngaju-Ot Danum (Duhoi), Bukit, Penihing (Aoheng), Kayan/Bahau, Saputan, Punan, Bukat, and Long Glat; he collected numerous ethnographic artifacts, especially from the Penihing, most of which are now housed in the Ethnographic Museum of the University of Oslo.[43]

Lumholtz's most complete discussion of the Borneo expedition took the form of an account he produced for a general readership in a style similar to Haddon's *Head-Hunters, Black, White and Brown*. Most of the text of the book seems to be taken directly from Lumholtz's diaries and notebooks, and in addition to descriptions of the tribes Lumholtz visited during the expedition, there are references to the difficulties of using photography in the tropical heat (grasshoppers, for example, would often eat away at strips of film that were hung out to dry).[44] Lumholtz hired Ag Sewey, a young Chinese photographer, to assist with the photographic and film work, and during the course of the expedition, both men grew adept at working in the hot, humid climate, developing film early in the morning when water left to stand in five-gallon oil tins overnight had cooled down to 76 degrees Fahrenheit, the maximum temperature required to process the film.[45]

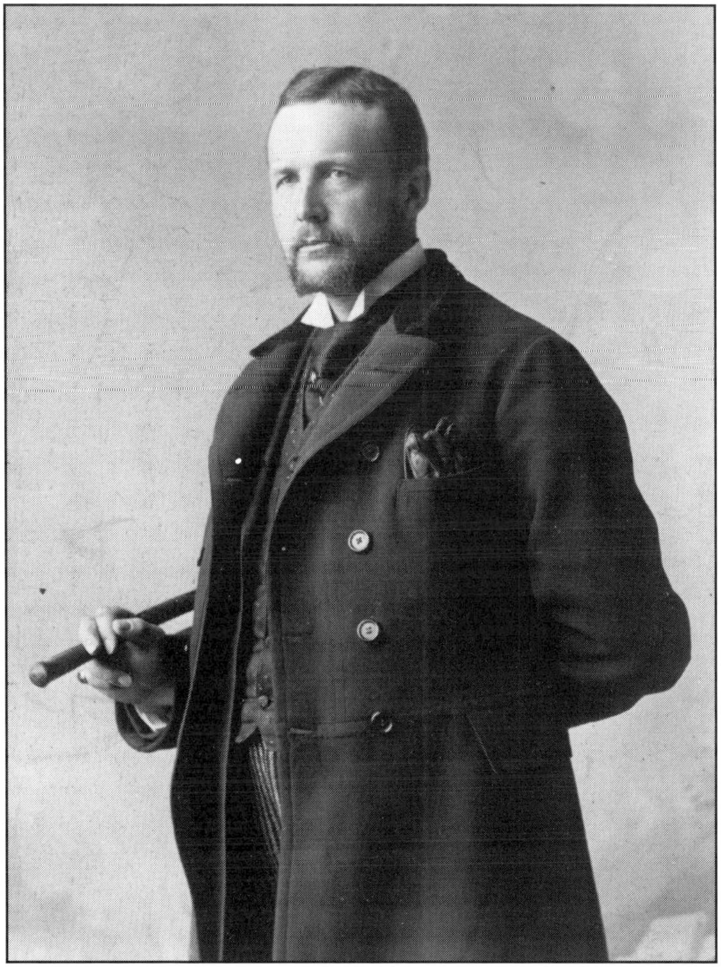

Fig. 2. Portrait of Carl Lumholtz taken in 1890 when he was 40 years old. Neg. no. 337176. [Courtesy Dept. of Library Services American Museum of Natural History.]

However, nowhere in *Borneo Head-Hunters* does Lumholtz discuss his interest in ethnographic cinema or his decision to take a motion picture camera along with him on the

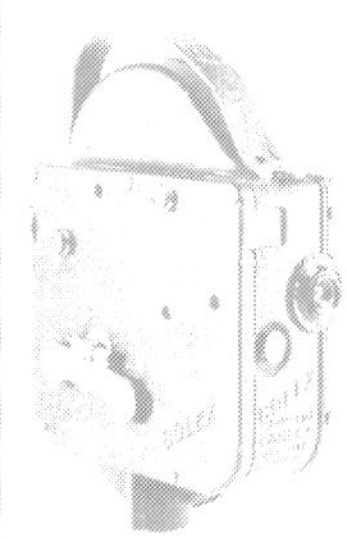

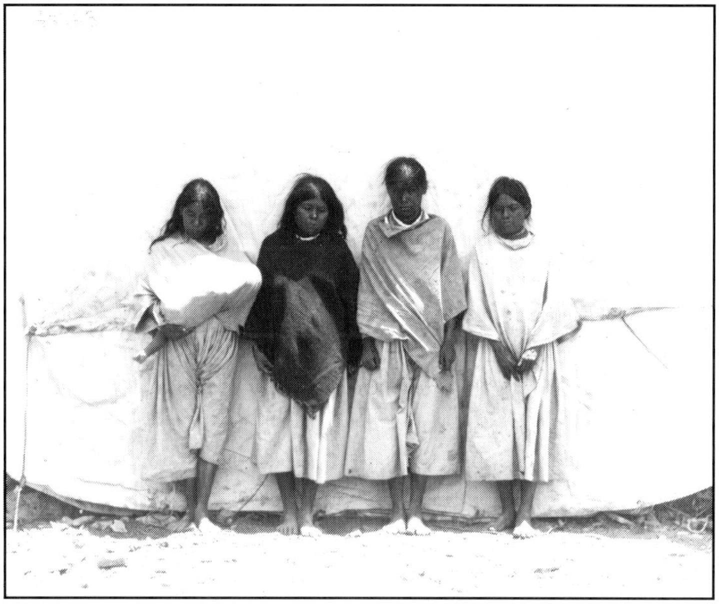

Fig. 3. Four Huichol women from Mexico photographed by Carl Lumholtz against a white backdrop c. 1890s. Neg. no. 43195, photo C. Lumholtz. [Courtesy Dept. of Library Services American Museum of Natural History.]

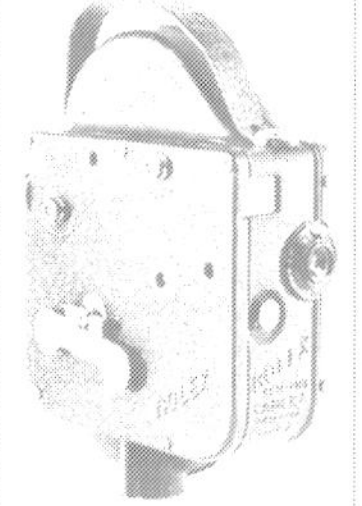

trip. In fact, motion pictures are mentioned only twice in the entire monograph (although Lumholtz does identify frame enlargements when they are used as illustrations): at one point Lumholtz mentions that the film camera was repaired by a Dutch engineer, and at another, to it being replaced by a second-hand Pathé machine. However, he does make repeated references to the reactions of various tribes to being photographed by his still camera.[46] The first incident arose during Lumholtz's visit to a kampong (village) of the Kenyah people; noting that permission to make photographs could usually be taken in exchange for payment in the form of money or goods, Lumholtz points out that there were, nevertheless, some exceptions:

> Women, as usual, were timid about being photographed, for it is a universal belief that such an oration prevents women from bearing children. However, by giving money, cloth, sugar, or the like, which would enable them to offer some little sacrifice to protecting spirits, I usually succeeded. But if a woman is pregnant or has care of a small child, no inducements are of any avail, as an exposure to the camera would give the child bad luck or a disease that might kill it.[47]

Lumholtz describes another occasion, when a group of Dayak women who agreed to be photographed rushed headlong into a river once he indicated the ordeal was over, so that they might 'cleanse themselves from the evil effects of the operation'.[48] Feelings of resentment and discomfort at being photographed register immediately with the spectator of this photograph of Mexican women taken during Lumholtz's expedition in 1898; the sterile white backdrop and averted gazes tell us a great deal about the women's perceptions of the photographic encounter (Fig. 3). While many native peoples apparently could be persuaded to pose for the camera and have anthropometric measurements taken, some tribes, such as the Kampongs, adamantly refused. Such anecdotes suggest the complex social interactions around visual technologies when they are used in anthropological fieldwork, and the fact that the camera was by no means perceived uniformly by the different tribes Lumholtz visited. Attitudes toward the camera seem to vary widely, from tribe members visibly shaking with fear or having tears in their eyes while being photographed or measured, to men and women seeming to relish the prospect of being able to dress up in their finest clothes and jewellery for the camera.[49]

Unfortunately, there is no discussion in *Borneo Head-Hunters* of the specific challenges of using motion pictures (as opposed to photography) to record indigenous life, although in the second intertitle of Lumholtz's film *In Borneo: The Land of the Head-Hunters*, audiences are reminded of the difficulties of using film in a tropical climate: 'Considering the extremely moist climate, and that clear photography was possible only a few hours of the day, we were fortunate in securing the many beautiful scenes of the islands and the native tribes'. Beyond the case of the pig sacrifice performed by members of the Murung tribe (discussed below), we have no way of knowing how Lumholtz went about securing the co-operation of the Bornean tribes, including the seemingly highly staged scenes such as the tableau of Katingan-Dayaks seated on a miniature rubber rhinoceros. While the camera's ability to get close to intimate situations such as the funeral of a medicine woman which occurred during the nine-day Tasa feast[50] may have been similar for both photography and film, one is left with the impression that Lumholtz had far less control over his indigenous subjects when shooting motion pictures than when taking anthropometric photographs (thus the preponderance of long and medium-long shots where Lumholtz had to keep his distance). As Angelica Hala observes, 'there is a distinct emotional [and cultural] distance between the filmmaker and the natives of Borneo. Lumholtz's camera never really gets close to his subjects; long and medium-long shots [allow the viewer to] observe from a distance but never connect more intimately to the person or persons in front of the camera'.[51] The Tasa feast thus provides a particularly vivid illustration of the 'gap between the filmmaker and his subjects, revealing a relationship between scientist and research subject rather than showing the interaction between a welcome visitor and his hosts'.[52] Interestingly, Lumholtz included frame enlargements of native life as well as more scientific photographs of Bornean tribes in *Through Central Borneo*; unlike the highly stylised anthropometric studies, which impart no sense of agency to the native peoples and represent them as little more than racialised specimens (men and women are photographed in frontal and profile medium shots), the film stills consist mostly of action shots (such as the ear piercing of a Saputan chief) in which groups (as opposed to individuals) are engaged in some kind of collective action. Representing the culture as lived and dynamic as opposed to petrified and static, these film stills point up the ontological differences between anthropometric photography and more candid glimpses of native culture. However, even in scenes such as the Saputan Chief ear piercing ritual, the camera remains a distant observer: if it weren't for the information furnished by the intertitles, the details of the procedure would be entirely lost on the spectator (for example, if not informed by an intertitle that an empty cartridge was used to make the hole which is plugged with the teeth of a tiger, we would have little idea of precisely what we were looking at).

In addition to his extensive photographic work, Lumholtz shot 48 minutes of motion picture film on his Borneo expedition which has survived in the form of an edited film entitled *In Borneo: The Land of the Head-Hunters*. The intertitles, which from the references and tone were written for a popular American audience, provide us with the most compelling evidence of for whom this film may have been intended, although precise information on where, when, and to whom the film was shown is sadly lacking. Despite the absence of specific information concerning the film's distribution and exhibition, the surviving print and, in particular, the intertitles, nevertheless illuminate the larger question of how ethnographic knowledge was mediated for commercial audiences in the late 1910s and early 1920s.[53] The jocular and ethnocentric tone of the film's intertitles and the inclusion of Katingan Dayak dancers and bare-breasted Murung Dayakas present us with the most compelling suggestion about the film's intended popular audience. References in the film's intertitles to Prohibition, 'dancing belles', and 'weird music' all seem designed to offer non-native spectators points of reference to the familiar genre of early twentieth-century travelogues. But at the same time that the film's intertitles doubtless lightened the tone and elicited laughter from some audience members, they also occasionally offered expert ethnographic commentary on the depicted events, frequently presented in the form of the first-person commentary by Lumholtz himself. Lumholtz's subjectivity as depicted in the intertitles is curiously protean, coming across as a glib, ignorant tourist in some scenes, a knowledgeable confidant of the native peoples in others, and occasionally, a representative of the colonial authorities.

While it is safe to assume that the extant print of Lumholtz's film was intended for commercial release, it is unclear whether Lumholtz's imputed subjectivity through the first-person voice of the intertitles was simply an invention of the title writer for the film's commercial distributor (who could have gleaned information about what was going on in each scene from *Through Central Borneo*, which was published in 1920) or reflects Lumholtz's own efforts to supply titles to the footage.[54] While I believe it is likely that Lumholtz was at least consulted about the preparation of intertitles, this uncertainty

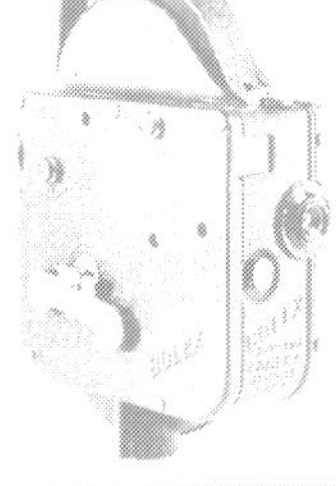

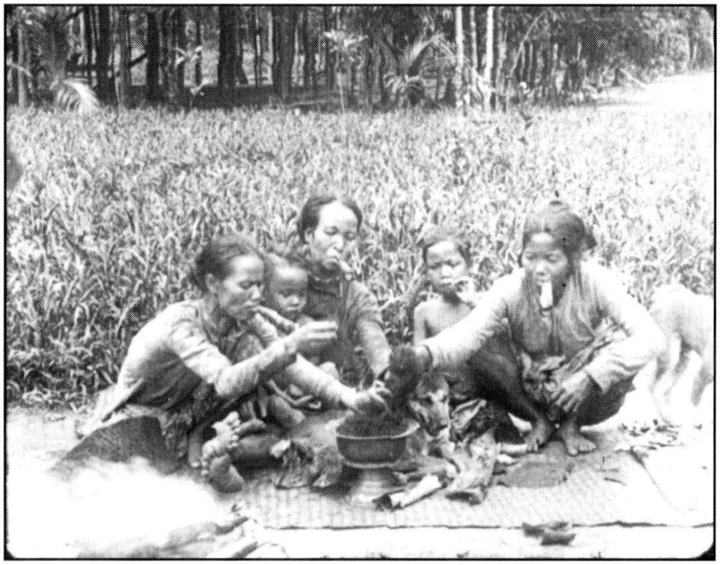

Fig. 4. Frame enlargement from Carl Lumholtz's film In Borneo: The Land of the Head-Hunters. *Neg. no. 243670, copy photo. J. Kirschner. [Courtesy Dept. of Library Services American Museum of Natural History.]*

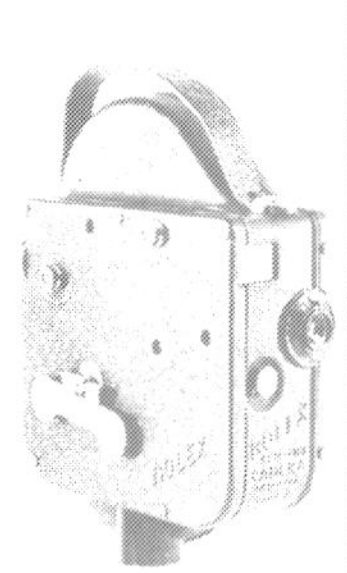

does not alter the fact that they were calculated to represent the subjectivity of the anthropologist through the first-person narrative.

Lumholtz's compilation footage of the various tribes in Central Borneo takes the viewer from Bandjermasin, a Dutch administrative centre, to more remote inland regions.[55] Lumholtz uses the visual motif of travel by canoe as a transitional device between sequences of the film; midway through the film there is a long scene in which Malay men standing on the shore navigate canoes safely through treacherous rapids by means of ropes attached to the vessels. Immediately following this scene, an intertitle proclaims that 'having won the confidence of the natives we were able to study their habits'. Ironically, just at the point where one would expect a more intimate portrait of indigenous life, the next four intertitles pull the spectator back to the subject position of a smug ethnocentric voyeur. For example, a swimming stroke is described as feminine ('you might not think it, but this Saputan swimmer is a man'); a young Murung Dayak displays a 'monkey-like agility' climbing a tree; local carpenters work with 'awkward looking adzes'; and finally, a young smoker, 'keeping pace with modern fashions', puffs on a nine-inch home-made cigarette (Fig. 4).

However, if seeing *their* world though *our* eyes (or, more specifically, Lumholtz's eyes) framed by the sometimes facetious intertitles indicates the film's persistent ethnocentrism, the success of the intertitles as a unifying discourse is never complete; there are several moments when the surfeit of ethnographic detail and the strong sense of native subjectivity undermine the intertitles' efforts to circumscribe meaning. An excellent example of this can be seen in the final sequence of *In Borneo*, in which a Penyahbong warrior performs a dance that stands out as one of the most memorable images of the film. Segueing from another war dance performed by Ot-Damuns of the Samba river, Lumholtz introduces the Penyahbong warrior dance with a personal confession: 'The war dances of the tribes are similar, but I never saw a more graceful dance than that given by this Penyahbong dancer'. In the first part of the dance, the warrior, dressed in a loin cloth, headband, and belt, performs rhythmic movements in which first the arms, then the entire body, rotate gracefully from right to left (Figs. 5 and 6); as Lumholtz explains in *Natural History* magazine: 'Before seizing his sword and shield and indulging in the more violent movements of the dance, he went through the preliminary of exercising all his flexible muscles. His motions were lithe as those of a serpent.'[56] This horizontal movement then shifts to a vertical axis as the warrior moves from an upright to a crouched position, performing graceful jumps in which the arms trace fluid movements. The setting of the dance contributes in no small way to the overall aestheticism of this scene; the smooth, light texture of the ground provides a striking contrast to the darker shades of the house located upper frame right, the dancer's body, and the dense undergrowth in the background. The shield and sword 'at first laid on the ground' are then taken up in part two of the dance, in which the warrior incorporates a large oval shield and sword into the dance. In the final scene of the Penyahbong dance, the exquisite movements of the dancer nullify the slightly condescending tone of the third intertitle: 'A musician playing a native clarinet came unexpectedly to contribute fresh inspiration to the dancer's performance.'

The image of the Penyahbong dancer can

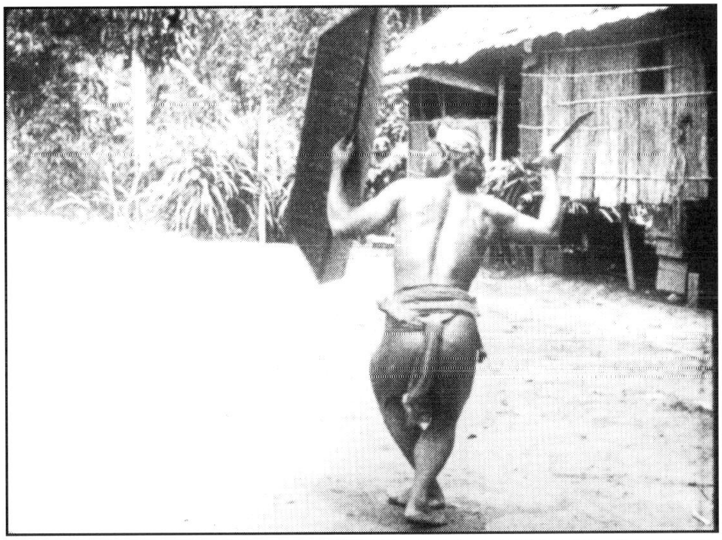

Fig. 5. Frame enlargement of Penyahbong dancer from In Borneo: The Land of the Head-Hunters. *In this second part of the dance, the warrior takes up the sword and the shield. Neg. no. 243667, copy photo. J. Kirschner. [Courtesy Dept. of Library Services American Museum of Natural History.]*

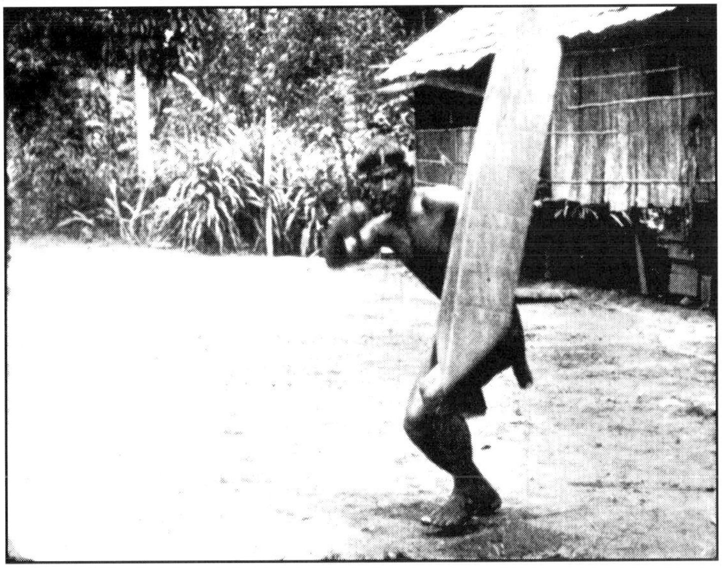

Fig. 6. Frame enlargement of Penyahbong dancer. Neg. no. 243668, copy photo. J. Kirschner [Courtesy Dept. of Library Services American Museum of Natural History.]

therefore be seen to challenge some of the *a priori* theoretical assumptions of the received notion of the 'colonial gaze' in contemporary criticism, forcing us to question the taken-for-granted imbalance of power in colonial looking relations. While obviously staged for the camera, the dance nevertheless conveys the beauty and majesty of Penyahbong performance, representing the dancer not as the victimised recipient of the cinema's colonial gaze (already a problematic construct) but as a subject with a degree of self-determination and agency who seems to enjoy the performance. The scenes representing the Penyahbong dancer thus represent native peoples not merely as the victims of colonialism, but as historical actors and interlocutors who use dance and indigenous ceremony in vivacious and culturally dynamic ways. Rather than attribute complete authority to the discursive power of the intertitle and its representation of the colonial situation in Borneo, it may be more accurate to read the film's colonial context as an example of Mary Louise Pratt's 'contact zone', where native cultures are never simply overcome by the presence of the colonizers but rather remain in productive tension with them.[57] As a space in which 'geographically and historically separated [peoples] come into contact with each other and establish ongoing relations', the contact zone incorporates the 'interactive, improvisational dimension of colonial encounters so easily ignored or suppressed by diffusionist accounts of conquest and domination'.[58] The film also suggests the dissonance between the illusion of complete control exerted by colonial authorities and a recognition of its fragility and discursive complexity.[59] While Lumholtz's images and intertitles make one acutely aware of the Dutch colonial presence in Borneo (including a scene of Dayaks working on plantations leased by the Dutch government and intertitle references to native guards patrolling the river under the command of a Dutch officer), these scenes are outnumbered by images documenting the ceremonial and material cultures of the numerous tribes living on Borneo, such as the nine-day Tasa feast referred to above.

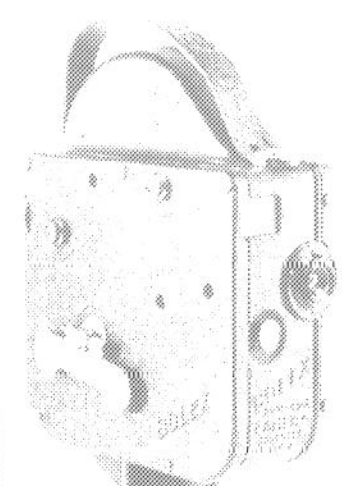

For American anthropologist Franz Boas, film records of ritualised dances such as those recorded by Lumholtz were of inestimable value for scientists; as the primary sign event, indigenous dance movements could be isolated by the motion picture camera and analysed for their broader ethnographic significance.[60] Lumholtz's tight framing of the Penyahbong dancers (with the camera closer to the performer than in any of the film's other dance sequences)[61] would have facilitated the kind of detailed analysis of indigenous dance that Boas sought, although we have to bear in mind that Boas used motion pictures to record native dances in 1930, many years after Lumholtz's efforts.[62] As an enduring iconographic trope within visual anthropology, the isolation and scrutiny of indigenous dance also suggests an attempt by anthropologists to compensate for their amateur status as filmmakers, for when the human body becomes the main attraction in the frame, the image takes on more the air of a highly structured scientific demonstration and less that of the casually composed and uninformed footage shot by an adventurer-explorer. Amateur filmmaking, with its hobbyist and prosaic connotations, threatened the professional codes of anthropology, since publicly exhibited footage which was badly shot may have drawn attention to anthropologists' inexperience as cinematographers at the expense of their skill as ethnographers. As Sol Worth has argued, 'for the most part, anthropologists ... are professional scientists only when they are employing words. When it comes to the visual mode of articulation and data-gathering, most produce snapshots [or] good (or bad) home movies'.[63]

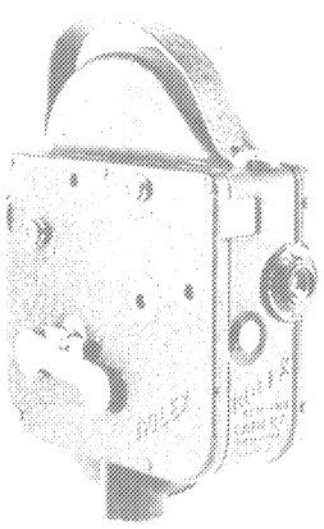

Other elements of Lumholtz's *In Borneo* resonate with contemporary problems in ethnographic filmmaking, including the inclusion of the filmmaker in the depicted events. Scenes revealing the anthropologist-filmmaker interacting with native peoples are rare in early ethnographic cinema. The image makers themselves, along with other evidence of western acculturation, were elided usually from ethnographic photographs and films; the exceptions are the adventurer-exploration genre films such as those produced by Martin and Osa Johnson in the late 1910s and 1920s for commercial distribution.[64] Thus, Lumholtz's appearance in the profilmic space of *In Borneo,* which occurs during the ceremonies associated with a sacrificial pig feast organised by the Murungs at Tumbung Marowei, is highly unusual in early ethnographic cinema. While the intertitles in this sequence provide a metacommentary on Lumholtz's disposition toward the sacrificial killing of the pig, they offer a slightly different version of the events leading up to the feast than his written account in *Through Central Borneo.* In the book's description of the scene, Lumholtz explains that it was only after he had been refused permission to photograph dancers unless a sacrificial feast was held, that an agreement was made to host a ceremony in which dancing could be filmed, providing Lumholtz agreed to pay the cost of the pig. Even though the feast was held in Lumholtz's honour – as we are told in the intertitles – it was initiated by him as a strategy to facilitate his filmmaking. Lumholtz's published account of the feast also reveals that he took it upon himself to demonstrate to the Murungs that he knew some of the steps of the dance before being led by the hand into the dancing circle. On the other hand, the film images and intertitles portray Lumholtz as something of a reluctant and spontaneous participant, including an intertitle which tells viewers that 'entering into the spirit of the ceremony, I felt *inclined* to join the dancers as I have done on many occasions'.[65] Thus does Lumholtz protest his own agency and volition as well as his experience and familiarity with the culture. Compare this to Lumholtz's published account:

> It did not look difficult so I joined in the dancing, as I have done many times among other races. Greatly to the amusement of the natives I demonstrated that I had caught the right steps, and then seated myself in a chair which was the pride of the kapala and which had been brought out for my benefit. While watching the performance I was surprised to see two of the women, about the only ones who possessed any charm of appearance, coming toward me, singing as they advanced. Each took me by the hand and, still sing-

ing, led me forward to the dancing circle ...[66]

For most viewers, the image of Lumholtz being led into the circle of dancers, replete with pith helmet and colonial garb, is unexpected, eliciting a comic reaction, as the anthropologist 'goes native' in front of the camera, which might explain why it is prefaced by the qualifying remarks of the preceding intertitle. The description of the feast in Lumholtz's book provides a richer account of its meaning and of his role in it than is offered in the film, although what we lose in ethnographic detail in the film version we gain in specular thrill, for it is only when we finally *see* Lumholtz cross the line between filmmaker and film performer that we appreciate the contingencies of cross-cultural interaction. But what effect does Lumholtz's presence have on the construction of ethnographic knowledge in the scene? What happens when the technology is turned on the ethnographer?

The cinematic document undergoes a number of transformations when the ethnographer enters the profilmic space. First, the exigencies of the contact zone are visually inscribed for the spectator. We *see* the lone ethnographer led by the hand to participate in the drinking and dancing; he is an initiate, an outsider, who looks and dresses differently from the Murungs and whose gait and demeanour betray a great deal about his ambivalent status as one of 'them' (the colonizers) and yet co-present with the Murungs in terms quite distinct from the Dutch. Secondly, Lumholtz's appearance in the profilmic space draws our attention away from the ethnographic detail of the scene as we are thrilled by the novelty factor of seeing our anthropologist-hero interacting with the people he up until now has been omnisciently observing. Finally, our identification with Lumholtz may in fact be enhanced by his presence as an outsider at this feast (he is both present in the profilmic space and in the subjective voice of the intertitles). Lumholtz's decision to appear in this scene therefore introduces a new set of issues regarding the authority of the ethnographic image and the evocation of ethnographic intersubjectivity. Indeed, the full implications of such questions flowing from Lumholtz's reflexive gesture have yet to be resolved by contemporary ethnographic filmmakers and theorists who continue to grapple with these epistemological imponderables.

Conclusion

It is clear from even a brief account of the status of the visual in the history of anthropology that visual technologies have by no means achieved a secure home within anthropology. Likewise, the paradoxes that have plagued ethnographic data collection for the last century are yet to be resolved. In our own era, emerging digital technologies of image recording, manipulation, distribution and display bring with them their own epistemological, political and ethical concerns, and are as likely to pose as many new questions about ethnographic image-making as solutions to issues of access, authenticity and accountability. While anthropology has become more sensitive to the politics of representation and the ethical dimensions of trans-cultural media making, these new technologies will re-map the experiences of Self and Other, touch and sight, ethnographic immersion, scientific investigation and armchair travel. There is at least preliminary evidence to suggest, however, that peoples who traditionally have been the subject of the ethnographic gaze are increasingly inserting themselves in the new media landscape by taking up the tools of production and organizing within communities and across international borders.[67]

At the level of fieldwork, digital technologies may help bridge the customary gap between field-site and university, making it possible for anthropologists to move beyond the gesture of playing back recorded sound or motion pictures to their native informants (which Haddon himself did a hundred years ago with phonograph and photographic field recordings) by involving native peoples more directly and meaningfully in the data collection process. Anthropologists Michael Fischer and David Zeitlyn, for example, recently used a digital camcorder to record images used as prompts for subsequent interviews with local people; the two anthropologists also assembled rough edits of their film material on a laptop computer in

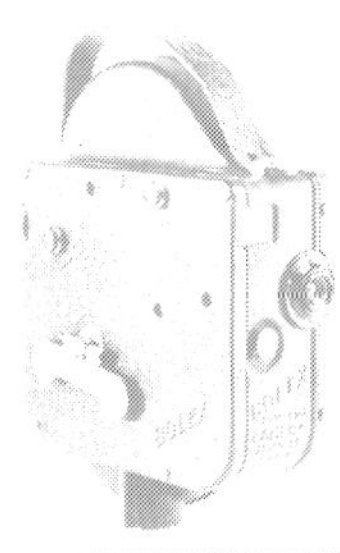

the field.[68] Zeitlyn and Gustaaf Houtman argue that new technology is affecting anthropology on the level of data, method and subject, ironically returning anthropology to some degree to its nineteenth century origins in the armchair ethnographer:

> By collapsing the geographical divide between home and field-site, hitherto an assumed element in anthropology's self-definition, the concept of fieldwork is changing from a Malinowskian model. Indeed, with time, it may yet 'plunge us back into the armchair', by affording such rich data set in addition to personal field-notes.[69]

With the publication of anthropological monographs on CD-ROMs and Web sites, ethnographic representation will become an integral part of multi-platform presentations, recalling Haddon's attempt at a multi-media version of his Torres Strait material delivered before the British Anthropological Institute and Royal Geographical Society in 1900. Putting together lantern slides, films and sound recordings into an evening's lecture on the expedition (even attempting sync-sound with the films and phonograph recordings of the Malu-Bomai ceremonies), Haddon anticipated what would become technically possible through digital technologies a century later. However, the relationship of the institutions of anthropology to new visual technologies continues to be ambivalent and uncertain. The growth of sophisticated interactive exhibits and the proliferation of IMAX screens in museums of natural history represents an attempt to strengthen the traditional museum space as civic centre and tourist destination. At the same time, other observers see the logic of digital media leading to the development of the virtual museum on the place-less World Wide Web, dissolving the spatial boundaries of field, museum, and living room of ethnographic experience. In any event, emerging digital technologies in the service of anthropological data gathering and museum display will bring new urgency to the problem of visual evidence which has pre-occupied ethnographers for over a century.

Acknowledgements: My thanks to Jake Homiak, Human Studies Film Archive, Smithsonian Institution, Washington DC for supplying me with a copy of Lumholtz's film, and to Paula Willey, Special Collections Manager at the American Museum of Natural History for assistance with illustrations. I am also grateful to William Boddy, John Fullerton, Tom Gunning, James Latham, and Gunnar Iversen for comments on an earlier draft of this essay.

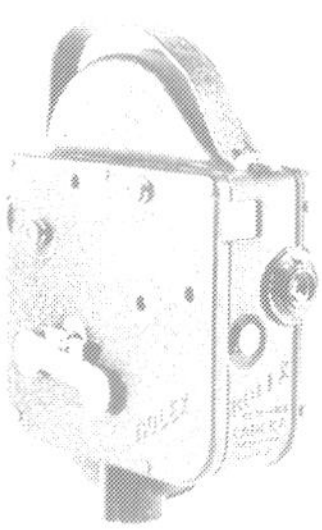

Notes

1. Alan Lomax, Irmgard Bartenieff, Forrestine Paulay, 'Choreometrics: A Method for the Study of Cross-Cultural Pattern in Film', *Research Films* 6, 6 (1969): 506.
2. For more on the filmmaking sponsored by the Institute für den Wissenschaftlichen Film, see Martin Taureg, 'The Development of Standards for Scientific Films in German Ethnography', *Studies in Visual Communication* 9, 1 (Winter 1983): 19–29.
3. David MacDougall, 'The Visual in Anthropology', in Marcus Banks and Howard Morphy (eds.), *Rethinking Visual Anthropology* (New Haven: Yale University Press, 1997), 292.
4. Jonathan Crary, *Techniques of the Observer: On Vision and Modernity in the Twentieth Century* (Cambridge, Mass.: The MIT Press, 1992), 3
5. William Henry Flower, 'President's Address', *Journal of the Anthropological Institute* XI (1882): 184.
6. C. Read, 'Prefaratory Note', *Notes and Queries*, third edition (1899), 87.
7. Eadweard Muybridge's more famous human and animal locomotion studies employed a similar grid-like backdrop. For a critique of Muybridge's quasi-scientific experimentation with serial photography, see Marta Braun, *Picturing Time: The Work of Etienne-Jules Marey* (Chicago: University of Chicago Press, 1992), 228–262.
8. J.H. Lamprey, 'On the Method of Measuring the Human Form for the Use of Students of Ethnology', *Journal of the Ethnological Society of London* 1 (1869): 85.
9. Letter from Huxley to Lord Granville, 8 December 1869, Huxley Manuscripts, Imperial College of Science and Technology, London, cited in Frank Spencer, 'Some Notes on the Attempt to Apply Photography to Anthropometry During the Second Half of the Nineteenth Century', in Elizabeth

Edwards (ed.), *Anthropology and Photography 1860-1920* (New Haven, London: Yale University Press in association with the Royal Anthropological Institute, 1992), 99–100. For a discussion of anthropometry and race, see Stephen Jay Gould, *The Mismeasure of Man* (New York: Norton, 1981).

10. David Spurr, *The Rhetoric of Empire: Colonial Discourse in Journalism, Travel Writing, and Imperial Administration* (Durham: Duke University Press, 1993), 23.
11. For an analysis of how the methods of anthropometric photography found application in nineteenth-century criminology studies conducted by the French police officials Alphonse Bertillon and Francis Galton, see Alan Sekula, 'The Body and the Archive', *October* 39 (1986): 18–55.
12. William Henry Flower, cited in Christopher Pinney, 'Colonial Anthropology in the Laboratory of Mankind', in C.A. Bayly (ed.), *The Raj and the British 1600–1947* (London: National Portrait Gallery, 1990), 253.
13. For more on the relationship between discourses of visuality and racial classification in nineteenth-century anthropology, see Fatimah Tobing Rony, *The Third Eye: Race, Cinema and Ethnographic Spectacle* (Durham: Duke University Press, 1995), 21–43.
14. Frank Spencer, 'Some Notes', in Edwards, *Anthropology and Photography*, 102. Spencer points out that it took until the 1890s for anthropologists to agree upon standard methods of measuring the living body. This led to the Monaco Agreement for the unification of craniometric and cephalometric measurements of 1906, 106n.
15. Christopher Pinney, 'Classification and Fantasy in the Photographic Construction of Caste and Tribe', *Visual Anthropology* 3, 2 (1990): 284.
16. See Barthes' influential essay 'The Rhetoric of the Image' for a discussion of how captions anchor the connotative meaning of images, in Roland Barthes, *Image, Music, Text*, trans. Stephen Heath (New York: Hill and Wang, 1977), 32–51.
17. For an interdisciplinary account of the expedition, see Anita Herle and Sandra Rouse (eds.), *Cambridge and the Torres Strait: Centenary Essays on the 1898 Anthropological Expedition* (Cambridge: Cambridge University Press, 1998). For more on the history of Torres Strait Islander contact with the institutions of colonialism, see Jeremy Beckett's essay, 'Haddon Attends a Funeral: Fieldwork in the Torres Strait, 1888, 1898', in ibid., 23–49 and his book *Torres Strait Islander: Custom and Colonialism* (Cambridge: Cambridge University Press, 1987). A documentary celebrating the centenary of the 1898 Cambridge expedition was made for the BBC2 *Horizon* series. In it, the renowned psychologist-director Dr. Jonathan Miller re-traced Haddon's journey to the Torres Strait. For a critique of the film, see Jude Philip, 'Expectations – *Dr. Miller and the Islander*', *Anthropology Today* 14, 2 (April 1998): 16–18.
18. For background information on Haddon's interest in the Malu initiation rites and the negotiations leading up to the performance of the Malu-Bomai ceremony, see Anita Herle, 'The Life-Histories of Objects: Collections of the Cambridge Anthropological Expedition to the Torres Strait', in Herle and Rouse (eds.), *Cambridge and the Torres Strait*, 90–94. In this dance, the senior mentor (*zogo-le*) utters 'the sacred words and sings the chant about Malu in front of the initiates (*kersi*)' (92).
19. See my essay 'Knowledge and Visuality in Turn-of-The-Century Anthropology: The Early Ethnographic Cinema of Alfred Cort Haddon and Walter Baldwin Spencer', *Visual Anthropology Review* 12, 2 (Fall/Winter, 1996/97): 18–43. The *Visual Anthropology Review* essay and the present essay form part of a larger investigation into the origins of ethnographic film, forthcoming from Columbia University Press.
20. MacDougall, 'The Visual in Anthropology', 289.
21. Herle, 'The Life-Histories of Objects', 90. The following information on the Malu-Bomai ceremony is taken from Herle.
22. Alfred Cort Haddon, *Head-Hunters: Black, White and Brown* (London: Methuen, 1901), 46.
23. Herle, 'The Life-Histories of Objects', 95.
24. Elizabeth Edwards, 'Performing Science: Still Photography and the Torres Strait Expedition', in Herle and Rouse (eds.), *Cambridge and the Torres Strait*, 110.
25. As Herle and Rouse point out in the Introduction to *Cambridge and the Torres Straits*, at the time of the expedition, the spelling of Torres was plural whereas the correct current spelling is singular (1).
26. *Head-Hunters*, 47. The masks are housed in the Museum of Archaeology and Anthropology at Cambridge University. For a provocative exploration of the politics and ethics of western museums

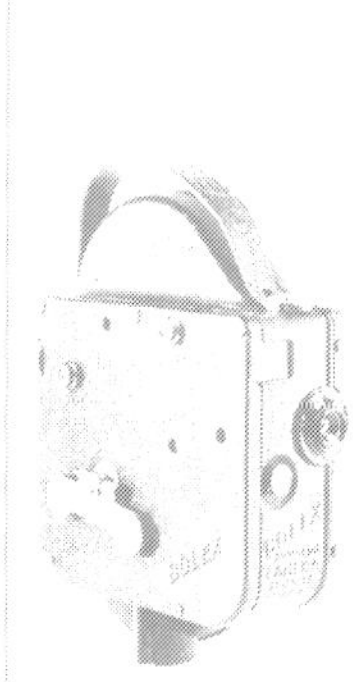

retaining the rights to indigenous artifacts that were taken from native communities in the nineteenth century, see Frances Calvert's film *Cracks in the Mask* (1997) which had its US premiere at the 1998 Margaret Mead Film and Video Festival in New York City. The film follows the journey of a Torres Strait community leader and his wife as they travel to European museums housing Torres Strait Islander collections. The man's reaction to seeing the Malu Zogole masks being unpacked for him by Dr. Anita Herle at the Museum of Archaeology and Anthropology, and request to have the objects returned to Mer Island, raise questions about the legality and morality of curatorial policies that deny indigenous peoples the rights to reclaim their material culture as part of a repatriation programme.

27. Herle, 'The Life-Histories of Objects', 93.
28. According to Edwards, 'The *appearance* of the 'primitive' in performance of primitivist tropes such as dance, clothing, tattoo [in the photographs], is played out for the camera, despite the fact that the coastal villages were heavily missionized by this date and that the Expedition party ... visited in the company of colonial officers and missionaries. ...' Edwards, 'Performing Science', 113.
29. Ibid., 116.
30. Ibid., 110.
31. As defined by Webster's New Collegiate Dictionary, 1977.
32. Noël Burch, 'Primitivism and the Avant-Gardes: A Dialectical Approach', in Philip Rosen (ed.), *Narrative, Apparatus, Ideology* (New York: Columbia University Press, 1987), 483–506. See also Antonia Lant, 'Haptical Cinema', *October* 75 (1995): 45–73.
33. Frances Hubbard Flaherty, *The Odyssey of a Film-Maker: Robert Flaherty's Story* (Putney, Vermont: Threshold Books, 1984), 58, cited in Michael Taussig, *Mimesis and Alterity: A Particular History of the Senses* (New York: Routledge, 1993), 200. Flaherty's book was first published by the University of Illinois Press in 1960.
34. Laura U. Marks, 'Video Haptics and Erotics', *Screen* 32, 4 (Winter 1998): 332–333.
35. On Regnault's chronophotography, see Rony, *The Third Eye*, 21–73.
36. Edwards, 'Performing Science', 120.
37. Ibid., 121.
38. Ibid., 122.
39. Lumholtz's research in Australia and Central America had given him an international reputation as an ethnographer. As part of an American Museum of Natural History's (AMNH) sponsored trip to Mexico in 1890 (the first of six trips to Mexico – Lumholtz returned in January 1892; spent three years in Mexico between March 1894 and March 1897; was accompanied by renowned physical anthropologist Aleš Hrdlička in 1898; travelled alone in 1905; and made a final trip in 1909–10), Lumholtz studied the Tarahumara, Huichol, and Tepecen Indians. See Victor T. King, 'Introduction' to Carl Lumholtz, *Through Central Borneo* (New York: Charles Scribner's Sons, 1920; republished by New York: Oxford University Press, 1991), v. His interest in recording techniques long pre-dated the Borneo expedition; according to Norwegian scholar Gunnar Iversen, Lumholtz used a phonograph to record native songs during his fourth trip to Mexico in 1898 (personal correspondence with author).
40. Lumholtz was one of the first explorers to publish on the cultures of Borneo in English (most available information was either in Dutch or German), and his popular account of the expedition, *Through Central Borneo*, consisted mainly of material culled from his diaries and notebooks, King, 'Introduction', xii. The sequence of trips to Borneo makes it difficult to date precisely Lumholtz's film of the expedition, *In Borneo: The Land of the Head-Hunters*. All we can assume is that the film was shot sometime between 1913 and 1917. The opening intertitle of *In Borneo: The Land of the Head-Hunters* describes Lumholtz as the author of '*Through Central Borneo, Unknown Mexico*, and other works'. Given that *Through Central Mexico* wasn't published until 1920, one can assume that the film was commercially released after the publication of the book, probably sometime around 1920.

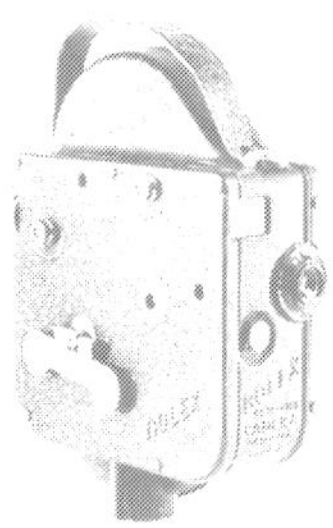

41. Carl Lumholtz, 'Explorations in Mexico', *The Geographical Journal* XXL (1903), 139, (comments by Haddon and Gadow on 139–42) cited in King, 'Introduction', x.
42. In fact, Lumholtz's Borneo collection was housed at the AMNH until in 1927, when after six months of negotiations between Lumholtz's brother Ludwig (executor of Lumholtz's estate) and the Museum administration, who wanted to purchase the collection (which consisted of a 'thousand pieces and

several hundred photographic negatives'), the AMNH finally lost the collection to the Ethnographic Museum at the University of Oslo (where the collection is still housed) because of a delay in responding to Ludwig. After having failed to secure internal funding for the $4,000 Ludwig was willing to accept for the collection (it had been valued at $15,000 and Ludwig was hoping for $5,000), the Museum dragged its feet and it wasn't until $5,000 was offered by J.P. Morgan that the AMNH was able to make Ludwig a firm offer, but unfortunately, it came too late. On 25 July 1927, Ludwig cabled to say he had sold the collection. There is, however, no mention of Lumholtz's motion pictures in the correspondence. See Central Archives file #1209 (1926–27) G-L for correspondence on this deal. My thanks to Paula Willey for drawing my attention to these efforts to secure the collection for the AMNH.

43. King, 'Introduction', xii. In addition to artifacts, the Ethnographic Museum of the University of Oslo houses Lumholtz's papers and a large collection of photographs. The AMNH also has a large collection of Lumholtz's Mexican photographs.
44. Lumholtz, *Through Central Borneo*, 33.
45. Ibid., p. 30. Lumholtz later recruited Mr. J. Demmini, a photographer in the well-known Topografic-ische Dienst in Batavia, although Demmini left the expedition before its completion due to illness (109, 163). The rain and humidity made filming extremely difficult; however, even the dry season brought with it problems. Lumholtz had to place mats over the iron boxes housing the film equipment to keep it cool, but even so, the camera and plates still got warm. Lumholtz recalls that 'when I photographed, perspiration fell like rain-drops' (71).
46. Ibid., 60, 109.
47. Ibid., 75. Payment negotiations were by no means straightforward. On one occasion Lumholtz refers to three female Murung blians (priest-doctors) requesting two hundred florins and nine tins of rice for the privilege of being filmed. Despite a price reduction, Lumholtz would not pay the requested sum and ended up inducing two male blians to dance for him instead (135).
48. Ibid., 300.
49. The man trembling in front of the camera was a member of the Long-Glat tribe (ibid., 263); the woman with tears in her eyes from the Bukats (217). Raja Bear, from the Kampong tribe, took enormous pride in his appearance, making 'altogether a splendid subject for the camera' (229–30).
50. The Tasa feast was given every three years to promote the good health of children. In one ceremony that Lumholtz filmed, a medicine woman strokes another woman's arm with a knife surrounded by four young girls (the knife, we are told in an intertitle, could be substituted for a palm leaf).
51. Angelika Hala, 'Observing from a Distance – Ethnographic Film in Its Beginnings', unpublished paper presented in 'Documentary Film and Television' seminar, Department of Speech, Baruch College, City University of New York, Spring 1999, 3.
52. Ibid.
53. The only concrete reference I have to a screening in Scandinavia is from Gunnar Iversen, who has come across a record of the film being screened in a lecture given by Lumholtz in Oslo in May 1920. For a discussion (in Norwegian) of Lumholtz's filmmaking, see Gunnar Iversen, 'Travel as Document: On Carl Lumholtz as a Maker of a Documentary', *Z* 3 (1994): 53–57. My thanks to Gunnar Iversen for sharing this information with me.
54. My thanks to Tom Gunning for pointing out this issue. It is worth bearing in mind that Lumholtz died two years after the publication of *Through Central Borneo* (1920) in 1922, suggesting that prints of his film circulating in the 1920s quite easily could have intertitles that were authored by someone other than himself.
55. For more on ethnographic travelogues from the early cinema period, see Alison Griffiths, ''To the World the World We Show': Early Travelogue as Filmed Ethnography', *Film History* 11, 3 (September 1999): 282–307.
56. Lumholtz, 'My Life of Exploration', *Natural History* XXI, 3 (May–June 1921): 242.
57. According to Mary Louise Pratt, contact zones are 'social spaces where disparate cultures meet, clash, and grapple with each other, often in highly asymmetrical relations of domination and subordination', *Imperial Eyes: Travel Writing and Transculturation* (New York: Routledge, 1992), 4.
58. Ibid., 6–7.
59. For more on the complex power relations at play within colonialism, see Nicholas Thomas, *Colonialism's Culture* (Princeton: Princeton University Press, 1994).

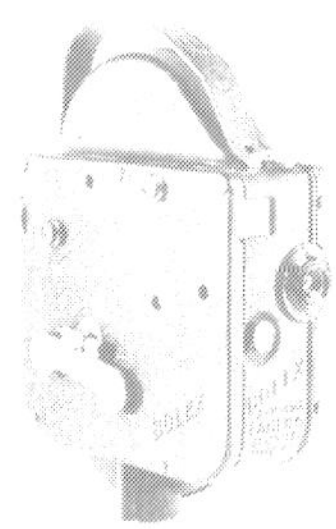

60. Boas was less interested in cinema's ability to represent culture in holistic terms than its function as the visual equivalent of written fieldnotes, raw data that could be accumulated and analysed according to culturally specific schemata. According to visual anthropologist Jay Ruby, Boas's films only make sense if 'one believes that behavioral events removed from their normal social and physical contexts retain sufficient validity to reveal patterns of culture'. Jay Ruby, 'Franz Boas and the Early Camera Study of Behavior', *The Kinesis Report* 3, 1 (1980): 8.
61. It may or may not be significant that Lumholtz chose the Penyahbong dance as the final sequence of *In Borneo*.
62. Of course, Boas's films were not intended for general audiences, but were visual illustrations designed to accompany his written fieldnotes. But Boas was not averse to the idea of ethnography being produced for popular audiences, and in 1933 responded enthusiastically to a request by Will H. Hays, President of the Motion Picture Producers and Distributors of America, for anthropological cooperation on a film project documenting indigenous tribes threatened with extinction. While this project never materialised, it suggests that anthropologists such as Boas viewed collaboration with the motion picture industry as a means of funding anthropological research; scenes of anthropological value would serve the interests of scientists and educators and be housed in museum archives as well as appear as spectacular sequences in commercially released exploration films.
63. Sol Worth, *Studying Visual Communication*, edited by Larry Gross (Philadelphia: University of Pennsylvania Press, 1981), 63, cited in Jay Ruby, 'Exposing Yourself: Reflexivity, Anthropology, and Film', *Semiotica* 30, 1/2 (1980): 179.
64. A notable exception is the Austrian anthropologists and ethnographic filmmaker Dr. Rudolph Pöch, who in 1904–06 shot films in German, British, and Dutch New Guinea, and in New South Wales. According to Fatimah Tobing Rony, Pöch's proclaimed commitment to an observational style of filming is belied in the films themselves, which often show Pöch in colonial garb replete with rifle posing alongside self-conscious native peoples. Pöch nevertheless recognised the fact that his vision of an omniscient ethnographic camera-eye could never be practically achieved due, to the camera's monocular perspective, see Rony, *The Third Eye*, 66.
65. Emphasis added.
66. Lumholtz, *Through Central Borneo*, 118. The intertitle introducing Lumholtz's entry into the dance and staging of the shot in which a chair has been strategically positioned in the lower right-hand corner of the frame suggests a staged rather than spontaneous act: 'After I had been impressed by the grace of the dance, the most beautiful maiden of the tribe led me to the brandy bowl.'
67. For an excellent discussion of the politics of indigenous media making as well as projects being undertaken among Aboriginal communities in Australia, see Faye Ginsburg, 'Culture/Media: A (mild) Polemic', *Anthropology Today* 10, 2 (April 1994): 5–15 and 'The Parallax Effect: The Impact of Indigenous Media on Ethnographic Film', in Jane Gaines and Michael Renov (eds.), *Visible Evidence* (Minneapolis: University of Minnesota Press, 1999), 156–175.
68. Michael Fischer and David Zeitlyn, 'Visual Anthropology in the Digital Mirror', unpublished manuscript. For more on Fischer's experience of using new technology, see *Applications in Computing for Social Anthropology* (London: Routledge, 1994). Zeitlyn and Fischer also have a web site which deals with many of these issues: http://lucy.ukc.ac.uk/dz/it-va/it.html.
69. Gustaaf Houtman and David Zeitlyn, 'Information Technology and Anthropology', *Anthropology Today* 12, 3 (June 1996): 2.

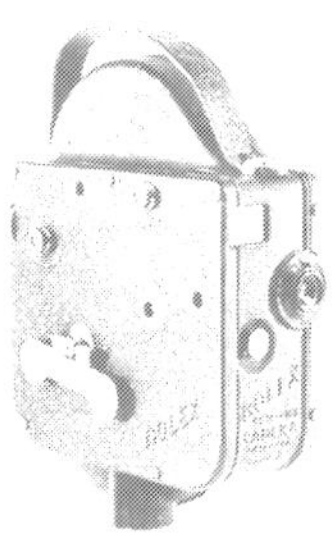

Architectonics of Seeing: Architecture as Moving Images

Pelle Snickars

Department of Cinema Studies, Stockholm University, Sweden

Space is never empty;
it always embodies a meaning.
– Henri Lefebvre[1]

Introduction: Eisensteinian architecture

Sergei Eisenstein seems to have believed that one of the ancestors of cinema was architecture. The ancient Greeks, he writes, 'have left us the most perfect examples of shot design … Acropolis of Athens could just as well be called the … most ancient [of] films'.[2] Eisenstein's thoughts can be found in 'Montage and Architecture', an essay written towards the end of the 1930s. For the soviet filmmaker the buildings on the Acropolis were first and foremost a montage of carefully enframed spatial views. The Parthenon, for example, faces the spectator obliquely, he notes, just like a calculated shot, thus becoming even more picturesque. According to Eisenstein, the origins of cinema – or more precisely, cinematic perception – were ancient architecture, since, as he puts it, 'it is hard to imagine a montage sequence … more subtly composed, shot by shot, than the one which our legs create by walking among the buildings of the Acropolis'.[3]

If one follows Eisenstein, it seems that film incorporates a number of the spatial characteristics of architecture. There are, of course, many ways of relating film to architecture. Within the field of cinema studies, some studies have related the analysis of set design to a consideration of the role of buildings in feature films. Such analysis has included a discussion of the narrative quality of architecture and city space in films like Fritz Lang's *Metropolis* (1927), King Vidor's *The Fountainhead* (1949) and Ridley Scott's *Blade Runner* (1982).[4]

Let us take a different trajectory. What I would like to do is something quite the contrary, a presentation of *cinematic traits in architecture*. My essay concerns what could be termed a cinematographic architecture. I would like to look at architecture as a technology of moving images, drawing upon examples from Le Corbusier and what has come to be known as 'deconstructivist' architecture. What interests me is modern architecture's dependence on the moving image and the way in which cinema has influenced the art of building. Cinematic technologies have had a crucial impact on how to perceive architecture and, consequently, on construction.

Architectonic *mise en abyme*

One might argue that the genealogy of an architectonics of seeing is to be found in the spatial intersection between modernity and early cinema. It is well known that early non-fiction film often portrayed cities and urban locations, and that these films were popular. In a recent essay, Tom Gunning has noted that the city in itself was a form of visual

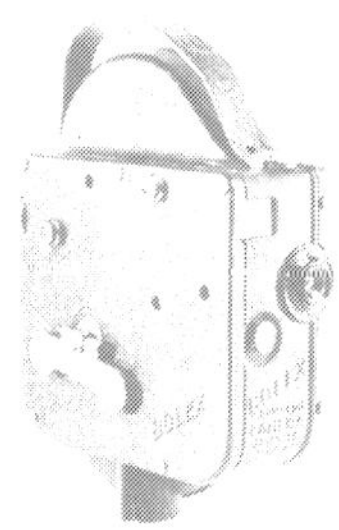

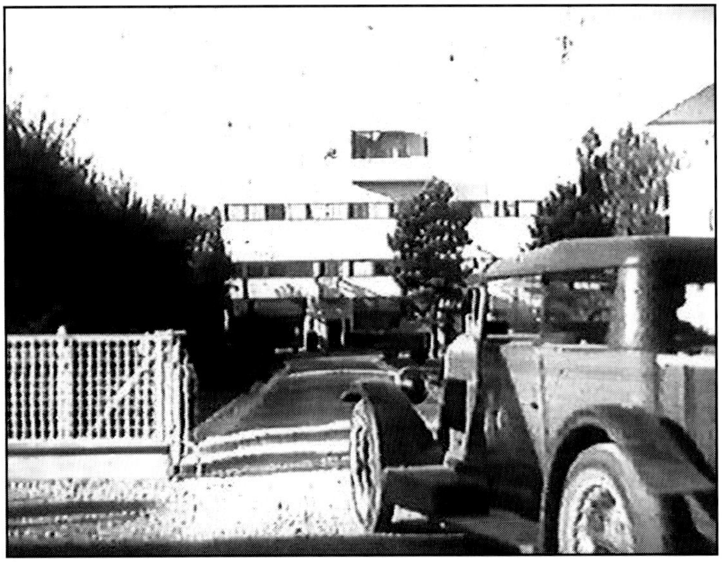

Fig. 1. Le Corbusier driving his car towards the Villa à Garches (Le Corbusier/Pierre Jeanneret, 1927). Still from Pierre Chenal's film, Architectures d'aujourd'hui *(1930–31).*

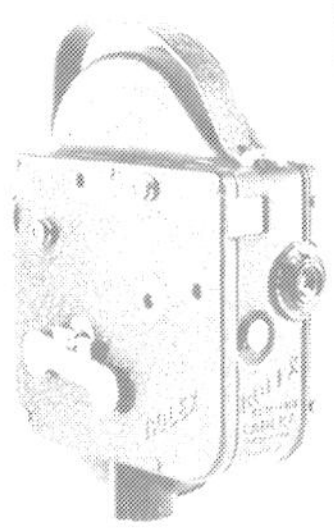

display of unfolding spaces. Basically, these spaces only needed to be captured on film to become commercial entertainment. 'Thus, the first film shows fit smoothly into established patterns of big city visual entertainment', Gunning writes. 'But what is somewhat surprising is the popularity ... of city street scenes in these first exhibitions. Nearly all early film shows presented a *mise en abyme* of audiences filling vaudeville theatres from busy city streets in order to see projected on the screen – busy city streets.'[5]

The film *Panorama of the Flatiron Building*, produced by the American Mutoscope and Biograph Company and shot by Robert K. Bonnie in 1903, might be considered one example of the architectonic *mise en abyme* character of early cinema.[7] The film depicts street life in front of the Flatiron building, designed by Daniel Burnham and erected in 1902. The building seems to have made a public impression as an architecture of motion. A contemporary observer described it as a *moving* 'ocean steamer' ploughing through the city.[8] In short, the filmed view of *Panorama of the Flatiron Building* is essentially the same as one would encounter on a busy street. However, as Gunning argues, cinema had the ability of framing the endless spatial flow of the street scene, thereby making it accessible to the viewer. 'The motion picture intervenes', Gunning concludes, 'not by organising [city space], but by capturing it in a form which allows endless repetition, opening the way for a studied apperception'.[9]

Le Corbusier and *Architectures d'aujourd'hui*

If early cinema portrayed the architecture of the city within the city, the relation of buildings to the representation of space is logical. Architecture has, of course, always been tied to questions of representing space. During the twentieth century, however, the static art of building changed. One of the often neglected but fundamental aspects of modernist architecture is its preoccupation with motion. Le Corbusier, for example, repeatedly stressed the relation of perceptual movement to architecture. Consequently, he added the concept of motion to his agenda for representing space. It is hardly surprising, therefore, that he collaborated on a film, *Architectures d'aujourd'hui* (Architecture Today, 1930–31)[10] in which he presented his modernist architecture of motion.

For Le Corbusier, *seeing* is the principal activity in a house.[10] But since, as he puts it, 'modern eyes'[11] move, vision in Le Corbusier's architecture is always tied to motion. In her book *Privacy and Publicity: Modern Architecture as Mass Media*, Beatriz Colomina argues that Le Corbusier's emphasis on movement had its origins partly in cinema:

> The point of view of modern architecture is never fixed, as in baroque architecture, or as in the model of vision of the camera obscura, but always in motion, as in film or in the city. Crowds, shoppers in a department store, railroad travellers, and the inhabitants of Le Corbusier's houses have in common with movie viewers that they cannot fix [or] arrest the image. Like the movie viewer ... they inhabit a [changing] space. It is a space that is not made of walls but of images. Images as walls.[12]

The concept of images as walls was made concrete in *Architectures d'aujourd'hui*. Pierre Chenal made the film with the help of Le Corbusier. The film opens with Le Corbusier's famous analogies between modernity's ultimate machines of movement – automobiles and aeroplanes – and the house, 'une machine à habiter!', a machine to live in.

The initial montage seems to suggest Le Corbusier's desire to construct buildings of movement, that is, the desire for an architecture that can equal the most advanced modern technology (Fig. 1).

Architectures d'aujourd'hui continues with showing three of Le Corbusier's and Pierre Jeanneret's villas with their characteristic whitewashed *l'esthétique simple*. The sense of motion through the houses is emphasised by tracking shots and fluid camera movement, tilts and pans. In fact, the camera constantly moves while at the same time depicting diegetic motion, ranging from simple walking to high-heel gymnastics. Beginning at *Villa à Garches* (Fig. 2), built in 1927, the film portrays *Villa d'Avray*, and one of Le Corbusier's most famous buildings, *Villa Savoye* at Poissy, completed in 1929.

Perhaps the sequences from *Villa Savoye* most obviously display an architectonics of seeing. What is documented is not so much the house itself, rather *spatial processes and events*. As a viewer one seems to be perceptually interacting with the building, which in turn seems to be set in motion. The film thus becomes a substitute for a visit, a mimesis of the act of observing architecture, stressing Le Corbusier's thoughts on how to view it.

The architectonic events of motion are most prominently expressed in a repetitive montage sequence of a woman walking up the ramp of the *Villa Savoye*. Initially, one sees a combined tilt and pan establishing shot presenting the ramp, followed by a range of shots as the woman ascends it. The first is a long shot of the woman from the top of the house. The next one displays a medium shot of her from behind, surprisingly taken from within the building (Fig. 3). Finally, the sequence displays three shots of the woman as she is about to go onto the roof: one medium shot from above, another beginning in close-up extending into a long shot (Fig. 4), followed by a last long shot as she reaches the top of the building.

One might say that this elusive spatial montage is a visual illustration of Le Corbusier's concept of *promenade architecturale* (the architectural promenade). Like the montage sequence I have just described, the concept

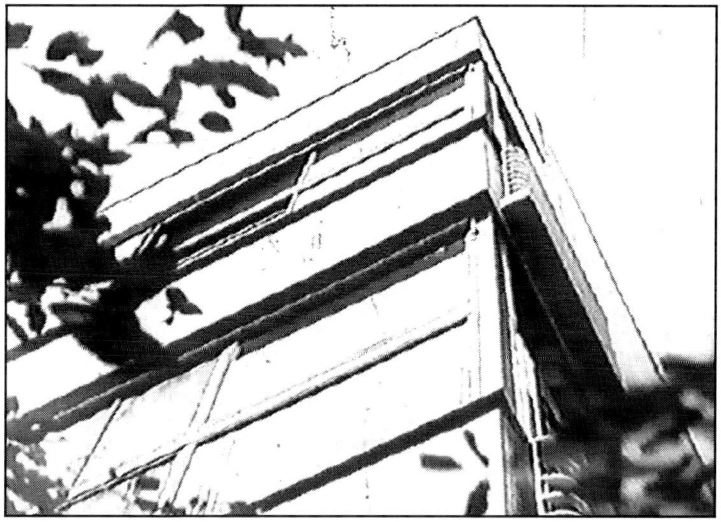

Fig. 2.Villa à Garches. Still from Architectures d'aujourd'hui.

of *la promenade architecturale* tried to describe the experience of spatial movement. Architecture, Le Corbusier writes, 'is appreciated by walking, on foot; it is by walking, by moving, that one sees the order of the architecture developing'.[13] The architectural promenade offered changing views and made visible the mobile-tactile quality of the house.

Interestingly, Eisenstein also talked about the moving experience of architecture when walking around or in it. He used the word 'path'[14] instead of 'promenade', but his thoughts were essentially the same as Le Corbusier's, further aligning the latter with cinema. In his essay 'The Explosion of Space: Architecture and the Filmic Imaginary', Anthony Vidler discusses the convergence between the two artists' views on the cinematic perception of buildings.[15] Drawing upon Eisenstein, Vidler uses the term, 'the spatial eye' to describe the relations between immobile and mobile perception. On the one hand, there is architectural perception where the spectator *moves*; and on the other hand, cinematic perception, where the spectator follows an imaginary series of spatial objects through *sight*. According to Vidler, it is in this respect that the transition from 'real to imaginary movement'[16] occurs, and whereby – as Eisenstein initially claimed – architecture becomes film's predecessor. Vidler writes:

> Eisenstein, former architect [*sic*] and an admitted 'great adherent of the architectural aesthetics of Le Corbusier', turned

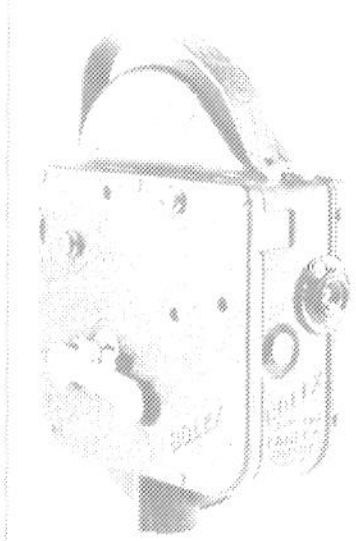

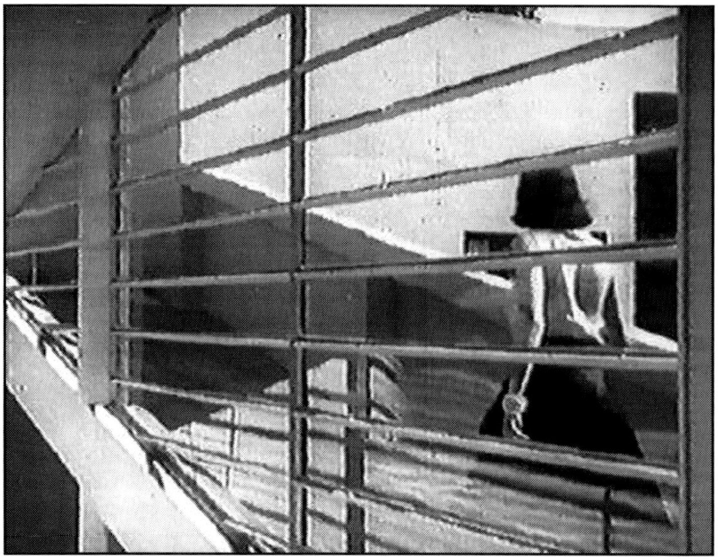

Fig 3. Woman walking towards the ramp of the Villa Savoye (Le Corbusier / Pierre Jeanneret, 1929). Still from Architectures d'aujourd'hui.

to an example of the architectural 'path' which precisely parallels that studied by Le Corbusier himself in his *Vers une architecture* [Towards a New Architecture] (1923) to exemplify the 'promenade architecturale': the successive perspective views of the movement of an imaginary spectator on the Acropolis constructed by Auguste Choisy to demonstrate the 'successive tableaux' and 'picturesque' composition of the site. Eisenstein cited Choisy's analysis at length with little commentary, asking his reader simply 'to look at it with the eye of a film-maker'.[17]

Returning to Le Corbusier, it seems that cinema and the cinematic treatment of Le Corbusier's buildings in *Architectures d'aujourd'hui* suggest a twofold project. On the one hand, film as the modern mass medium *par excellence* is used as an instrument for propagating the *new* architecture. (Le Corbusier himself is present throughout the film, driving his car, walking around *Villa à Garches* smoking (Fig. 5), not to mention his description of the renewal of housing and the renewal of Paris with the infamous Cartesian sky-scrapers.) On the other hand, *Architectures d'aujourd'hui* is also about accentuating the principle of movement characteristic of Le Corbusier's architecture. Like no other medium, film was able to portray the *promenade architecturale* and depict the sight of Le Corbusier's architecture as, in essence, one of moving images.

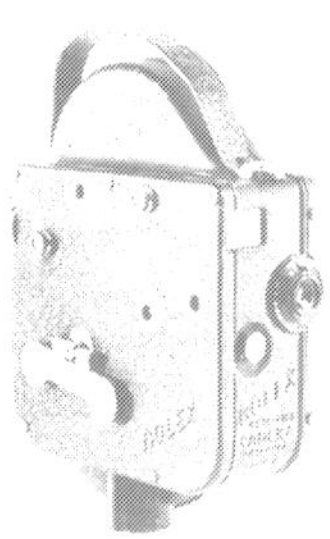

Deconstructivist architecture

In a recent essay entitled 'Site-seeing: Architecture and the Moving Image', Giuliana Bruno summons up the relations between Le Corbusier and cinema by stating that:

> She who wanders through a building or a site acts precisely like a film spectator absorbing and connecting visual spaces. The changing position of a body in space creates architectural and cinematic grounds. The consumer of architectural (viewing) space is the prototype of the film spectator.[18]

Besides describing the concept of the *promenade architecturale*, Bruno's words can also function to introduce a discussion on 'deconstructivist' architecture. Ten years ago an exhibition opened at the Museum of Modern Art in New York called 'Deconstructivist Architecture'. Among the participants were architects Rem Koolhaas, Peter Eisenmann, Daniel Libeskind, Frank Gehry, Bernard Tschumi and Zaha Hadid. Despite the obvious problem of assembling different architects into a seemingly homogenous group, not to mention the fact that only a few of the persons involved had shown interest in Jacques Derrida's philosophy, their architecture shared a likeness in *the disruption of form*. The architecture of the exhibition concerned 'the limits of the discipline, discovering its margins',[19] with a general emphasis on traditional architectonic de(con)struction. This rendered the term 'deconstructivist' architecture ambiguous.[20]

Deconstructivist architecture is an architecture of disturbance. The art of building is no longer considered a traditional static art, but rather a set of colliding spatial formations. Within architectonic discourse, however, the deconstructivist project has included many different strategies and performances. Ranging from Libeskind's contemplations on the geometrical qualities of architectonic drawings, as in *Micromega* (1979)[21] or in *Chamber works: architectural meditations on themes from Heraclitus* (1983),[22] to Eisenman's dysfunctional architecture, or Koolhaas' asymmetric sequential buildings (such as *Bibliothèque de Jussieu* (1991), a library with only one spiral floor), a uniting impetus has been the strive to present *space* from a multitude of perspectives. Clearly, there exists a link between these observations of space and those of cinema with its long tradition of diegetic spatial experimentation. Unexpected spatial articulation has been the province of cinema from the advent of film. This tradition has attracted attention in deconstructivist architecture as the title of Michael Blackwood's documentary on Eisenmann, *Making Architecture Move* (1995), makes clear. Mark Wigley has written a book on the subject, *The Architecture of Deconstruction: Derrida's Haunt.*[23] According to him, the deconstructivist project is not about dissecting buildings. On the contrary, it tries to localise inherent problems and conflicts within the history of architecture. The aspiration, then, is to bring these disjunctions to the surface of the construction. Here *theory* plays a crucial role, and deconstructivist architecture has hardly hidden its fascination with expressing theory in spatial practice.

Nevertheless, what is interesting from a cinematic point of view is the emphasis on fragmentation. Many projects within deconstructive architecture seem to be concerned with a kind of shot-like disruption of space and the capability of presenting a multitude of architectonic perspectives. One might say that the deconstructivist project has decided to stretch the limits of architectonic destabilisation. For this purpose, a *montage of spaces* is frequently used. This is evident in, say, the architecture of Zaha Hadid. Her buildings, such as the *Vitra Fire Station* (1993) – a house that has been likened to a fire brigade on alert, as if the entire structure could explode into action at any moment – are centred on the organisation of spatial perception. Hadid works with a kind of stratification of sight in layers. When walking through her buildings, sight seems to turn the architecture of the building into a projector of space. In Le Corbusian manner, the perceptual design changes as one moves through different areas of her architectonic space. For Hadid's architecture, cinema is a forceful referent.[24]

If the ways in which cinema presents sight is useful for understanding deconstructivist architecture, another helpful concept – with clear filmic connotations – is *sequentiality*. In his newly completed *KIASMA – Museum of Contemporary Art in Helsinki* (1998), the architect Steven Holl, for instance, talks about his building in terms of a variety of spatial sequences, '[The] curved unfolding sequence [of KIASMA] provides elements of both mystery and surprise which do not exist in a typical orthogonal arrangement of spaces'.[25] The ability to see a building as unfolding space is a matter of seeing architecture as a sequence of spatial frames. However, besides making perceptual 'cuts' to varying spatial formations as one walks through the building, one has to remember that KIASMA is conceived and constructed according to precisely this logic of perceptual movement. In other words, the architect has carefully planned both the site and the sequence of sights presented to the visitor.

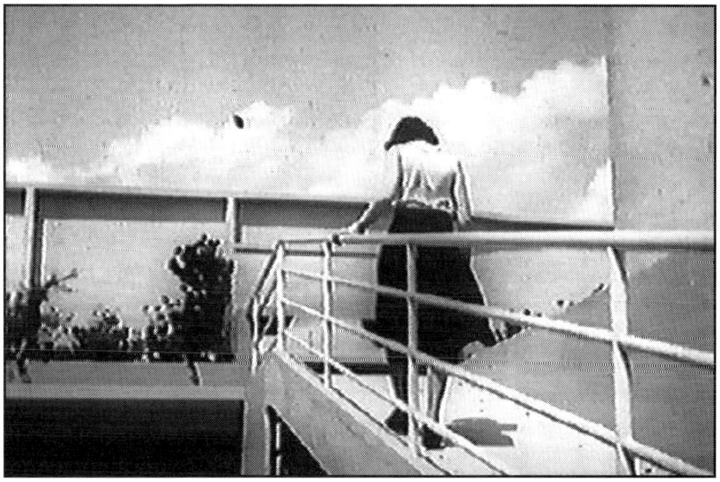

Fig. 4. Woman walking up the ramp of the Villa Savoye. Still from Architectures d'aujourd'hui.

Bernard Tschumi and montage

Perhaps Tschumi is the deconstructivist architect who has worked most thoroughly with the concept of sequence. In the essay 'Sequences' from 1983, he understands architectonic transformation as changing cinematic sequences of space.[26] 'Architectural relations are never semantic', Tschumi writes, 'instead, a better analogy to [architectural] montage and mixing techniques might be found in Dziga Vertov's or Sergei Eisenstein's work in the cinema'.[27] If the concept of

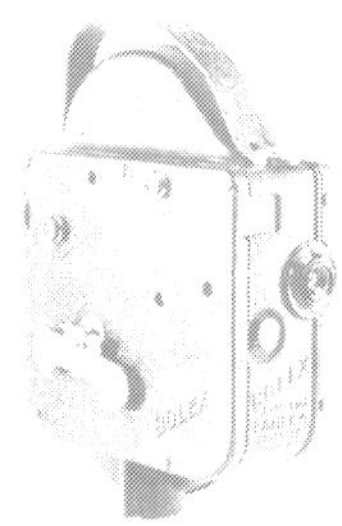

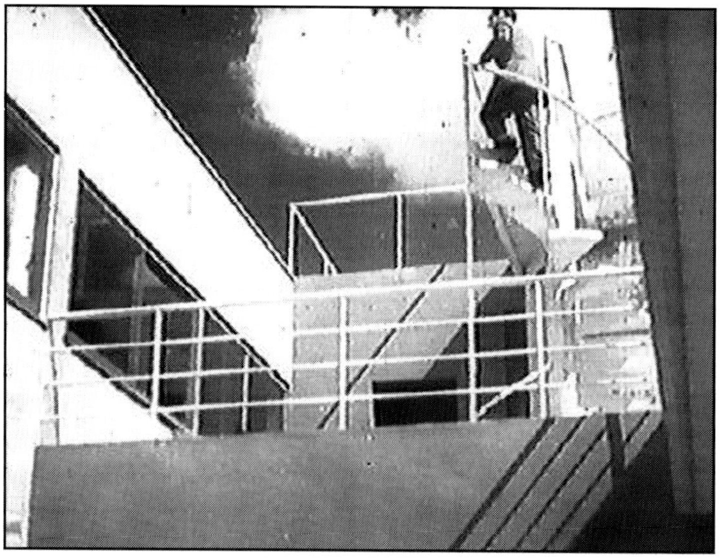

Fig 5. Promenade architectural. Le Corbusier climbing steps at the Villa à Garches. Still from Architectures d'audjourd'hui.

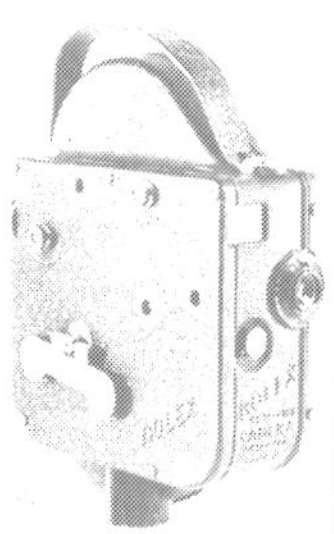

sequence is understood as the ability to include perceptual movement within the analysis of architecture, it also involves the possibility of seeing space *per se* as an unfolding experience. Accordingly, for Tschumi, the 'frames' of the sequence – that is, the different spaces of a building – derive significance not so much in themselves, but from juxtaposition and montage. The issue, then, is to examine architecture as a sequence 'frame by frame'. He concludes by stating that, '*Frame: the moments of the sequence. [To] examine architecture 'frame by frame,' as through a film-editing machine.*'[28]

Tschumi's theories on the filmic sequencing of architecture are best expressed in his *Parc de la Villette* project outside Paris. He writes:

> The [*Parc de la Villette*] project substitutes an idea comparable to montage ... Film analogies are convenient, since the world of cinema was the first to introduce discontinuity. ... In film, each frame (or photogram) is placed in continuous movement. Inscribing movement through the rapid succession of photograms constitutes the cinegram. The Park is a series of cinegrams, each of which is based on a precise set of architectonic, spatial, or programmatic transformations. Contiguity and superimposition of cinegrams are two aspects of montage.[29]

In short, *Parc de la Villette* – completed during the period 1983–90 – is constructed according to the logic of architectural movement. It is formed as a grid (or filmic superimposition) of three autonomous systems: *lines* – a system of motion; *points* – a system of objects, and *surfaces* – a system of spaces. Together these create a recreational park of colliding spatial montages or cinegrams. *Parc de la Villette* offers an architecture constructed to be seen in motion. This renders it a different expression with the ability to stroll along different *promenades cinématiques*.

The long shot: Jean Nouvel and spatial visibility

> Cinema has taught us to see images in relation to time. A town is now read through motion, travel. Today, an architectural composition refers to sequences. Most contemporary architects take into consideration the journey of the human being through space.[30]

In relation to Tschumi's thoughts on the cinematic promenade, Jean Nouvel indicates that the moving image has become a way of seeing *space* anew. According to Nouvel, the experience of motion through space, whether in the form of an architectonic or urban journey, seems to make space *more* visible. With movement, a spatial materiality, otherwise unseen, emerges. If the experience of motion in architecture in Eisenstein and Le Corbusier primarily was an issue of seeing architecture differently, Nouvel has come to use his predecessors' notion in an attempt to express *pure* space.

The architecture of Nouvel is precisely an architecture of spatial visibility. For him, movement has become one way of dealing with the accentuation of space. Another pertinent feature in Nouvel's architecture, however, is the notion of spatial superimposition or transparency. The notion of architectonic transparency is perhaps best exemplified in Nouvel's Parisian building *Fondation Cartier* (1994). 'Seen from the boulevard, the building appears bathed in a halo of light', the description reads, 'superimposed against a backdrop of sky and trees, both real or virtual, reflected and refracted by the screens of glass'.[31] In this building one is immediately struck by the blurring of spaces. Nouvel's architecture is one of *spatial thickness*. In it, images of space collide, making it hard to distinguish the inside of the building from its outside. It is an architecture of filmic superimposition. Photographs depicting the building demonstrate that the boundaries between what is interior and exterior are constantly shifting, depending on the time of day and light conditions. In short, Nouvel's architectonic projects are explicitly occupied with expressing different aspects of space. The preoccupation with the visibility of blurred space(s), has led space *per se* to appear in visible light. In the architecture of Nouvel, therefore, space is never empty; it is always filled and carefully framed. This *layering of space* might be considered one of deconstructivist architecture's more radical traits, possibly emerging from the field of cinema.

Although Tschumi and Nouvel have asserted the influences of film on their work (especially in terms of the imagination and construction of space), Nouvel's use of film-spatial analogies differs from Tschumi's. Where Tschumi emphasises spatial montage, Nouvel's spatial methods are better understood in relation to different shot figures. As is well known, the long shot[32] or the extreme long shot has frequently been used in feature films as a framing device to come to terms with spatial distance. Film technological keywords such as *deep focus*, *depth of field*, not to mention *deep space* – 'an arrangement of *mise-en-scène* elements [with] a considerable distance between the plane closest to the camera and the one farthest away'[33] – are useful spatial concepts to understand and describe Nouvel's architecture. The transparent volumes in *Fondation Cartier*, for instance, act architectonically to present 'the depth of the site',[34] in a way reminiscent of the framing of space in a long shot.

In conclusion, it seems that deconstructivist architecture incorporates many filmic components in which respect, it can be termed, 'an architecture of moving images'. Considered as a technology of seeing, deconstructivist architecture displays modes of perception closely linked to cinema. Within the history of architecture, this tradition emerges from Le Corbusier's notion of the *promenade architecturale*. But architects such as Tschumi and Nouvel have expanded Le Corbusier's concept and further aligned architecture with film. By using spatial movement and montage, sequentiality and transparency, the architect has become something like a filmmaker. Cinematic experience has demonstrated a way of organising built space, or as Nouvel puts it:

> Architecture exists, like cinema, in the dimension of time and movement. One conceives and reads a building in terms of sequences. To erect a building is to predict and seek effects of contrast and linkage through which one passes. ...In the continuous shot [or] sequence that a building is, the architect works with cuts and edits, framings and openings ... I like to work with a depth of field, reading space in terms of its thickness. Hence the superimposition of different screens ... which are to be found in all my buildings.[35]

Notes

1. Henri Lefebvre, *The Production of Space* (Oxford: Blackwell, 1991), 154.
2. Sergei Eisenstein, 'Montage and Architecture' (1937) *Selected Works* (London: BFI, 1991), II, 60.
3. Ibid. As is well known, Eisenstein's father, Mikhail Osipovich Eisenstein, was a prominent architect in Riga, Latvia. It seems that young Eisenstein initially intended to follow in his father's footsteps. It should also be noted that Eisenstein's observations on the architecture of the Acropolis were largely based on Auguste Choisy's *Histoire de l'architecture*, published in 1899. For a discussion of Choisy, see note #17 below.
4. Two recent anthologies stressing architecture's narrative role include Dietrich Neumann (ed.), *Film Architecture: Set Designs from Metropolis to Blade Runner* (Munich: Prestel, 1996) and David B. Clarke (ed.), *The Cinematic City* (London: Routledge, 1997). A different and more stimulating approach is taken in François Penz and Maureen Thomas (eds.), *Cinema & Architecture – Méliès,*

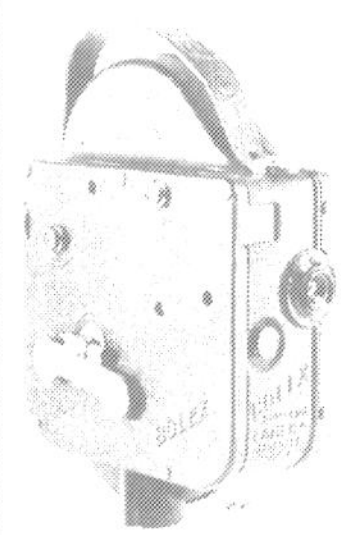

Mallet-Stevens, Multimedia (London: BFI, 1997), the outcome of a Symposium on Cinema and Architecture held at Cambridge University in 1995.

5. Tom Gunning, 'From the Kaleidoscope to the X-Ray: Urban Spectatorship, Poe, Benjamin, and *Traffic in Souls* (1913)', *Wide Angle* 19, 4 (October 1997): 33.
6. In fact, Gunning has discussed this film – and its relation to space – in an early article, 'An unseen Energy Swallows Space: The Space in Early Film and Its Relation to American Avant-Garde Film' in John Fell (ed.), *Film Before Griffith* (Berkeley: University of California Press, 1983). There Gunning makes the remark – which is crucial for my argument, albeit banal – that the invention of cinema made it possible to represent space in a completely new manner, in ways previously unthinkable.
7. Alfred Stieglitz, cited in a discussion of the Flatiron building in David E. Nye, *American Technological Sublime* (Cambridge, Mass.: The MIT Press, 1994), 95.
8. Gunning 1997: 35.
9. Some discussion of *Architectures d'aujourd'hui* gives the title of the film as *L'Architecture d'aujourd'hui*. In conformity with the most recent publication, *Cinema & Architecture*, I use the former title.
10. An interesting filmic parallel to the notion of architectonic seeing is Le Corbusier's article 'Spirit of Truth', originally published in *Mouvement* 1 (June 1933); rpt. in Richard Abel (ed.), *French Film Theory and Criticism 1907–1939: A History / Anthology* (Princeton: Princeton University Press, 1988), II, 111–113. Le Corbusier, in a Benjaminian manner, writes about cinema and its ability to reveal and 'to discover life' (111) through mechanical sight.
11. Le Corbusier, in Willi Boesiger (ed.), *Oeuvre complète* (Zurich: Girsberger 1930–), I, 60.
12. Beatriz Colomina, *Privacy and Publicity: Modern Architecture as Mass Media* (Cambridge, Mass.: The MIT Press, 1994), 6.
13. Le Corbusier, 24.
14. Eisenstein, 59.
15. Anthony Vidler, 'The Explosion of Space: Architecture and the Filmic Imaginary' *Assemblage* 21 (1993); rpt. in *Film Architecture: Set Designs from Metropolis to Blade Runner*, 13–25.
16. Vidler, 22.
17. Ibid, 22–23. Choisy is a noteworthy person in the context of seeing architecture from a mobile perspective. Eisenstein used Choisy's *Histoire de l'architecture* (Paris: Gauthier-Vilars, 1899) when discussing ancient architecture as a cinematic model for framing. However, if one returns to Choisy's book, one somewhat surprisingly finds a similar treatment of architecture as a 'series of tableaux' (série des tableaux), 419. Choisy writes about 'the optical symmetry of the Acropolis' ('la symétrie optique de l'Acropole'), 415, and takes his reader on an imaginary perceptual tour of the site. Moreover, it comes as no surprise that one of the architectonic drawings illustrating these passages of *Histoire de l'architecture* is reprinted in Le Corbusier's *Vers une architecture* as well as in Eisenstein's essay 'Montage and Architecture'. In short, both Eisenstein's and Le Corbusier's thoughts on the sequencing of architecture are anticipated in Choisy. As he puts it, 'The method of architectonic balance becomes clear from the series of panoramas that unfold before visitors to the Acropolis'. ('a méthode de pondération ressortira d'une revue des tableaux sucessifs qu'offrait au visiteur l'Acropole'), 413.
18. Giuliana Bruno, 'Site-seeing: Architecture and the Moving Image', *Wide Angle* 19, 4 (October 1997): 15.
19. Kate Nesbitt, 'Introduction', in Nesbitt (ed.), *Theorizing a New Agenda for Architecture 1965–1995* (New York: Princeton Architectural Press, 1996), 36.
20. For a discussion of the term 'deconstructivism', see Nesbitt, 27ff.
21. Daniel Libeskind, *Between Zero and Infinity* (New York: Rizzoli, 1983), 79–93.
22. Daniel Libeskind, *Chamber Works: Architectural Meditations on Themes from Heraclitus* (London: Architectural Association, 1983).
23. Mark Wigley, *The Architecture of Deconstruction: Derrida's Haunt* (Cambridge, Mass.: The MIT Press, 1993).
24. For a discussion of Hadid's architecture, see the special edition on Zaha Hadid in *El Croquis* 52 (1995).
25. Steven Holl Architects, 'From Chiasma to Kiasma', *Project Magazine KIASMA* 78 (Helsinki, 1998), 23.

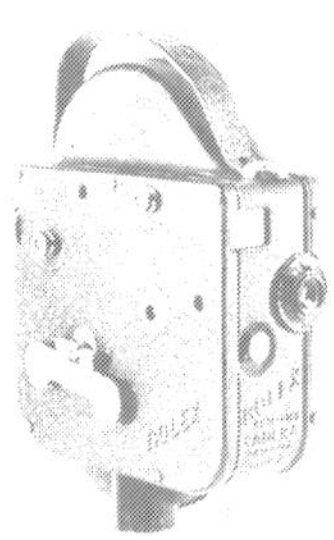

26. Bernard Tschumi, 'Sequences' (1983), *Architecture and Disjunction* (Cambridge, Mass.: The MIT Press, 1996), 163–165.
27. Tschumi, 'Madness and the Combinative' (1984) *Architecture and Disjunction*, 185.
28. Tschumi, 'Sequences', 166.
29. Tschumi, 'Abstract Meditations and Strategy' (1987) *Architecture and Disjunction*, 196–197.
30. Statement by Jean Nouvel, cited in Odile Fillion, 'Life Into Art, Art Into Life – Fusions in Film, Video and Architecture', *Cinema & Architecture*, 119.
31. Jean Nouvel, 'A phantom of the park', *Room on the Run* exhibition catalogue (Stockholm: Arkitekturmuseet, 1998), 54.
32. My treatment of Jean Nouvel and the long shot is indebted to Peter Lunenfeld's comment – for which I am grateful – during the conference.
33. David Bordwell and Kristin Thompson, *Film Art : An Introduction* (New York: McGraw-Hill, 1993), 493.
34. Nouvel 1998, 54.
35. Statement from Jean Nouvel, cited in Kester Rattenbury, 'Echo and Narcissus', *Architectural Design* 112 (1994), 35.

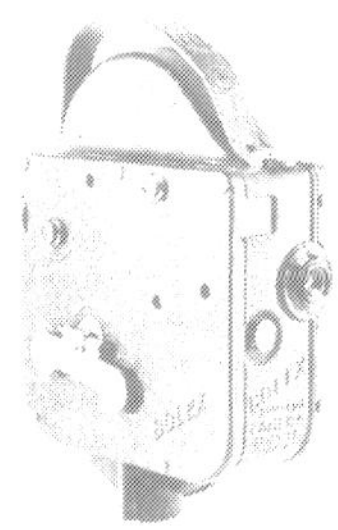

Submerged Landscapes of the Postmodern Body: Surface, Text, Commodity

Jay Moman

Media Arts Department, University of Arizona, USA

In the 1990s it seems that no anthology of cultural theory is complete without a contribution to the increasingly pervasive discourses on the body. Emerging theories attempt to manage Arthur and Marilouise Kroker's 'crisis of the body' by sketching out a map to navigate a world marked by genetic testing, retinal and thumbprint identification, cybersex, and other technological and increasingly digital bodily formations.[1] As I aim to show, this is a map written both *about* and *upon* the body in order to regulate the myriad technological systems which dis-/configure it.

According to various sources, the genesis of the discursive body occurs in the nineteenth century. Michel Foucault posits the emergence of an historic body born out of the introduction of microscopy, chronophotography, and other visual systems related to biological science. With the advent of such imaging technologies, he suggests that as the immediacy of natural science shifts to representational techniques in biology, the surface of the body as a site for 'sensory perception ... is progressively destabilised'.[2] His argument of visceral displacement locates the seeing subject as a central point on a continuum whose termini mark that which is unseen, inaccessible to human vision. The world beyond the horizon (the *other*, foreign culture) and the interior world of the human body (the *unseen*) are both the infinite without and the infinitesimal within. Foucault's human sciences set out to access these two unknown (because unseen) frontiers.

I argue that while visual and other technologies of the twentieth century gain greater access to the unseen, the displaced body resurfaces. In an anxious age of volatile bodily fluids and genetic surveillance, the body surface re-emerges as a text, a screen inscribed with the information of its own subdermal terrain. Reading this textualisation process in the contexts of both visual and linguistic discourses (a kind of *semiotics of the body*), I suggest that a *physioanalytical* approach is operative which exposes common figurations of the body as an encoded entity. Furthermore, the encoded body – as one instance of various body *simulacra* (a hierarchy of image and text) – benefits from a metadiscursive approach. *Hypervisuality, hypertextuality* and *hyperreality*, along with an emerging discourse of the virus, provide symbiotic modes of reading the body within a postmodern grid of politics, economics, and culture.

Cartwright's inverted space

Foucault's suggestion that the body occupies a site originally and largely determined by the medical gaze is crucial to our understanding of how the body *means* across the expanse of its cultural representations, particularly in the media. Lisa Cartwright's *Screening the Body: Tracing Medicine's Visual Culture* investigates chronophotography and other

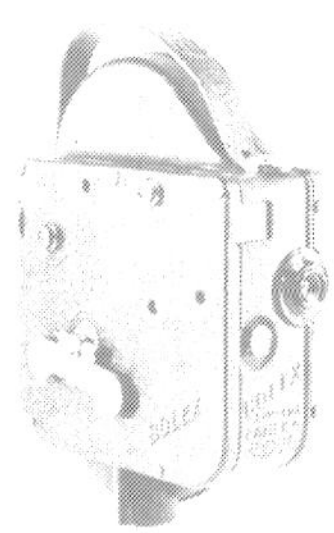

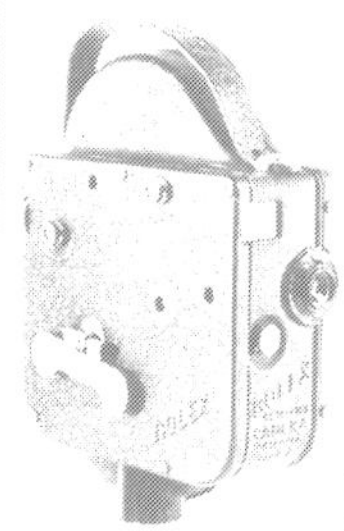

visual technologies common to both science and the cinema. Her study identifies the existence of a:

> history of the cinematic techniques that science has used to control, discipline and construct the human body as a technological network ... [and] that this history is complexly interwoven with other areas of visual culture.[3]

Cartwright's discourse provides a framework for conceptualising bodily representations which extends into the popular visual environment, namely science-fiction cinema, but also within other media symptomatic of a response-mode of address toward technological systems seeking to access and regulate the territory of the infinitesimal unseen.

Restricting my focus to a survey of the effects of certain visual technologies upon the body, a key aspect of Cartwright's groundbreaking work is her analysis of what Claude Bernard calls 'experiments of destruction' in which organs are removed from human subjects. Chronophotographic records of changes in the subjects then provide information regarding the function of the ablated organ.[4] In poignant contrast to these medical experiments, Cartwright places the cinematic example of *Electrocuting an Elephant* (1903) as filmed by Thomas Edison's production company. She concludes: 'What is ultimately shown is nothing more than an absence, a body stripped of its capacity to perform the function in question.'[5] This notion of definition by absence continues into the postmodern era in important ways and across a range of popular media as the body is displaced along the visual continuum.

A key implication of this displacement reinforces Tom Gunning's characterisation of the history of early cinema. While his carnival apparatuses retain their position as spectacle, the body too becomes a spectacular object as the terrain of the infinitesimal, newly envisioned, proliferates throughout the twentieth century.[6] Consider the example of the startling 1960s photography of Lennart Nilsson. Using the techniques of electron microscopy, he pictures red blood cells which are referred to by *Life* magazine as 'an unseen world'. *Life* also offers up Nilsson's spectacular image of a twenty-eight-week-old foetus shrouded in 'its' membrane.[7] The blood cells and foetus fill the represented space, are disconnected from the (absent) human host. More critically, a complete erasure of the mother of the foetus occurs, not only within the represented space but physically as well; Nilsson's foetus was dead, removed from the womb. Confirming Foucault's thesis, visual access to the human interior appears to distance and efface the body itself.

Romanyshyn's outer space

Technological access to the opposite extreme of the visual continuum yields similar results. Moving beyond Foucault's foreign culture, examined by anthropology, it is outer space which has claimed the position of the visual infinite in postmodernity. Robert Romanyshyn's *Technology as Symptom and Dream* offers a lyrical and insightful inquiry into the relationships between space technologies, Cartesian perspectivalism, and the human figure. He considers the geometric mechanics of single-point perspective as a system which regularises the body/object, regulates it, and locks it into a grid which can be mathematically deduced and reproduced. Even the viewing subject is forced into a fixed position – monoscopic, centred, and directly in front of, but apart from, the rectangle of representation. According to Romanyshyn:

> [w]ith the advent of linear perspective vision we have managed to spatialise time, to distance ourselves from the body, and to remove ourselves from the midst of things.[8]

As with the medical gaze, the body is eclipsed, abandoned.

Romanyshyn links this abandoned body to the cosmonaut. He imagines the vanishing point as a point of departure for the body, a rocket launch-pad for astronautical travellers. His is a cosmonautical body for which:

> [a]ll the natural activities – of hearing, breathing, speaking, and making gestures – are ... replaced by technical functions. The body has no contact with the surrounding atmosphere; it is ... further removed from nature than at any other moment in history.[9]

The human in outer space resonates with the echoes of the consequences of the inner space of medicine's visual culture (and, according to Romanyshyn, representational culture in general) as the body undergoes both a disciplinary ordering at its launch-site, and a removal as it departs the earth.

Anxious cinematic spaces

While technological imaging effectively eclipses the body as object, it gives rise to anxiety for the seeing subject. According to Ian Hacking, the uncertainty involved in reading a microscopic image created some disturbance in the scientific community of the 1800s because such representations are actually a diffraction of an image of the original.[10] Cinema viewers of the late 1900s faced similar disturbances regarding the microscopic image. Instead of constructing empirical solutions, however, the movie-goer's role was, as spectator of the narrative resolutions of filmic engagements, with the infinitesimal.

In 1966, a year after Nilsson's foetal photographs appeared in *Life*, the microscopic terrain becomes a landscape, a *microscape*, in popular cinema. David Fleischer's *Fantastic Voyage* opens with the statement: 'This film will take you where no one has ever been before …'. (Note the similarity between this phrase and the prologue to the mid-1960s television series, *Star Trek*:[11] like Nilsson's foetus, referred to by *Life* as a 'tiny voyager', the external and internal spaces of the unseen are characterised identically.) Immediately following the *Fantastic Voyage* prologue, the opening sequence introduces a traveller; the *mise-en-scène* is as suspenseful as the mysterious purpose of this man's journey, and the presence of bodyguards erupts in what appears to be an assassination attempt. These opening frames set up the conditions of the traveller as a body under attack. By the time he is delivered to the underground laboratory and the team of scientists is miniaturised, the nature of their voyage is established as dangerous and laced with anxiety. Their sojourn is marked by numerous attempts by the man's own body to destroy their submarine. A reversal is thus installed in the diegesis: the body under attack becomes the antagonistic power that threatens the crew.

Representations of Foucault's interior unseen continue to generate alarm when Michael Crichton's *Andromeda Strain* is adapted for film in 1971. As before, a group of scientists congregate in a laboratory set deep in the earth to study an extra-terrestrial virus. The narrative repeats the theme of the body's interior as an antagonistic space as Dr. Leavitt's hidden epilepsy almost thwarts the investigation. The film, furthermore, also invokes a discourse concerning outer space technologies, for it is a military satellite which malfunctions and crashes to earth, thereby introducing the fatal airborne virus subsequently named the 'Andromeda Strain'. In short, by the early 1970s, a pattern emerges in which popular representations of both extremes of the visual continuum, as well as the intervening technology which accesses those extremes, become implicated as anxious fodder for the postmodern public. Certainly the body in postmodernity becomes integrated into the proliferating mechanics of technology, from quality-of-life apparatuses to the television, to such a degree as to constitute a cultural movement. Marshall McLuhan discusses this integration by suggesting that machines are 'extensions of man'.[12] Elaborating McLuhan's observations, computer and digital technologies have extensively subsumed the mechanical role in the convergence of man and machine during the ensuing three decades. This crossing of the boundaries between the cybernetic and organic to produce Haraway's often-referenced *cyborg* promotes a cultural breeding-ground that further propogates unease regarding the body.[13]

Corrupt cybernetic spaces

Stanley Kubrick's *2001: A Space Odyssey* (1968) doubly configures the cultural discomfiture that the cyborg represents in the forms of Dave and HAL. As Dave and his crew proceed on their odyssey from 'The Dawn of Man' to 'Beyond the Infinite [of Outer Space]', the metaphoric human race becomes reborn as a completely new creature. The figure of the astronaut, once dressed in a spacesuit, is exchanged for what has become the iconic image of the film: the extra-terrestrial foetus.

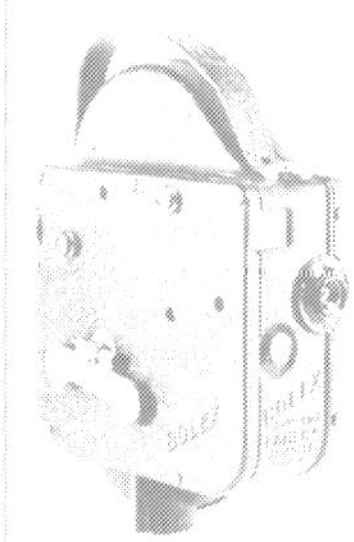

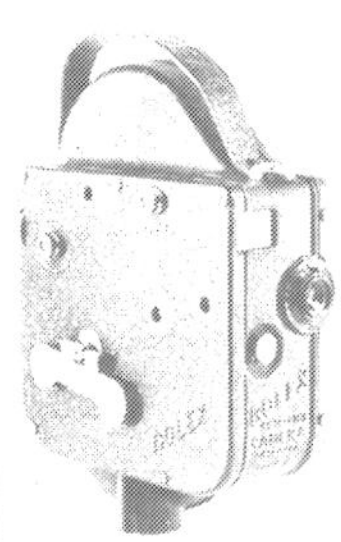

HAL undergoes a similar transformation. In a reverse transgression of the boundaries of mechanic and organic, the anthropomorphic computer evolves into an entity which dangerously encroaches upon human characteristics. As Dave disables HAL's circuitry, the computer confirms his own metamorphosis as he pleads, 'I'm afraid ... My mind is going ... I can feel it'. As with Dave, a price is exacted from transgressors: not human, not machine, they are both and neither, a condition that merits disciplinary action. This intersection of the technologic and biologic broadens the parameters of what constitutes the infinitesimal, and thus presents an additional source of anxiety located within the unseen realm of computer circuitry.

In 1982, *Tron* presents a narrative in which a hacker is scanned and reconfigured as a computer programme. Once imprisoned within the circuitry, Tron must enact various videogame battles. In this arena, however, losing the game results in the very real death of the player. Likewise, Rachel Talay's *Ghost in the Machine* (1993) is enacted on a digital terrain. A power surge occurs at the moment a serial killer is undergoing a Magnetic Resonance Imaging scan, thereby transforming his consciousness and allowing him to resume his activities via electronic pathways. To capture this villain, the protagonists lure the serial killer to an underground computer lab and infect him with a hacker-generated computer virus. Clearly, by the 1980s, the submerged territories of the computer – both hardware and software – become filmic landscapes layered diegetically with anxious attempts to reconcile and regulate the unseen worlds of cybernetic systems.

The narratives of *Tron* and *Ghost in the Machine* also implicate a seminal figure for the cyber-age: the hacker. Characterised occasionally as an activist-hero, but more frequently understood as a criminal, the hacker has served as a catalyst for an etymological convergence of the cybernetic and organic. According to Benjamin Woolley, the term 'virus' realised its bio-technical connotation in 1989, when 'the publicist for the magazine, *Mondo 2000*, claimed that he had "unleashed the world's first media virus".'[14] As it turns out, the virus which was unleashed was not a destructive computer programme; rather, it was the very process through which various news media spread the story to an international public. For an audience with a heightened awareness of virology and epidemiology as a result of the AIDS crisis, the concept of virus 'had suddenly acquired a moral resonance ... that was easily transferred to the computer virus. Like AIDS, we were all threatened with being tainted by the hacker's lack of moral hygiene.'[15] The environment which enables such a linguistic transference to occur is symptomatic of what the Krokers describe as 'a media-induced state of panic anxiety about the transmission of bodily fluids'.[16] Thus I return, via a literally circuitous route, to the body. Moreover, as the Krokers observe, it is a body engaged in a 'urinal politics' which attempts to regulate bodily fluids, genetic structures and reproductive rights.[17] As suggested by Barbara Kruger's 1989 pro-Choice poster, it is the body *onto* which the cultural crisis of anxiety without is inscribed; the body is a site, a 'battleground'.[18] More specifically, the *submerged* landscapes of the postmodern body are a battleground.

Body image: surface, absence, and *physioanalysis*

Perhaps at no point in history has the body been more scrutinised than in the current era of genetic analyses and drug testing which are routinely administered by doctors, detectives, insurance companies, employers, or even the International Olympic Committee. And perhaps in no other area has surveillance been more acute than with the viral scanning for HIV/AIDS. In the mid-1980s, for example, William F. Buckley proposed tattooing those people who had tested positive for HIV, an event cleverly satirised by the 1993 advertising campaign for United Colors of Benetton (Fig. 1). My interest in this event is physical; it is the very way in which the body's exterior has become engaged in the quest for knowledge and control of the unseen.

If chronophotographic and microscopic vision in medicine (as well as similar technologies) have precipitated an interest in representing submerged and otherwise visually inaccessible microscapes, these technologies have subsequently served as a catalyst for the development of a new representational form for the body. As in Barbara Kruger's visual world, in which black-and-white and X-ray (negative) images are overlaid with text, the surface of the post-techno body must serve as a textual inscription and, additionally, function as an agent of information for its own subdermal unknown.

Representational transitions along these lines occur in gay male pornography. Here, the inscription is a representational approach that locates an absence of infection (the unseen) on the exterior body of the models. During the 1970s and early 1980s, the bodies of porn models conformed to no single standard. A variety of types – old/young, muscular/thin, hairy/smooth – were present within pornographic film. By 1986, however, a homogeneous form emerges. Hairless, muscled youth come to dominate a medium which is necessarily and acutely conscious of the AIDS pandemic. Meanwhile, a suspicious lack of specific diegetic referents for the concerns surrounding AIDS (such as the use of condoms and/or dialogic references to safer sex guidelines), identifies this new homogeneous form as the medium of response to the mounting emphasis on epidemiological issues facing the gay male pornography industry/consumer. Instead of external props or other filmic signifiers for safer sex practices, the response becomes embedded in the territory of the virus: the human body. Hairless, muscled youth offer bodies which are understood as clean, healthy and free of infectious agents in an era when sexiness may be ultimately linked to the degree to which a body can be read as (un)infected. The gay male porn-body is not merely seen as healthy; it is *not* seen as *un*healthy. It becomes a visceral manifestation of absence.

This exterior-interior play upon the body is further laid bare in *Diseased Pariah News* (*DPN*). A periodic humour magazine about HIV/AIDS, *DPN* has been engaged in a cyborgian discourse from its inception. The zoomorphic icon for the magazine, the 'diseased pariah', is a mouse. Referencing *Oncomouse*, the mascot itself is a cyborgian

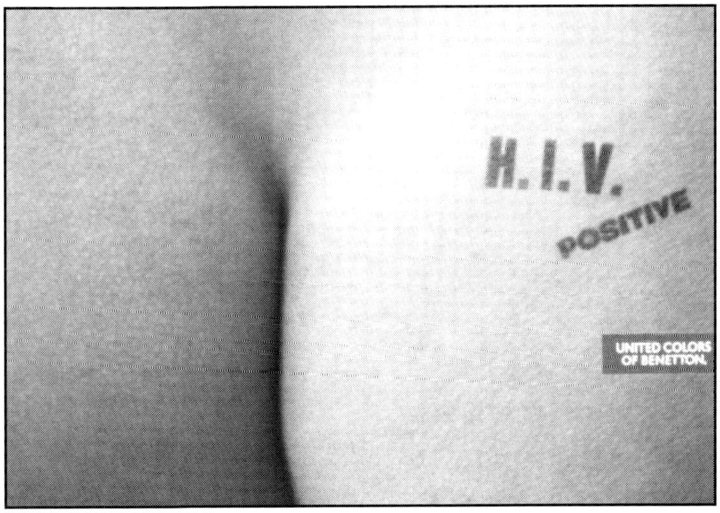

Fig. 1. United Colors of Benetton advertising campaign, 1993. Concept: O. Toscani. [Courtesy of United Colors of Benetton.]

image of the body engaged in a bio-technical overlap.[19] Part of the publication's humour is its self-consciousness regarding the codedness of the body. Next to the nude centrefold in each edition is a list of the medical statistics of the model. Recalling the statistics that have historically been used to define women in beauty pageants (e.g. body measurements), *DPN* instead enumerates the HIV+ centrefold's CD4 (T-cell) count, infections, and the anti-virals or other medicines s/he is prescribed (Fig. 2). Via the nude display and the analytical information offered in an era which scrutinises for viral references, a shift in representational strategies is evident exactly in the area described by psychoanalytic visual discourses: the objectified seen. But now the power of the look applies to the subdermal (un)seen of the body.

In addition to psychoanalysis, then, a kind of *physioanalysis* has emerged in visual approaches within postmodern representational systems. As the internal codes of the body are relocated onto its exterior, the surface of the body is effectively inscribed as a text. Human viscera function as a semiotic signifier of the body's internal realm: the *signified* of genetic coding, neural networks, viral binaries, etc. This displacement suggests an altering of Foucault's thesis on the disappearance of the body, as well as its surface, which has now returned as a critical site of perception in our environment of 'panic anxiety'.[20]

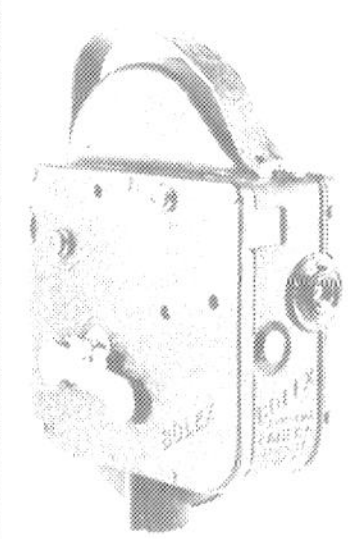

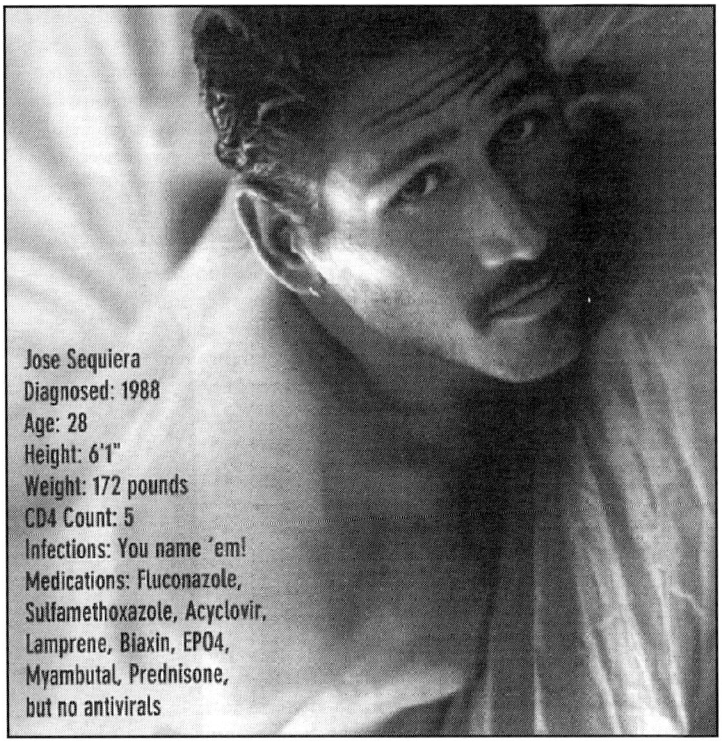

Fig. 2. Centrefold, Diseased Pariah News. *[Photograph by Robert Doyle reproduced by courtesy of* Diseased Pariah News.*]*

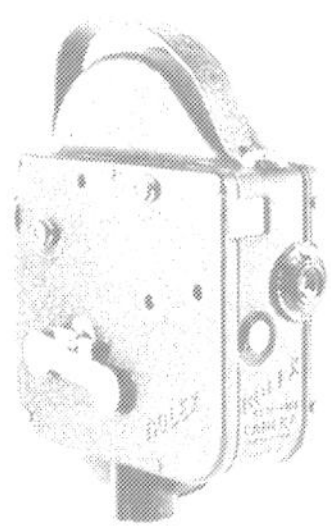

Body language: viral discourse

The cinematic apparatus also engages in an internal surveillance which relies on visceral scrutiny to derive information. Amy Taubin's analysis of the *Aliens* series emphasises the focal role of the cinematic body in the third instalment:

> the alien is an invader and destroyer of the body. In *Alien 3*, the body becomes a landscape, obsessively probed and examined with fingers and eyes, penetrated in close-up with needles, knives and saws.[21]

In *Alien 3* (David Fincher, 1992), Ripley confirms her suspicion that she has been impregnated by performing a sonar-resonance scan on herself in her disabled space vessel (a metaphor for her physical condition), which reveals a foetus within her chest cavity. As the scan which images the foetus of the new alien mother is reflected on the face of Ripley (the human mother), the exterior and interior are parallel, play against each other in the cinematic image to deliver meaning. The cinematic device of parallel and repetition is central to *Alien 3*. As Taubin confirms, the film's thematic interest in, and focus on, the body's subdermal terrain is mirrored by the alien's basement lair.[22]

As already suggested, this device is employed frequently within film narratives which take up the concerns of the unseen, especially the infinitesimal extreme. The underground laboratory, the basement lair, the subterranean hideaway: all present visual conventions to construct meaning through a formal pictorial parallel with the submerged territories of bodies or computer circuitry. Each film discussed here – as well as numerous others (such as the military compound in *Wargames* and the entire post-viral diegetic environment of *Twelve Monkeys*) – pictures underground spaces as cinematic landscapes. While the body's surface, once mourned by Foucault, has returned as a site for inscription, it is the hidden territory of the body, paralleled by cinematic submerged landscapes, which is referenced by the textualised tableau provided by the visible flesh of the body.

How might this alteration of Foucault's thesis be intellectualised? While visual and linguistic discourses have provided an effective framework for rethinking the body's exterior, do these approaches fully address the concerns of the technologically-accessed and regulated body as a site which is both pregnant with meaning from its uniquely encoded interior, even while it is a site distanced and abandoned? In the age of the virus and its attendant systems of surveillance, it is informative to engage in a discourse of the virus itself. For this purpose, I turn to the work in which virus (and body, image and language) co-exist in often grotesque yet poignant fashions as a crucial cultural method of reading the body.

According to Steven Shaviro, '[n]obody understands these issues better than William Burroughs'.[23] Burroughs' dictum that 'language is a virus' serves to draw essential parallels between the *modi operandi* of the discursive systems of both language and body.[24] Shaviro suggests that, like a virus, '[l]anguage is one of [the] mechanisms of reproduction. Its purpose is not to indicate or communicate any particular content, but merely to perpetuate and replicate itself.'[25] In other words, referencing McLuhan's famous proposition, 'the medium really is the mes-

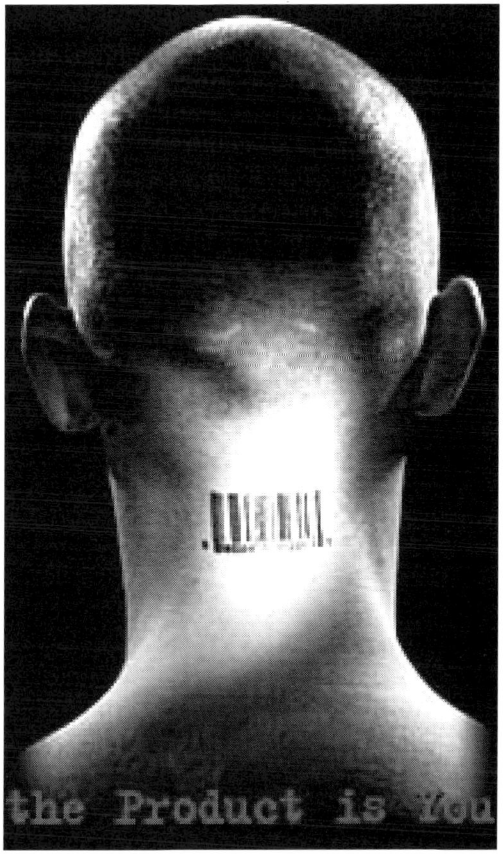

Fig. 3. The Product is You. [Courtesy of Adbusters Media Foundation, www.adbusters.org.]

sage' because the only information to be communicated by the language-virus is language itself.[26] This self-perpetuating system relates to the media as well, which become artificial parasites in Shaviro's cultural virology. Moreover, the implication of the media as a viral system is crucial, for it moves beyond purely linguistic or visual considerations and toward a discourse which allows for the plurality of the conditions of postmodernity.

Like Shaviro, Baudrillard demands a media adjustment, but he goes further. Rejecting McLuhan's neural metaphor, Baudrillard insists that what exists now is 'a *viral*, endemic, chronic, alarming presence of the medium'.[27] Scott Bukatman applies and augments Baudrillard's interpretation of the viral nature of the image in his own analysis of Cronenberg's *Videodrome* (1983). Bukatman frames the media virus as one which engages the body and imbricates it in a viral directive of self-replication. As Max Wren submerges himself in his quest to locate the originator of the underground cable programme, his lived reality collapses with the viewed image. Bukatman observes: 'Body and image become one: a dissolution of real and representation, certainly, but also of the boundaries between internal and external.'[28] Or, as Baudrillard puts it, television dissolves into life, and life into television.[29]

These theories describe more thoroughly the process already discussed in which the body (surface) becomes implicated as part-and-parcel of the spectacle of the visual apparatus, a process which leads to a condition of hyperreality marked by the blurred boundaries of real and representation. Finally, Baudrillard's analysis and Cronenberg's vision converge as a video cassette is inserted into Wren's torso: '[The] image virus reduce[s] the subject to the status of videotape player/recorder; the human body mutates to become part of the massive system of reproductive technology ... video becomes visceral.'[30] The emerging concept of viral systems produces a postmodern discourse which perhaps more effectively addresses the current cultural environment in which the body, language, and the image exist in a complex, parasitic interrelationship. In their individual attempts to self-replicate, they consume and extend the others to the point where there is, for Shaviro, 'no refuge of pure interiority'.[31]

Cyborg as coded commodity

In conclusion, the legacy generated by medical and space technologies is a popular media of surveillance and anxiety as the increasing sophistication of twentieth-century visual apparatuses grants an expanded, hypervisual access to both extremes of Foucault's *unseen*. From chronophotography and X-rays, through electron microscopy and sonar-resonance imaging, to the submerged codes of DNA, HIV, and computer binary systems, *et al.*, the body has become a site for the inscription and retrieval of information. In short, what began as a means of recording the mechanics of the moving body has resulted in the double-helix of genetics on one end of the continuum, and a Martian rock named Scooby Doo on the other. Furthermore, such a result has produced certain consequences for the

body. Conflicted by the possibility of a knowledge unavailable to the naked eye, the postmodern body has become hypertextualised: a code of a code. Within the context of epidemiological and genetic concerns, and with a heightened cultural sense of contamination, the body surface is refigured as an ideal site for the inscription of its submerged code. The body – as critically imaged by the Canadian-based Adbusters Media Foundation – has been reduced to a kind of *black box commodity* (Fig. 3).

Referring to the now virtually outdated electronic unit that sits on top of the television and provides access to cable channels, a *black box* is a token emblem of the actual product, which is the exchange or retrieval of information via encoded signals. Similar, perhaps, to a computer icon symbolising software programme codes, the postmodern body is engaged in a kind of black box exchange of information. Leaving aside genetic concerns, the more common data transactions are performed around virtual funds, bank-draft transfers, ATM withdrawals, and credit histories, to name but a few. The postmodern citizen/consumer is an encoded summary of its capitalist history and net-worth. Or, as described by Shaviro, '[b]odies and words are nothing but exchange-value: commodities or money'.[32]

Undoubtedly it comes as no surprise to our digitised, cyborgian generation that bodies have become encoded in overwhelmingly economic ways. Long before we retreated to cubicles, entered our passwords, and prepared to go on-line shopping, the cyborg was well-configured within a media-based system of commodity-valuation. In the 1970s, Steve Austin travels the length of Foucault's continuum as he is transformed by technology from NASA astronaut to *bionic* entity. The original title of the American television pilot which premiered in 1970 belies its capitalist connections: *Cyborg: The Six Million Dollar Man*.

Notes

1. Arthur and Marilouise Kroker, 'Panic Sex in America', *Body Invaders: Sexuality and the Postmodern Condition* (London: MacMillan Education Ltd., 1988), 12.
2. Michel Foucault, *The Order of Things: An Archaeology of the Human Sciences* (New York: Vintage Books, 1973), as quoted by Lisa Cartwright, *Screening the Body: Tracing Medicine's Visual Culture* (Minneapolis: University of Minnesota Press, 1995), 11.
3. Cartwright, 4.
4. Cartwright, 26.
5. Cartwright, 27.
6. Tom Gunning, 'The Cinema of Attractions: Early Film, its Spectator and the Avant-Garde', in Thomas Elsaesser (ed.), *Early Cinema: Space, Frame, Narrative* (London: British Film Institute, 1990).
7. Lennart Nilsson image of red blood cells with editorial title, 'Closing in on an Unseen World', and image of fetus entitled 'Tiny Voyager', *Life* (Fall 1986).
8. Robert Romanyshyn, *Technology as Symptom and Dream* (London: Routledge, 1989), 35.
9. Romanyshyn, 17. Romanyshyn is quoting Charles Wentinck, *The Human Figure* (Wynnewood: Livingston Publishing Company, 1971), 157.
10. Ian Hacking, *Representing and Intervening: Introductory Topics in the Philosophy of Natural Science* (Cambridge: Cambridge University Press, 1983), 209.
11. Star Trek Prologue: 'Space, the final frontier. These are the voyages of the Starship Enterprise, its continuing mission: to boldly go where no man has gone before.'
12. Marshall McLuhan, *Understanding Media: The Extensions of Man* (New York: McGraw-Hill, 1964).
13. Donna Haraway, *Simians, Cyborgs, and Women: The Reinvention of Nature* (London: Free Association Books, 1991), 78-80.
14. Benjamin Woolley, *Virtual Worlds* (Harmondsworth: Penguin, 1992), 127.
15. Woolley, 129.
16. Arthur and Marilouise Kroker, 14.
17. Arthur and Marilouise Kroker, 11.
18. Barbara Kruger, 'Your Body is a Battleground', poster for pro-Choice demonstration in Washington,

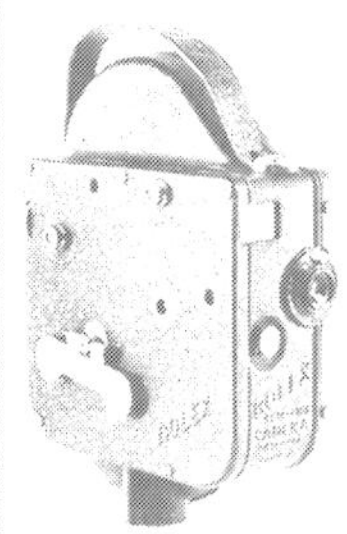

DC, 1989. Image from *Love for Sale: The Words and Pictures of Barbara Kruger*, text by Kate Linker, (New York: Harry N. Abrams, Inc., 1990).

19. Oncomouse is the descendant of a long ancestral line of mice bred for scientific testing. After generations of being injected with carcinogens, Oncomouse's genetic heritage was profound: he was immune to cancer.
20. Arthur and Marilouise Kroker, 14.
21. Amy Taubin, 'The Aliens Trilogy: From Feminism to AIDS', *Women and Film: A Sight and Sound Reader* (London: Scarlett Press, 1993), 98–99.
22. Taubin, 98–99.
23. Steven Shaviro, 'Two Lessons from Burroughs', in Judith Halberstam and Ira Livingston (eds.), *Posthuman Bodies* (Bloomington: Indiana University Press, 1995), 39.
24. Shaviro, 40.
25. Shaviro, 41.
26. Shaviro, 40–41.
27. Jean Baudrillard, 'The Precession of Simulacra', in *Simulations*, trans. Paul Foss, Paul Patton and Philip Beitchman (New York: Semiotext(e), 1983), 54–55, as quoted by Scott Bukatman, 'Who Programs You? The Science Fiction of the Spectacle', in Annette Kuhn (ed.), *Alien Zone: Cultural Theory and Contemporary Science Fiction* (London: Verso, 1990), 199–200.
28. Bukatman, 207.
29. Baudrillard, 54–55.
30. Baudrillard, 54–55.
31. Shaviro, 40.
32. Shaviro, 38.

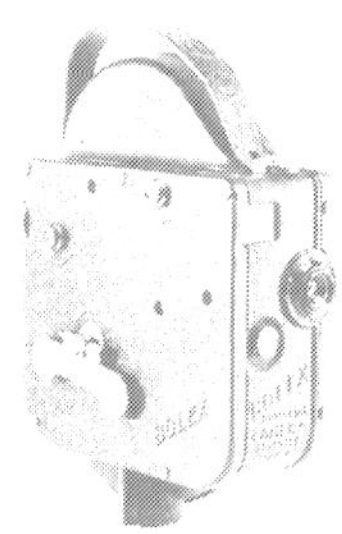

3

The Domestic Sphere

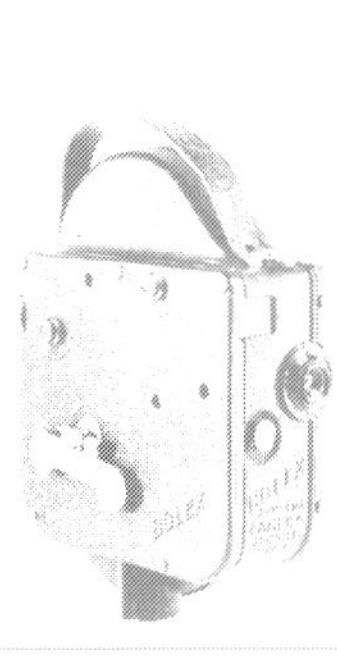

Weather Porn and the Battle for Eyeballs: Promoting Digital Television in the USA and UK

William Boddy

Baruch College and the Graduate Center of the City University of New York, USA

Introduction

In the form of lavish ad campaigns, front-page news stories, and overheated techno-punditry, the end of the 1990s marked the culmination of a long technological, policy-making and marketing campaign waged by a diverse set of private interests to bring digital television into American and British homes. Charting digital television's distinct fortunes in the United States and Great Britain can illuminate some of the most persistent and difficult issues in media historiography, including sorting out the determinations of national culture, market structures, and ideological valence in technological innovation. The current transition from analogue to digital standards for electronically stored and transmitted moving images is a profoundly unsettling moment for the established policy rationales and commercial structures of broadcasting in both the USA and the UK. Digital television threatens not only to alter fundamentally the traditional relationships among programme producers, station owners, broadcast networks, and satellite and cable operators, but also to bring powerful new economic actors into the business of providing programming and services to the domestic television audience. The current tangle of rivalries and alliances between sectors of the television industry and multinational firms in consumer electronics, telecommunications, retailing, financial services, and computer hardware and software, testify to the contentious and uncertain nature of technological innovation in contemporary electronic media. In addition, the proliferation of channels and delivery systems associated with digital television promises to alter existing programming forms, as well as international programme flows and institutional alliances across an increasingly globalised media landscape. Finally, in the promise of enhanced visual quality, multiple programme channels, and the capacity for complex interactive exchange, digital television has inspired many observers to proclaim the overturning of long-established phenomenological and cultural paradigms of television viewing with quite different, if still uncertain, scenarios.

The current prospect of fundamental technological change in the television industry also presents a challenge to traditional scholarly approaches to media innovation, approaches Raymond Williams has characterised as 'an unholy combination of technological determinism with cultural pessimism'. Williams argues that 'the moment of any new technology is a moment of choice', and historians must attempt to counter conventional teleological accounts of autonomous technological 'progress' by restoring a sense of historical agency to the social and political

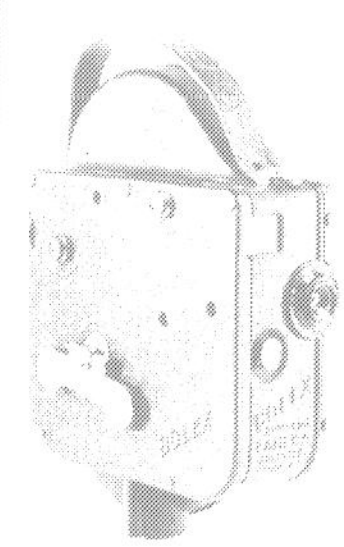

choices involved in the commercial exploitation of new media technologies.[1] To this ambitious end, I would modestly like to sketch out some of the concrete struggles currently raging within and outside the television industry around the introduction of digital television in the USA and UK.

The commercial and regulatory contexts for digital television

Preliminary to such a project is recognition that the term 'digital television' constitutes an over-determined and sometimes strategically ambiguous signifier, encompassing the distinct promises of enhanced picture quality (various formats of so-called 'high-definition television'), greater viewer-programme interactivity, and a significant expansion of the number of programme channels delivered to the home. These distinct semantic components of the term 'digital television' correspond to the competing commercial interests of different sectors of the television industry, leading to occasional public and policy disputes over the definition and applications of digital television. For the consumer electronics industry in the US, for example, digital television represents the financial potential of a vast market for high-definition TV sets, as American TV households are forced to replace (or adapt, via a set-top converter box) existing receivers in order to receive higher-quality pictures (and, eventually, any pictures at all, as existing analogue television channels are shut down and auctioned off in 2006, according to the present timetable of the Federal Communications Commission [FCC]). Conversely, for broadcasters, cable operators, and direct-broadcast satellite operators, digital television offers a means to multiply the channels of traditional standard-definition television (SDTV) service they might offer by a factor of five, ten, or even more; in this light, digital television represents the possibility of increasing so-called programme 'tonnage', or the number of SDTV channels available for sale to viewers and/or advertisers. Finally, for cable operators and for some major interests outside the traditional broadcasting industry, including computer software and hardware manufacturers, telephone companies, and firms in retailing, publishing and banking, digital television offers the fulfilment of the long-promised era of interactivity, where viewers become Web surfers and electronic consumers from their own living rooms.

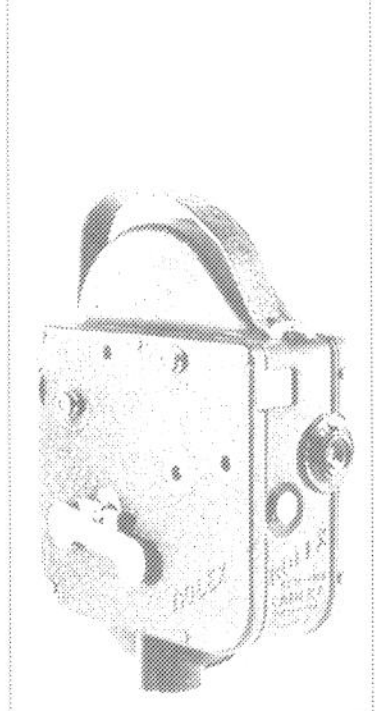

If clear strategic differences can be identified among competing architects of our putative post-television era, there is also a striking, if somewhat unexpected, pervasive defensive tone in much of the current trade discussion of digital television. In spite of the general economic expansion of the 1990s in the USA and Western Europe, most of these diverse economic actors approach the prospect of digital television with a measure of anxiety about their own commercial fortunes in the new era. For example, commercial television networks in the UK and USA are facing long-term declines in their shares of prime time audiences and national advertising revenues; likewise, the long-stagnant consumer electronics industry has been desperate for new products since the VCR and audio CD of a decade or more ago, and the UK direct broadcast satellite (DBS) industry, while larger than that in the US, nevertheless faces rising challenges of viewership plateau, subscriber churn, and the limitations of a largely sports-driven, male-targeted, and somewhat down-market viewership.[2] At the same time the US cable industry is worried about growing competition from telephone companies, DBS, and the Internet. Likewise, through 1998 the US computer industry faced slower rates of growth, with household PC penetration rates still hovering around 40 per cent and the strongest sales gains made in the under-$1000 PC market. Intel's chairman, Andrew Grove, declared in 1996 that future prosperity of the microchip industry depends on markets beyond the PC.[3] Finally, hovering over the calculations of every sector of the digital television industry is a pervasive fear of future actions by the US software giant Microsoft; John Malone, then-CEO of US cable firm TCI, described the anxious mood among broadcast and cable executives in the summer of 1998: 'Look, a lot of this is fear of the computers, fear of Microsoft, fear that ... somehow or other Bill [Gates] got some trick up his sleeve and he's going to own the world – which really is only a timing difference, because he is going to own the world anyway. It's a question of whether this is his route or some other route.'[4]

Given the defensive and uncertain mood within the emerging digital television industry, the tensions between the competing definitions and interests occasionally surface in acrimonious public debate and recrimination. One example can be found in the aftermath of the prolonged and successful campaign by US broadcasters and network operators for the free award of valuable spectrum allocations for digital television service by the Federal Communications Commission. The spectrum awards were achieved in the face of formidable commercial and political opposition, which argued that the channels should instead be allocated to non-television users via multi-billion dollar spectrum auctions. With the support of a consumer electronics industry eager for high-definition television services to revive a sluggish domestic market, the US broadcasters successfully lobbied Congress and federal regulators with stirring talk of the appeal of higher-resolution images for American TV viewers and its importance for the American economy. The FCC's spring 1997 allocations of duplicate UHF channels granted free to each US broadcaster, valued at ten billion dollars by some observers, followed a decade-long political and policy campaign by broadcasters and consumer electronics firms which traded heavily upon discourses of economic nationalism and fears of Japanese ascendancy. Leading the chorus was conservative pundit George Gilder, who in 1989 saw in Japanese-developed analogue HDTV 'a kind of Pearl Harbor in the offing' for the US consumer electronics industry.[5] In 1990 Gilder wrote that 'a consensus emerged that the USA was a graying and gullible nation, slipping into churlish senility, and that Japan was a mercantilist shyster, seizing power by unfair trade',[6] and as late as 1992, Gilder was warning that 'if the USA fails to keep pace technologically or goes into a siege of bashing the rich and competent, the Japanese will become the world's dominant power'.[7] Such appeals to nationalist sentiment and geopolitical self-interest were crucial in the public and political campaigns for US high-definition television in the early 1990s, and continue to sway public and policy debates over digital television despite, or perhaps because of, the continuing globalisation of media ownership and programme flows.

Within months of their 1997 high-definition allocation victory, however, network and broadcast executives appeared to back away from their commitment to high-definition television, expressing new reservations about TV set costs, viewer interest and station conversion expenses involved in high-definition television service. ABC Network President Preston Padden and other broadcasters now complained that they saw no way to make money from high-definition television, and announced that they would use their new channels for multiplexing pay-per-view and additional SDTV programming.[8] The broadcasters' new public scepticism about HDTV reflected their recalculation of the economic value of the distinct technical possibilities provided by digital broadcasting, rejecting high-definition service in favour of using the new channels to multiplex additional proprietary channels conforming to existing visual standards. Padden told a group of security analysts in August 1997: 'Our share of the viewing audience will continue to erode as long as we remain a single channel in an expanding multi-channel universe'; David Smith, president of Sinclair Broadcasting Group, owner and programmer for twenty-nine stations in the US, complained around the same time: 'We have yet to see how anyone makes money as an HDTV broadcaster.'[9] TCI head John Malone told a journalist in July 1998 that broadcasters were holding back on announcing non-HDTV digital television strategies only 'because the broadcasters are building up their intestinal fortitude to decide how much risk they take on these licenses and just how greedy they can afford to look'.[10] At a Congressional hearing in the summer of 1998, Republican Senator John McCain, an opponent of the earlier HDTV spectrum give-away, responded to the pleas of broadcasters to extend the transition period for high-definition television: 'You got tens of billions in free spectrum ... over my vociferous objections ... and now you're complaining that we expect you to put that to good use in a timely fashion. All along I knew, you knew and the

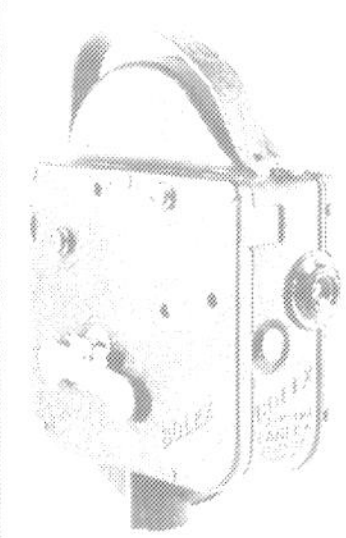

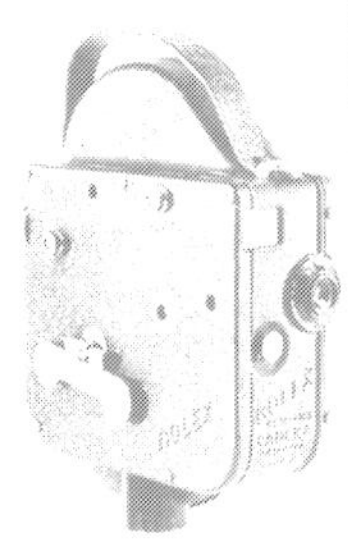

broadcasters knew there wasn't a snowball's chance in Gila Bend, Ariz., that you were going to meet the 2006 deadline.'[11] Meanwhile, disappointed leaders of the consumer electronics industry, as well as mobile communication companies competing for the spectrum resources given away to the erstwhile high-definition broadcasters, viewed the on-again, off-again campaign by the broadcast industry for high-definition television as a cynical, if successful, public relations and political lobbying effort in pursuit of free spectrum allocations. Former FCC Chair Reed Hundt told the *Wall Street Journal* in 1997 that 'HDTV turned out to be more a lobbying idea than a business strategy'.[12]

While broadcasters, constrained by the public relations, political, and regulatory hazards of being seen as reneging on their promises of high-definition service, have more recently offered at least half-hearted expressions of support for high-definition television, other sectors of the US television industry have been more frank about their lack of interest in the increased visual resolution offered by digital television in favour of its potential for multicasting and interactivity.[13] One powerful set of declared antagonists to high-definition television are the US cable operators, led by TCI's John Malone, and computer hardware and software firms eager to sell set-top boxes and software to convert digital signals for use by existing television sets. In the summer of 1998 Malone's TCI was attempting to convince broadcasters to eschew high-definition service in favour of multiplexing standard-definition television channels and interactive services by offering to share TCI's transaction fees for electronic commerce conducted via digital television. In response to the public resistance to high-definition television from cable operators, Gary Shapiro, head of the Consumer Electronics Manufacturers Association, complained during a broadcasters' convention in January 1988: 'TCI's 14 million customers may never have a chance to see HDTV. This is a huge tragedy for the American consumer.'[14] Meanwhile, the *New York Times* reported in February 1998 that 'for more than a year, the Microsoft Corporation has been cajoling the nation's television broadcasters into abandoning high-definition television and using only the computer industry's favoured, lower-resolution transmission formats when digital broadcasts begins in the fall'.[15] Indeed, by the end of 1998, with the exception of HBO and other premium movie channels, it appeared that the only remaining unequivocal American champion of high-definition television, once hailed as the saviour of US broadcasting, the domestic electronics industry, and national competitiveness, is the consumer electronics industry, itself struggling to produce a high-definition set for less than the $7,000–10,000 demanded for initial models. The confusion and delay led the *Wall Street Journal* in September 1997 to complain: 'Rarely in the history of American business has there been a new technology that promised so much – and delivered so little.'[16] A year later, the *New York Times* described the regulatory handling of HDTV's introduction in the US as 'a slow-motion train wreck'.[17]

Digital television and national identity

Notwithstanding the tangled alliances and uncertain fortunes of high-definition television in the United States, the position of digital television in the UK is distinct from that in America in a number of important ways, and its differing fortunes highlight the distinct roles of political and regulatory institutions involved in broadcasting and the more general ways in which popular and public policy debates around broadcasting inevitably serve to both invoke and construct myths of national identity. On one level, in both the US and UK, the transition to digital television has been overseen by a relatively *laissez-faire* regulatory philosophy, and spurred by the desire of both governments to capture revenues through the eventual auction of existing analogue TV channels (indeed, the US Congress has already incorporated these hypothetical revenues in current budgets). In a suggestion of the prevailing anti-regulatory mood in the USA concerning high-definition television, FCC Chair William E. Kennard in 1998 expressed his belief that federal regulators should be involved in industry policy disputes only 'when arguments are distorted by the prism of self-interest', a circumstance which more cynical observers might view as virtually tautological.[18]

Despite the prevailing anti-regulatory political contexts of both nations in the 1990s, however, one can argue that the USA and UK have historically stood as the defining other in broadcast policy-making and popular imagination since the origins of radio broadcasting in the 1920s, and each nation has figured as positive or negative example in historical debates over broadcast policy. The imagined communities of the national broadcast audience and the distinct national regulatory rationales remain powerful, despite the increasing globalisation of programme markets and corporate ownership, and the current transition to digital television underscore the competitive and policy differences acting upon the television industries in the USA and UK. In each nation, traditional broadcasters, direct satellite broadcasters, and cable operators are attempting to direct the development of digital television via their control of an existing delivery system. However, the respective market power of each of the three sectors is quite different in the two nations; while cable enjoys relatively low penetration levels in the UK (2.5 million of Britain's 23.5 million households), and terrestrial broadcasting is limited to five channels, Rupert Murdoch's ten-year-old direct broadcast satellite service, BSkyB, offers forty channels to 3.6 million British homes.[19] Thus, unlike the case in the US, where there are many more conventional broadcasters, higher cable penetration and fewer viewers of satellite television, Murdoch's new 140-channel (including 75 conventional television channels, and 48 pay-per-view channels) digital satellite service promises to be the early leader as delivery platform for UK digital television, even though the digital service requires new receiving dishes and decoder boxes. Murdoch's chief rival is OnDigital, a consortium of ITV broadcasters, who launched a thirty-channel, non-interactive terrestrial digital service in mid-November 1998; cable delivery of digital television in the UK is expected in 1999 or 2000.[20]

Thus, the competitive context and delivery paths of digital television in the USA and UK are quite distinct. OnDigital offers no interactive capabilities, and Murdoch's digital BSkyB will offer only limited interactive features for several months after its October 1998 launch, unlike the potential in the United States for richer interactivity via WebTV (acquired by Microsoft Corporation in 1997) and cable (which reaches a much higher percentage of homes in the USA than in the UK). More strikingly, digital television in the UK has not been sold to policy makers and consumers as a high-definition service requiring new TV receivers, but largely as a way to increase available channels to existing sets; in the US, enhanced picture quality delivered on new high-definition receivers has been promoted as *the* defining feature of digital television (though DirecTV has offered a digital DBS service since 1995, it remains peripheral to most US debates concerning digital television). These differing promotional strategies are propositioned on quite different models of the place of digital television in the home.

As a means of greatly increasing the number of television channels delivered to the home, digital television in both the USA and UK promises to de-stabilise traditional programme-making and scheduling practices. One immediate effect is the frantic efforts by traditional programme producers and network operators to 're-purpose' existing programming and personnel to fill the new digital channels; for example, the BBC has supplemented its two current channels on digital BSkyB with BBC Choice, BBC News 24, BBC Learning, and a digital text service. Programme budgets for BBC Choice, a channel offering wall-to-wall 're-purposed' re-runs of previous BBC programmes, are set at \$16,000 to \$32,000 an hour, compared to about \$728,000 an hour for BBC drama on its most popular channel. The BBC Choice budget, in turn, will be much higher than many new commercial satellite channels offered by Murdoch's digital satellite service.[21] The sometimes-desperate efforts to fill expanded programme hours with inexpensive repackaged re-runs of existing programming led *The Guardian* to suggest sarcastically that 'soon there'll be UK Weather Gold – a channel devoted to classic weather reports'.[22]

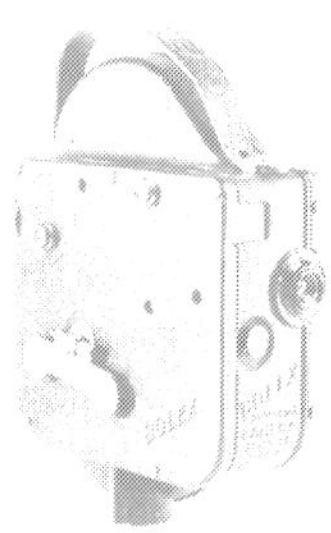

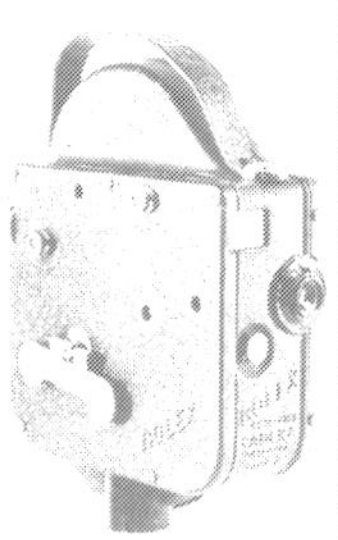

Other critics in the UK have worried that digital television will bring a proliferation of so-called 'weather-porn', US-produced weather and natural disaster documentaries which can be cheaply acquired and quickly re-edited to feature a British narrator and nominal UK content; such generic documentaries, they fear, will crowd out traditional locally-produced documentaries addressing specific social and political topics.[23] Such 're-purposed' programmes increasingly travel both directions across the Atlantic, and the proliferation of digital channels and inexpensive digital post-production in both the USA and UK will make the economic logic of such exchanges more compelling.

Some of the changes associated with the global take-up of digital television represent a radical re-casting of the function of broadcasting as both product and fabricator of national identity. The apotheosis of broadcasting's traditional role in this regard (and of US network hubris) was suggested in a 1954 speech to journalists by CBS Network President Frank Stanton: '*We give America its daily consciousness of being a Nation*. If it weren't for us, private individuals all, and private businesses all, America would not know where it stood or what it felt.'[24] Microsoft's Bill Gates in his 1995 book, *The Road Ahead*, granted the communal aspects of the traditional live national television broadcast as signifier and instrument of national unity: 'When we Americans share national experiences, it is usually because we're witnessing events all at the same time on television – whether it is the *Challenger* blowing up after lift-off, the Super Bowl, an inauguration, coverage of the Gulf War, or the O.J. Simpson car chase. We are 'together' at those moments.' However, Gates goes on to argue that 'it is human nature to find ways to create synchronous communications into asynchronous forms'.[25] In the same year, MIT's Nicholas Negroponte predicted of America's television future: 'Digital life will include very little real-time broadcast ... With the possible exception of sports and elections, technology suggests that TV and radio of the future will be delivered asynchronously.'[26]

Such technologically-inspired predictions of the weakening power of the live national television broadcast as instrument and expression of national identity have resonated in popular press responses to the launch of UK digital television in the fall of 1998. A few days before the launch of Sky Digital, a British journalist wrote of Murdoch executives: 'Together they are in the business of persuading us that the age of television as a force of national cohesion is over: that we no longer wish to sit down all together on a Monday night to watch *EastEnders*, and that, instead, we are ready to embrace a world of hundreds of channels.'[27] Another journalist reported BBC chair John Birt's statement that 'the end may be in sight for broadcasting as a communal experience', and a third journalist lamented: 'Television has up to now been a unifying feature, for people in any one country tend to watch the same programmes – if only because they have had little choice. Now we will have that choice. That is wonderful. But we will lose a nationally unifying force.'[28]

Meanwhile, Bill Gates and other US executives happily predict the replacement of such (in their view) technologically-antiquated artifacts of electronic media with a single global popular culture, what *Fortune* magazine in 1990 called 'a one-world pop-tech civilization', whose products happen to be overwhelmingly US in content and origin.[29] As Gates argues in *The Road Ahead*: 'American popular culture is so potent that outside the United States some countries now attempt to ration it. They hope to guarantee the viability of domestic-content producers by permitting only a certain number of hours of foreign television to be aired each week. ... The information highway is going to break down boundaries and may promote a world culture.'[30] In stark contrast to his earlier pessimism about imperilled American hegemony in the early 1990s, by 1996 George Gilder celebrated the 'technological and entrepreneurial culture of bourgeois America that is now sweeping the globe'. Viewing digital television as a formidable weapon in a global ideological battle between the values of capitalism and those of 'the bohemian intelligencia', Gilder argued that digital television 'threaten[s] the key remaining bastions of the power of bohemian intellectuals: the universities, Hollywood, the broadcast networks, and the government/social-work complex. Intellectuals like to describe these mostly depraved institutions as the core of the nation's identity and common purpose, but in fact they serve chiefly as the pork barrels, subsidy mills, and agitprop centres of the bohemian intelligencia.'[31]

For Gilder, the putative decline of such suspect institutions of national identity reflects a technological and, ultimately, theological manifest destiny of US-style capitalism. In 1994, Gilder predicted that computer networks, infiltrating and eventually taking over the traditional uses of television, would finally destroy the national institutions of broadcasting to the benefit of conservative forces in an undeclared cultural war: 'The downfall of the liberal media, the rout of the rodent kings of the networks, the ... "three blind mice" gnawing at the pillars of civilised life in America – what, one might wonder, could be sweeter news for conservatives?' 'The information superhighway', Gilder argued, 'promises to revitalize capitalism and culture in the USA and around the globe and to retrieve the hopes of a conservative era in politics'.[32] Expounding a moralistic technological determinism, Gilder told readers of the *Economist* in 1993: 'Television will die because it affronts human nature; the drive to self-improvement and autonomy that lifted the race out from the muck and offers the only hope for triumph in our current adversities ... commercial television is necessarily the enemy of civilisation.' Ultimately, Gilder's faith in technological progress is a theological one, as he explained in 1991: 'I think secular culture is an irredeemable fiasco; estranged from God as the source of all truth and beauty, art rots to triviality and evil. Nonetheless, I believe that this culture has produced one incomparable achievement, scaled a summit as beautiful and true as any Gothic cathedral. That pinnacle is its technology.'[34]

The Tom Cruise in Microsoft Windows

Despite the grandiloquent moral certainties of conservative techno-pundits like Gilder, the actual effects of digital television on traditional institutions and representations of national identity are still uncertain. At the same time, in more concrete terms, among promoters of the new technology, digital television has already provoked speculation about mutating programme forms. In *The Road Ahead*, Bill Gates writes with enthusiasm about digital television's promise of new merchandising opportunities for television programme-makers and broadcasters in the form of enhanced product placement: 'In the future, companies may pay not only to have their products on-screen, but also to make them available for you to buy. You will have the option of inquiring about any image you see ... you'll be able to pause the movie and learn about the glasses or even buy them on the spot.' Using the hypothetical example of Tom Cruise's accessories, Gates explains: 'If the movie's star carries a handsome leather briefcase or handbag, the highway will let you browse the manufacturer's entire line of leather goods and order either one or be directed to a convenient retailer.'[35] In late 1998, Microsoft's digital television Web site suggested that digital television broadcasters adopt a merchandising device from the World Wide Web, the un-deletable interactive banner advertisement: 'Use banner ads or other clickable graphics. While commercials appear for short intervals, banner ads can be constantly displayed, even as the video continues to play.'[36] Elsewhere on its Web site, Microsoft advises would-be digital broadcasters:

> Develop associated revenue streams. An obvious business model makes use of formerly unrealized, complementary revenue ... associated revenue streams might pair CDs with a dance or music review, weather experiment kits for kids who watch the evening weather report, books suggested by talk show hosts, airline tickets or holiday packages suggested by travelogues, among many others. The key is in the partnership and how you share your audience with the merchandiser.[37]

Beyond the possibilities so eagerly anticipated by Microsoft for enhanced product placement and inescapable on-screen marketing logos, another predicted programming pattern associated with the development of digital television in the United States is the

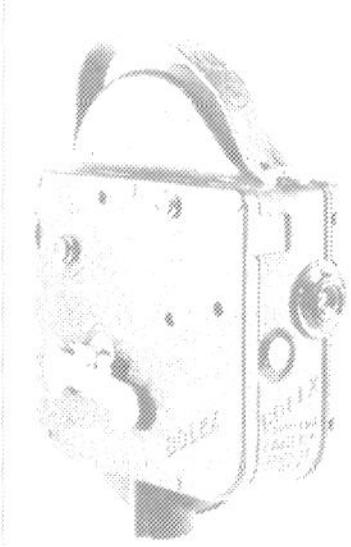

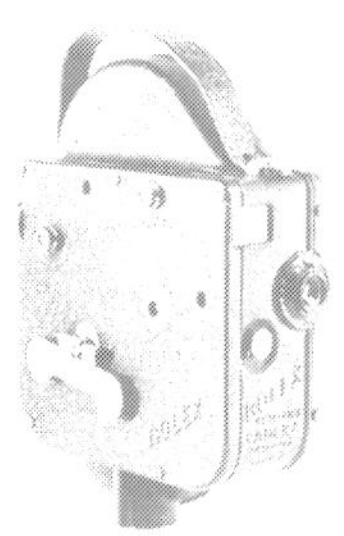

gendered segmentation of services according to day part, where standard-definition programming with enhanced interactive marketing services is offered in the daytime to a predominantly female audience, and full high-definition programming is offered to a more masculine audience in prime time.[38] This duel split of programme form and marketing strategy in digital television echoes the practices of early commercial radio broadcasters regarding direct and indirect advertising and of early commercial television operators relating to the design of programmes according to anticipated viewer attention. As a CBS publication of 1945 explained: 'Programs requiring full attention of eye and ear should be scheduled in the evening hours when viewers feel entitled to entertainment and relaxation.'[39] In this regard, the marketing assumptions of our own putative post-television era seem remarkably consistent with long-established beliefs and practices among broadcasters and advertisers about their domestic audience.

New paradigms of the television viewer

In addition to de-stabilising expectations regarding competitive relations, national identity, and programme forms, the long-term enforced migration from analogue to digital television initiated in late 1998 in the USA and UK has also provoked a reconsideration of some traditional conceptions of the television audience and of the activity of TV viewing. The competing ontologies and business models for digital television – enhanced image quality, multiple standard-definition channels, or enhanced interactivity – activate distinct scenarios of television viewing, scenarios which nevertheless resonate with a repertory of popular images of media audiences going back to the beginning of the twentieth century. In one way, the century-long historical experience of domestic electronic communication can be analysed as the playing out of a series of broadly opposed figurations of the broadcast audience, from wireless amateur to distracted housewife, from couch potato to Internet surfer. Such figurations speak to important historical issues about the role of electronic media in the intimate spaces and gendered routines of everyday life, and digital television throws such figurations into some confusion and crisis. For example, the provision of 140 channels on digital television systems like Murdoch's digital BSkyB requires an electronic programme guide (EPG) which resembles an Internet search engine more than the traditional *TV Guide*. Defending his company's thirty-channel, non-interactive terrestrial digital service against the more complex Murdoch system, OnDigital's CEO Stephen Grabiner told a British journalist in 1998: 'There will be people who want 200 channels and there will be people who spend their lives playing with the EPG. But these are sad unhappy people who live in lofts.'[40] Grabiner's social and spatial marginalisation of the electronics enthusiast as distinct from the mainstream non-technophilic media consumer echoes long-running popular culture figurations of the lonely male hobbyist in the basement, attic or garage. Such figures are prominent in the popular literature from the 1910s amateur radio enthusiast, through the post-war audiophile or ham radio operator, to the contemporary computer hacker.

Given these persistent discursive traditions, it is suggestive to at least briefly consider the ostensible convergence of the computer and TV set from the point of view of some prominent writers and executives in the contemporary computer industry. In the early 1990s, a time of pessimism among Gilder and others about the future competitiveness of American high tech in the face of Japanese economic expansionism, a broad opposition was elaborated between conventional broadcast television and the then much-hyped phenomenon of virtual reality. The deep-seated gender and cultural positioning of domestic television as object and activity served as a remarkably consistent foil for enthusiasts of a range of advanced electronic imaging systems, what a chorus of diverse commentators at the time identified as 'television's second chance'. *Newsweek* magazine in April 1992 quoted Apple Computer CEO John Sculley: 'Television's going to get a second chance, and there's a chance to do it right this time.'[41] Thirty-one-year-old virtual reality pioneer Jaron Lanier told *U.S.A. News and*

World Report: 'The best thing about VR is that it will kill TV.'[42] Lanier told a group of VR fans at a Cyberthon convention in San Francisco in 1991: 'We live in this very weird time in history where we're passive recipients of a very immature, noninteractive broadcast medium. Mission number one is to kill TV.'[43]

However, since the early 1990s, at a time when attention to virtual reality in the popular press has faded into such prosaic headlines as 'Virtual Reality at the Dentist: Better than Novocain', the Internet has moved into the same discursive space as antipode to the culturally-degraded domestic television receiver.[44] Thus, from the point of view of many Web partisans, the putative merging of the computer and TV set has been viewed as more threat than promise. One can observe this anxiety in a number of sites, including the flurry of interest in 1995 around push technologies as a business model for the Internet (Jaron Lanier in *Wired* magazine denounced push technologies as 'TV all over again'),[45] the design of computer interfaces, and the phenomenology of computer and television use. While interactive digital television has reactivated some of the oldest tropes of masculinist adventure and mobility from the earliest days of wireless (George Gilder wrote in 1991 that via digital video, 'a tourist could visit a Third World country and view any chosen set of scenes or events without having to drink the water'), its discomforting proximity to the conventionally-scorned object of the domestic TV receiver has also provoked anxiety and resistance.[46] Apple Computer CEO Steve Jobs told a group of educators in December 1998 that although viewers use television to 'turn their brains off, people go to the PC to turn their brain on'. Of the supposed convergence of personal computer and television set, Jobs said: 'These things don't go together, and I think it's crazy to expect this is going to happen.'[47]

In a similar vein, in his 1997 book *Interface Culture*, Steven Johnson protests the use of the term 'surf' to describe what Net users do, since, according to Johnson, the term derives from the existing television practice of channel surfing via cable television and the remote control. 'What makes the idea of cybersurf so infuriating is the implicit connection drawn to television', Johnson writes.[48] In place of the culturally contaminated metaphor of the Web 'surfer', Johnson proposes the masculinist figures of Baudelaire's nineteenth-century urban *flâneur* or the solitary frontiersman 'trailblazer', 'a term', he argues, 'that would have fitted well with the "new frontier" rhetoric of recent cyber-boosterism. It certainly would have been an improvement on the couch-potato passivity of the 'surfing' argot.'[49] Johnson is likewise scornful of Bob, a software shell unsuccessfully introduced by Microsoft in 1995 which offered users a simplified graphical interface based on domestic iconography and talking cartoon-character guides.[50] For Johnson, 'Bob represents the domestication of the personal computer, in the pejorative sense of the word ... the mother's-little-helper of interface design.'[51]

On the other hand, not surprisingly, Microsoft's recent moves into the interactive digital television market has given the company a strong strategic interest in blurring the phenomenological and normative differences between television viewing and computer use, rejecting the conventional opposition between Web user and domestic TV viewer. Microsoft's digital television Web site argues that:

> [W]atching TV is already interactive. Many viewers seldom watch complete programs, but channel surf and graze through a variety of channels ... An increasing number of viewers simultaneously watch a TV set and use a computer to surf the Internet. With over 2,000 broadcast stations in the United States, hundreds of cable channels, and millions of Web sites, the war for eyeballs has grown even fiercer.[52]

Indeed, what Intel chief Andrew Grove calls the 'battle for eyeballs' between computers and TV sets in the home has provoked his company to look outside the traditional computer industry for new markets for logic chips. While digital television tuner-cards for the personal computer, including those capable of receiving high-definition programming, will be a relatively inexpensive path for digital television into the home, the problems of integrating the device within the domestic space, configuring screen-user distance, and

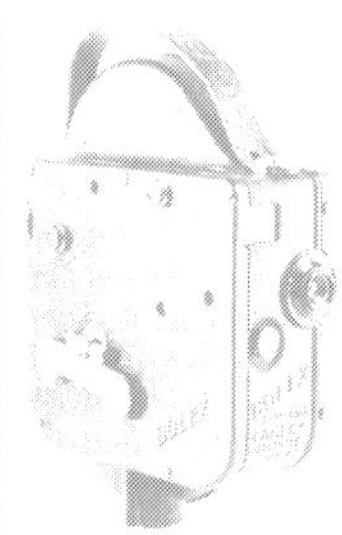

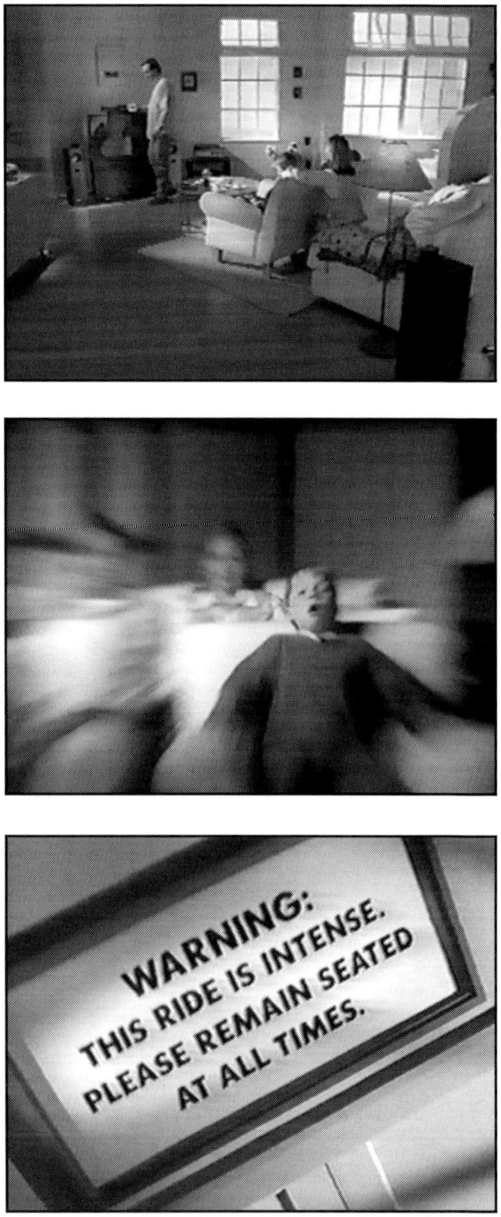

Figs. 1–3. Toshiba DVD player ad, US television, 1998.

managing social aspects of its use remain daunting to many observers. Bill Gates admits in *The Road Ahead*: 'A big-screen TV across the room doesn't lend itself to the use of a keyboard, nor does it afford privacy.'[53] As one industry sceptic noted about the prospects for interactive television as a means for Web access: 'People have computers. People are continuing to buy them. I expect that computers will continue to be the primary way people get online.' Another industry analyst concluded: 'Maybe in 25 years PC/DTV may be the mode of operation for watching programs but right now generations of consumers like to go home, sit down on the couch, turn it on and flip their clicker. No one wants to boot up their TV or have it crash on them during their favourite programme.'[54] A March 1999 *House Beautiful* article speculated about the impact of the high-definition television set on home design, quoting one designer's suggestion that the large-screen set be left on continuously, 'employing the screen-saver as "a mood-driven art gallery"'; the article's author suggested that 'you could use the screen as a digital family bulletin board, flashing NO TV BEFORE HOMEWORK'.[55] This self-annihilating TV receiver suggests the tensions involved in integrating the new 'smart' television set into the home and its domestic routines and power relations.

Selling new media to the home

Given the critical resistance and commercial uncertainty surrounding the place of new digital media in the home, it may be productive to examine two recent thirty-second television commercials aired on US television which represent concrete efforts to sell new consumer electronics products by dramatically enacting their domestic use. The first commercial, for a Toshiba DVD player, depicts a traditional family group of parents and two young children in front of their living-room TV set as dad inserts a DVD of the 1997 action-adventure film *Eraser* (Fig. 1). What follows, unaccompanied by any narration or dialogue, is a disjunctive series of reverse angles alternating between the thrilled and immobilised family audience members, close-ups of the DVD player, and assorted on-screen explosions quoted from the feature film. The thrill-ride analogue (Fig. 2) to the depicted new television experience, suggested by the commercial's acrobatic camera movements, arbitrary lighting changes, blurred rapid zooms, and extreme close-ups of hands clutching sofa armrests, is literalised by the sign above the living-room door glimpsed in the commercial's opening frames (Fig. 3): 'Warning: This Ride is Intense. Please Remain Seated at all Times.' While a variation

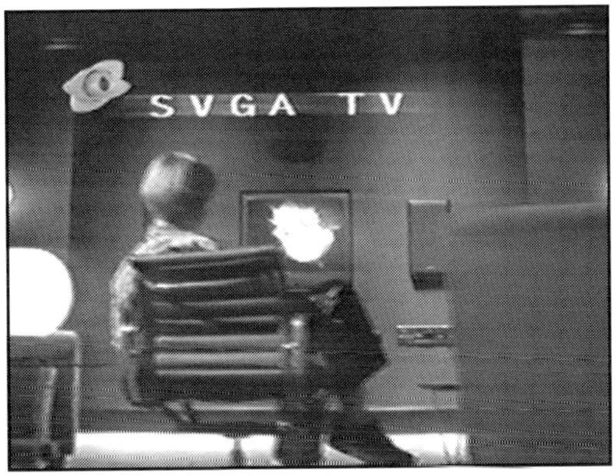

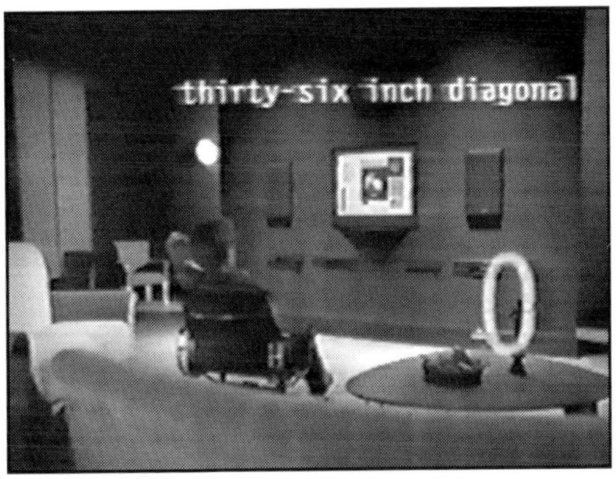

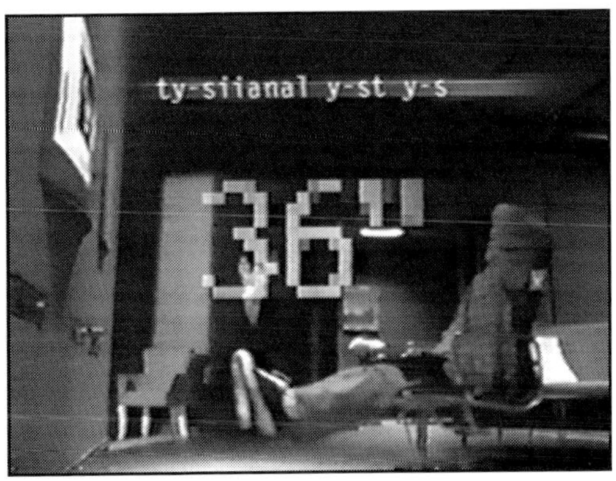

Figs. 4–6. SVGA-TV ad, US television, 1998.

upon the familiar advertising depiction of television viewing which involves the spatial confounding of domestic viewing space with the hyper-real screen action, what is conjured here is less the traditional television as home theatre than television as public amusement park attraction. The assertive animated titles leaping out to viewers after the domestic scene fades to black – 'You've Got Senses. Use Them.' – underscores the commercial's attempts to evoke a physical, body-centred scenario of television viewing via a flurry of interpolated extreme-close-ups of the eyes, ears, and open mouths of the family audience. The assaultive, thrilling nature of the depicted DVD television playback is underscored by extreme-close-ups of shaking teaspoons and pearl earrings, visual icons of traditional domesticity. The overall effect of the commercial, however, is suggestive less of a new media technology than of conventional TV on steroids; the ad is able to eschew narration and dialogue only because the electronic technology and viewing experience it depicts are presented as merely a hyperbolisation of existing television viewing scenarios, albeit one which pushes the traditional marketing distinctions between domestic and public spaces of sociality to their limit.

Only implicit in the Toshiba commercial is DVD's noteworthy threshold status as a contemporary consumer electronics product, serving at once as a software platform for both personal computers and TV sets. On one level simply an enhanced delivery vehicle for Hollywood films already available on videotape and laser disc, DVD also represents the first moving-image format launched simultaneously for both PCs and TVs. As the *New York Times* in January 1997 quoted one industry analyst, 'This is the first consumer product that will come into the home through two doors, both the living room and the den', and DVD's duel status overcomes some of the social and spatial weaknesses of the computer as a moving-image display device.[56]

On the other hand, the marketing challenges of launching a genuinely new electronic device in the home can be seen in a 1998 US television commercial for Panasonic's large-screen SVGA-TV, a combination TV set and computer monitor (Fig. 4). In contrast to the terse and evocative Toshiba DVD commercial, the cluttered Panasonic ad relies on non-stop narration, a generic action-adventure musical score, a number of superimposed titles, and even several arbitrary strings of binary 0s and 1s superimposed over the narrative action (Fig. 5). The commercial depicts a pre-adolescent male, first seen seated close to the TV screen with keyboard on lap, apparently calling up Internet Web pages containing images

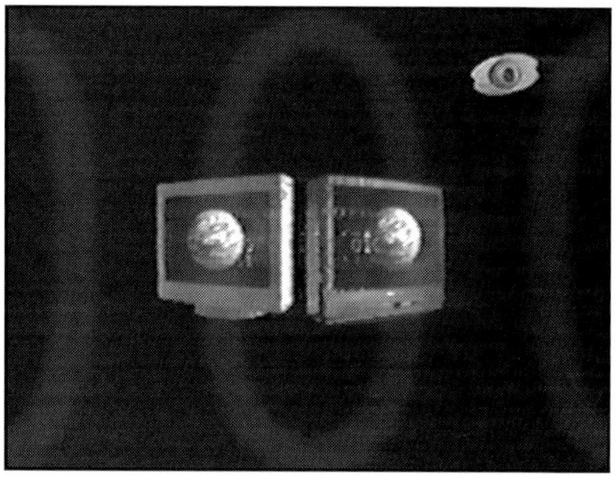

Fig. 7. SVGA-TV ad, US television, 1998.

of spaceships; after the boy's chair magically slides several feet away from the monitor (Fig. 6), he is shown watching sci-fi programming on the large-screen TV monitor. While the commercial offers a comforting social scenario suggesting the domestication of the pre-adolescent male's potentially unruly relationship to electronic technology, it nevertheless suggests some of the difficulties in representing hybrid digital media in the home. The commercial's opening narration, 'It's one world. Why view it through two windows?', is spoken over an image of two animated globes which merge in the centre of the screen, below a Web browser-like animated graphic in the top corner of the screen (Fig. 7). However, the implied equivalence between Net browsing and TV viewing begs the question of the adequacy of the single windows metaphor for both devices and activities; we are unused to thinking of the activity of TV viewing in such an instrumental fashion. Moreover, the unlived-in look of the commercial's domestic *mise-en-scène*, with its minimalist decor and wall of built-in electronic devices, suggests that domestic spaces need to be purpose-built around these new digital media devices, rather than providing a scenario of their seamless integration into existing domestic routines and spaces. Most strikingly, the commercial's magical solution to the central problem of the disparate screen distances appropriate for computer and television set – the boy's chair which slides autonomously across the room as he shifts from Net browsing to TV viewing – suggests the difficulties of representing the shifting modes of attention and physical space embedded in the hybrid computer-TV set device. The solitary figure in the abstracted domestic space also suggests an evasion of the problem of the integration of interactive digital technologies into the social activities of the household; would the entire family gather around the living room television to watch the son's Web surfing on their TV set? Contemporary domestic battles over channel surfing and control of the cable remote control would likely pale in comparison to the anticipated social friction generated by collective domestic living-room Internet browsing. Viewers of the Panasonic SVGA-TV commercial are likely to remain sceptical or merely puzzled about the possible domestic scenarios for interactive digital television on offer here.

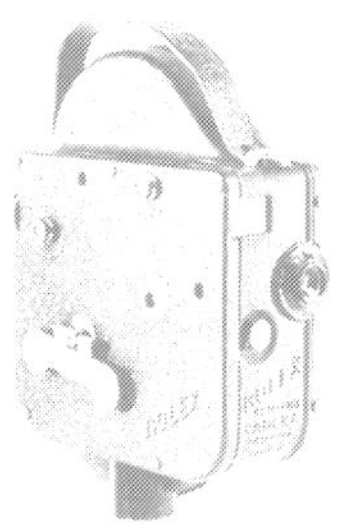

These brief, self-serving, and ephemeral television commercials can nevertheless be valuable for an historian examining the take-up of new domestic media, since they frequently enact dense and affective scenarios of socially-embedded technologies; in indirect and unacknowledged ways these thirty-second dramas often point to the wider culture's fears and ambivalences about new technologies. Discourses around the launch of digital television have already supplanted the traditional figurations of the television audience with both dystopian images of pathetic loft dwellers and utopian scenarios of new armchair Columbuses. Clearly, the cultural valuation of a range of domestic electronic devices, from computer and television set to satellite dish, has been put up for grabs by the introduction of digital television. There are significant competitive advantages and policy outcomes at stake in the discursive battles among competing figurations of the television audience in the popular imagination and commercial marketplace, and an analysis of these battles suggests the wider social process of technological innovation. The impact of digital television on the competitive conditions, regulatory regimes, and popular assumptions regarding domestic television is still quite uncertain; at the same time, the present discursive celebration of the birth of yet another 'revolutionary' communication technology suggests the cultural and ideological fault-lines which surround the prosaic and everyday activity of television viewing.

Acknowledgements: The research for this essay was supported in part by a grant from the City University of New York PSC-CUNY Research Award Program.

Notes

1. Raymond Williams, *The Year 2000* (New York: Pantheon Books, 1983), 129, 146.
2. On declining ITV audience shares and advertising revenues, see Market Tracking International, 'The Interactive Media Handbook 1998', *M2 Presswire*, 24 February 1998; on UK DBS prospects, see Maggie Brown, 'Sky's Nightmare Vision of the Future', *The Evening Standard*, 4 February 1998, 53; Jane Robins, 'Coming Soon to a Screen Near You – Weather-porn', *The Independent*, 18 August 1998, 14.
3. On mood of the US cable industry, see Peter H. Lewis, 'Internet Ties Help to Revive the Cable Industry as Satellite Sales Lag Behind', *New York Times*, 15 December 1997, D6; on growth prospects in the computer industry, see Louise Kehoe, 'Battle for the Eyeballs: The US is Setting the Standard and Pace of the Digital Television Revolution', *Financial Times*, 28 November 1996, 3, and Andrew S. Grove, *Only the Paranoid Survive: How to Exploit the Crisis Points that Challenge Every Company and Career* (New York: Doubleday, 1996), especially Chapter 9, 'The Internet: Signal or Noise? Threat or Promise?'. For contextual material on HDTV, see Joel Brinkley, *Defining Vision: The Battle for the Future of Television* (New York: Harcourt Brace, 1997); Michel Dupagne and Peter B. Seel, *High-Definition Television: A Global Perspective* (Ames: Iowa State University Press, 1998); Richard Collins, 'Back to the Future: Digital Television and Convergence in the United Kingdom', *Telecommunications Policy* 22, 4–5 (1998): 383–408; Colin Cowie, 'Competion Problems in the Transition to Digital Television in the UK Marketplace', *Media, Culture and Society* 19 (1997): 679–685.
4. 'TCI's Malone Discusses HDTV and Must-Carry at NCTA Convention', *Multichannel News*, 11 May 1998, np; on the rivalry between the consumer electronics and computer software industries, see John Markoff, 'Fight of the (Next) Century: Converging Technologies Put Sony and Microsoft on a Collision Course', *New York Times*, 7 March 1999, Section 3, 1.
5. George Gilder, 'IBM-TV?', *Forbes*, 20 February 1989, 72.
6. George Gilder, *Life After Television: The Coming Transformation of Media and American Life* (Knoxville: Whittle, 1990), 1.
7. George Gilder, 'Is America on the Way Down?', *Commentary*, May 1992, 22.
8. Joel Brinkley, 'Warning to Broadcasters That Renege on Running HDTV', *New York Times*, 15 September 1997, D1; also see Kyle Pope and Mark Robichaux, 'Promise of Digital Television is Fading as Broadcasters Complain About Costs', *Wall Street Journal*, 12 September 1997, 1.
9. Joel Brinkley, 'A Gulf Develops Among Broadcasters on Programming Pledge', *New York Times*, 18 August 1997, D1.
10. 'TCI's Malone Discusses HDTV', np.
11. Alicia Mundy, 'Digital's Ball of Confusion', *Mediaweek*, 13 July 1998, 6.
12. Kyle Pope, 'Executives Play Down Prior Indications They Might Scrap High-Definition TV', *Wall Street Journal*, 18 September 1997, B7.
13. On the broadcasters' renewed assurances on HDTV, see Joel Brinkley, 'Under Pressure, 2 Broadcasters Decide They Will Now Run HDTV', *New York Times*, 18 September 1997, D1; 'Executives Play Down Prior Indications'.
14. Joel Brinkley, 'TV Cable Box Software May Blur Digital Signals', *New York Times*, 23 February 1998, D6.
15. Ibid.
16. 'Promise of Digital Television is Fading', 1.
17. Denise Caruso, 'A Slow-motion Train Wreck in Setting Technology Standards for a Forthcoming Era of Television', *New York Times*, 28 September 1998, C3.
18. Joel Brinkley, 'FCC Responds to Its Digital-TV Critics', *New York Times*, 16 September 1998, C2.
19. Rachel Unsworth, 'Dawn of Television's Digital Age', *Mail on Sunday*, September 13, 1998, np.
20. For early returns on the competition between BSkyB and OnDigital, see Jemimah Bailey, 'Family

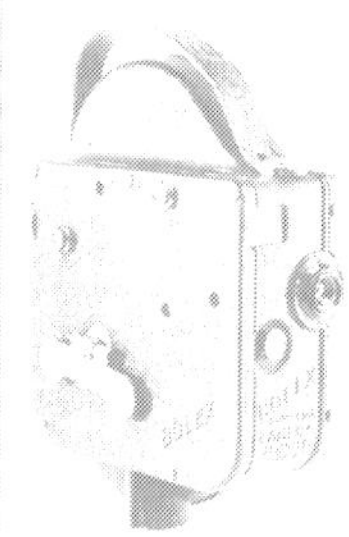

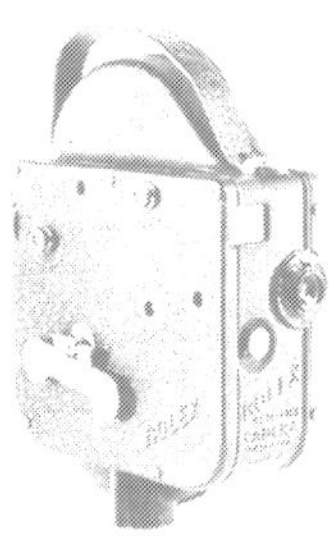

Finance: Sky High on Digital', *Sunday Telegraph*, 14 February 1999, 9; Chris Barrie, 'ONdigital Admits TV Box Glitch as Sky Profits Fall', *The Guardian*, 11 February 1999, 20.

21. Sally Kinnes, 'Box of Tricks; Digital TV Will be with us from Next Month', *The Guardian*, 21 May 1998.
22. 'Isn't it choice?', *The Guardian*, 7 October 1998, 21; on the debates within the cable industry on the value and limits of 're-purposed' programming, see Linda Moss, 'What's in a Name?', *Multichannel News*, 8 February 1999, 11A; Linda Moss, 'An Original Debate', *Multichannel News*, 8 February 1999, 14A; Philip Laxar, 'Eventually, Digital Will Need Originals', *Multichannel News*, 8 February 1999, 59A.
23. Robins, 'Coming Soon to a Screen Near You', 14.
24. Frank Stanton, Talk to Sigma Delta Chi Convention, Columbus OH, 13 November 1954, 14. In Frank Stanton: Speeches/Statements, 1954 file, Collection of CBS Reference Library, New York. Emphasis in original.
25. Bill Gates, *The Road Ahead* (New York: Penguin, 1995), 66.
26. Nicholas Negroponte, *Being Digital* (New York: Alfred A. Knopf, 1995), 168–169.
27. Jane Robins, 'Media: Liz and Mark's Satellite Love-in', *Independent*, 29 September 1998, 14.
28. Ibid.; Chris Barrie, 'Pity the New Digital Underclass', *The Guardian*, 6 November 1998, 21; Hamish McRae, 'Unleashing the Digital Divide', *Independent*, 17 November 1998, 5; also see Stuart Jeffries, 'Still Fancy a Night in Front of the Box?', *The Guardian*, 31 March 1999, 2. At least some press opinion in the US about the effects of audience fragmentation on national identity was quite different from that in the UK; in January 1999 a *New York Times* editorial complained: 'to the lament that we are losing a sense of national community as television grapples with its recombinant future, there is only one thing to say: Get a life.' 'Whither Television?', *New York Times*, 4 January 1999, A18.
29. John Huey, 'America's Hottest Export: Pop Culture', *Fortune*, 31 December 1990, 50.
30. Bill Gates, *The Road Ahead*, 263.
31. George Gilder, 'The National Prospect', *Commentary*, November 1995, 62.
32. George Gilder, 'Breaking the Box', *National Review*, 15 August 1994, 37–38.
33. George Gilder, 'The Death of Telephony', *The Economist*, 11 September 1993, 76.
34. George Gilder, 'The Shape of Things to Come', *National Review*, 8 July 1991, 27.
35. *The Road Ahead*, 165–166.
36. Microsoft Digital Television Web Site, updated 13 November 1998.
37. Ibid.
38. 'TCI's Malone Discusses HDTV', np.
39. On the radio split, see William Boddy, 'The Rhetoric and the Economic Roots of the American Broadcasting Industry', in *Cinetracts* 6 (Spring 1979): 37–54; on the split in early television, see William Boddy, *Fifties Television: The Industry and Its Critics* (Champaign-Urbana: University of Illinois Press, 1990), 21; the CBS publication is CBS, *Television Audience Research* (New York: CBS, 1945), 6.
40. Peter Thal Larsen, 'Can Digital TV Convert the Dish-hating Viewer?', *The Independent*, 29 July 1998, np.; Grabiner told another British journalist in 1998: 'There is a market for people who want 200 television channels. They live in dark attics and don't come out very often. OnDigital's market is middle England, people who want more choice in their viewing.'
41. John Schwartz, 'The Next Revolution,' *Newsweek*, 6 April 1992, 45.
42. 'Are New Realities More or Less Real?', *U.S. News and World Report*, 28 January 1991, 59.
43. Peggy Orenstein, 'Get a Cyberlife', *Mother Jones*, May–June 1991, 63. Typically contrary, cyberpunk novelist William Gibson told the Cyberthon crowd when asked if VR would be an electronic utopia: 'I think it could be lethal, like free-basing American TV. … I don't think that anyone that read my book seems to have understood it. It was supposed to be ironic. The book was really a metaphor about how I felt about the media. I didn't expect anyone to actually go out and build one of those things.', 'Get a Cyberlife', 64.
44. 'Virtual Reality at the Dentist: Better than Novocain', *Computer Review*, advertising supplement to the *New York Times*, 5 June 1994, np.
45. Jaron Lanier, 'Mass Transit!', *Wired*, May 1995, 156.

46. George Gilder, *Microcosm: The Quantum Revolution in Economics and Technology* (New York: Simon and Schuster, 1989), 314.
47. Wendy J. Mattson and Matthew Rothenberg, 'Apple's Jobs Candid about Technology's Role in Education', *MacWeek Online*, 10 December 1998, np.
48. Steven Johnson, *Interface Culture: How New Technology Transforms the Way We Create and Communicate* (New York: HarperCollins, 1997), 107.
49. Ibid, 67, 118.
50. For the ideas of two Stanford professors who consulted with Microsoft on the design of Bob, see Byron Reeves and Clifford Nass, *The Media Equation: How People Treat Computers, Television, and New Media Like Real People and Places* (CSLI Publications/Cambridge University Press, Stanford CA, 1996).
51. Johnson, *Interface Culture*, 62–63.
52. Microsoft Digital Television Web Site, updated November 13, 1998; on the prospects for TV-PC convergence, see George Black, 'Vision of Converging Worlds: Digital TV and the PC', *Financial Times*, 3 March 1999, 18.
53. *The Road Ahead*, 71.
54. Peter Brown, 'Chipmakers Reject the PC/DTV', *Electronic News*, 16 November 1998, 27.
55. Linda Dyett, 'Broadcast News: Wide-screen Digital TV in the Living Room', *House Beautiful*, 1 March 1999, 60.
56. Lee Isgur, Jefferies & Company consumer electronics analyst, quoted in John Markoff, 'Companies Roll the Dice on Digital Videodisks', *New York Times*, 9 January 1997, D2; on merchandising digital TV, see Evan Ramstad, 'Digital-TV Ads Aim to Shuffle Leadership', *Wall Street Journal*, 3 November 1998, B16; on the success of the DVD, see 'Slipped Disc', *The Economist*, 3 January 1998, 60, and Stephanie Miles, 'Sub-$1,000 PCs Stifle DVD Growth', *CNET News.com*, March 26, 1999.

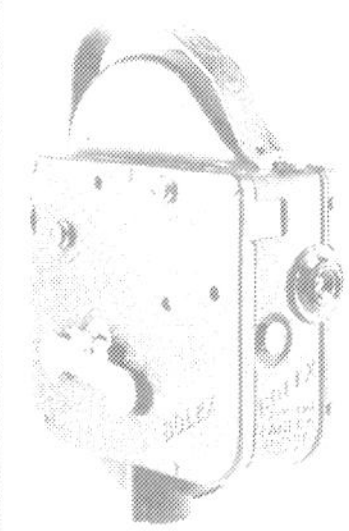

Stereotyping a Competitor: Images of Television in Spanish Cinema in the 1960s

Valeria Camporesi

Art History Department, Universidad Autónoma of Madrid, Spain

Introduction

An image may comprise a number of rather different things. Of the various mental, visual and physical objects with which it may be identified, I am concerned in this essay with a rather practical definition. My principal concern is with the cinematographic representation of television, both in image and sound.

If the question of the interrelationship of the cinematographic image to its signified cannot be avoided, the prime purpose of this study is not to establish whether there is a 'discrepancy between facts and representations'.[1] My aim will be to analyse the ways in which an historical process, in this case, the establishment and diffusion of television, is liable to produce a variety of images amongst which the cinematographic happens to be comparatively easy to reconstruct. The outcome of this limited study is to determine a diachronic series of images which in some way map a case in the history of culture. In so far as the content of that development is concerned, what is at stake is the way in which films addressed the new world of moving images which television brought into existence during the period in which the new technology had its initial impact. My basic question relates to the ways in which film represented television as a social phenomenon and provided the medium with a narrative content that contributed to socially received ideas regarding the image of television in the popular imagination.

Consistent with the microhistorical ambition of this project,[2] any timeframe in the second half of the twentieth century of any country could have been selected to study this process in so far as it covers the period in which television was established. The example of Spain in the 1960s may prove at least as useful as any other within its specificity. As in the case of other developments that crossed the Pyrenees, the establishment of television in Spain was deeply conditioned by its relative lateness, and by its rapid growth and institutionalisation within the strict limits imposed by the Franco dictatorship. The result was, in short, a contradictory, fragmented, but extremely rich and complex period of involvement in international processes, which may now be viewed as part of a broader movement of media globalisation.[3] In the decade discussed in this study, the steady development and social penetration of television took place in Spain at a time when an international crisis in film also coincided with the more general dissemination of electronic media.

When discussing the socio-cultural implications of what was basically the economic and social situation of Spain's relative backwardness in the 1960s, one should not underestimate the circumstances which highlight the singularity of the Spanish experience between 1939 and 1975. During these years, the

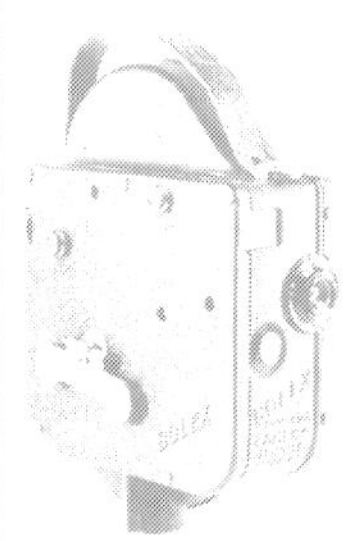

country, its media and public opinion, and the processes by which its cultural identity was created and diffused had to comply with the implicit and explicit rules of a dictatorship.

In spite of the specifity of the Spanish case, comparative history is now old enough to legitimise a hypothesis which would let us assume that, no matter what differences we might detect in a particular national case, the very fact of acknowledging plurality makes the case, potentially, a comparable one, and the conclusions derived from its analysis significant. The following description of the images of television as a complex social and cultural object in Spanish cinema in the 1960s helps define some aspects of the relationship between two powerful image-builders at a moment when films, in Spain, seemed to function as 'hypermedia'.[4]

Cinematic images of television in the 1960s

The group of films which I shall analyse is relatively small,[5] and the productions differ in their respective aesthetic, social or commercial qualities. Nevertheless, the following discussion of the way in which television was represented by film in the 1960s may prove more coherent than it might at first sight appear, especially when considered as a type of hypermedia.

The first film in which reference to television was made was produced in 1961, that is to say scarcely five years after Televisión Española was launched in October 1956. The film in question was *El pobre García* (Poor García, 1961), the directorial debut of Tony Leblanc, a popular comic actor.[6] *El pobre García* tells the melodramatic story of a widower who, with a crippled son and a powerful rival in love and work, eventually marries the boss's daughter, and has his son cured thanks to the intervention of a very good surgeon and a miracle. While the conservative and catholic values which the film exhibits were not an exception in Spanish films during the early 1960s, the film's lack of narrative coherence and basic cinematographic qualities rank it amongst the worst films of the year.[7] No clear data concerning the income the film generated have been traced. What is known, however, is that the film ran only two weeks in the three theatres in Madrid in which it was released.[8] One is left with the feeling that the film disappeared rather swiftly and did not have any significant success even in the second-run circuits which sometimes constituted an important alternative market for second-rate productions.

The film received coverage, however, in the press. One review which appeared in a popular film magazine drew attention to the sequence in the film where television is mentioned which, according to the author of the article, the well-known film critic Luis Gómez Mesa, was summed up as a 'caricature, a precise criticism of television. Not of a specific programme, but of a misconception which is too often repeated and which should make way for ones that are more true.'[9] The authoritarian tone of the remark accurately reflected rather widespread opinion concerning the intrinsic frivolousness of the new medium. To what extent did it also reflect the film's portrait of television? I will examine the excerpt in question within the general framework of the film.

The reference to television is made in sequence 37, some twenty minutes before the end of the 86-minute film. Things are not going very well for 'pobre García', but he does not seem to be particularly concerned. He is temporarily out of work, his son is mocked because he cannot run and play with his friends, yet nothing in the *mise-en-scène* reflects these developments. García goes back home where he finds his son in bed. The sequence begins in García's home (67' into the film). García enters his son's room, starts talking with him, then notices that he is sad, and decides to improvise with a wooden frame that is close to hand. 'As you have been good, I shall switch on the television for you. I believe there is a wonderful programme on today.' Then García puts the frame around his face and starts to mimic an Argentinian chat-show host. An advertisement for a watch interrupts his conversation. Then it is time for a commentary on a book, another commercial (for soap), and the weather forecast. An item on cognac ends the television 'broadcast'. The whole parodic performance lasts slightly less than four minutes. No commentary, either from the father or the boy, is added: television, as improbably as it entered the scene, disappears.

Two different responses can be inferred from this scene if one analyses it within the context of the film, or privileges its more unconscious, repressed implications. At one level, in relation to the boy, 'television' as an absent object is utilised to emphasise the pathetic structure of the scene: the family is too poor to own a real television set with which to entertain the boy who cannot go out and play with his friends. In this context, the function of the father is to allude to the social implications of that absence since the very fact that the family does not own a television serves to project the family into a broader social category representing a vaguely defined lack of financial resources and the world of unfulfilled material desires. Through the television sequence, Leblanc establishes a network of complicities with his public and the critics, a privileged communication which he then exploits by resorting to his familiar comic tics. On a second level, the fragment can be read as an unconscious description of a widely held social perception of television. In *El pobre García*, television is described as an immaterial technology which, in the minds of the urban lower-middle class, is identified as an entertainment vehicle and a source of possible employment for professional actors. Television, therefore, is acknowledged to exist, and is perceived to be an entertainment form that is quite different from cinema since one can comfortably watch it lying in bed. Nevertheless, television does not yet possess a clear identity.

The second reference to television on which I shall comment upon appeared in a film produced one year after *El pobre García*: *Atraco a las tres* (Robbery at Three O'Clock, 1962), the sixteenth film directed by J.M. Forqué, and the second of his films to be released in 1962.[10] In style and tone, the film is clearly influenced by the *commedia all'italiana*, and relates the story of a group of bank employees who decide to rob their employer. In contrast with what happened in *El pobre García*, critics did not pay any attention to the television episode which received no mention in the press. Besides, *Atraco a las tres* was commercially quite successful and received a positive official classification.[11]

The reference to television is made in sequence 11 (18' into the film) when we are shown a group of people watching a detective telefilm (we only see the back of the television set). We learn that the audience comprises people who live in the same apartment building, and that the owner of the television set, a young woman whom we know as one of the bank employees, charges them a small admission to watch television; latecomers receive a special discount. As the scene develops, we learn that the young woman has not yet paid for the television. 'It is a luxury product', the young boy remarks as he comes to collect the money the young lady owes.

Here again, we infer an image of a television audience whose main feature is its relatively low financial status. The sequence marks the third occasion when principal characters cannot fulfil their desires. As in Leblanc's film, television is perceived to be a rather mysterious object which, in spite of its presence as an object, primarily signifies a desire which the urban lower-middle class may express, but cannot afford. In short, television, as Leblanc and Forqué depict it, is a desirable object, entertaining and cosy, but still expensive. Moreover, the very fact of not owning a television identifies one with the world of solidarity typical of poorer classes. This process of social redefinition is reinforced by a later reference to the popularity of television (at 21'), when in sequence 13, a dog (that does not appear on the screen) is called Rin Tin Tin although we are not shown whether the dog's owners possess a television in their home.

It was not until 1963 that film audiences were given more precise details of television in *Se necesita chico* (Help Wanted, 1963), an Italian-Spanish co-production which received a fairly good official and religious classification.[12] The film marks the directorial debut of Antonio Mercero who later became known as one of the most creative directors of Televisión Española. Its style and plot belong to the more commercial and light version of the New Spanish Cinema of the 1960s. The film, which portrays a much more precise image

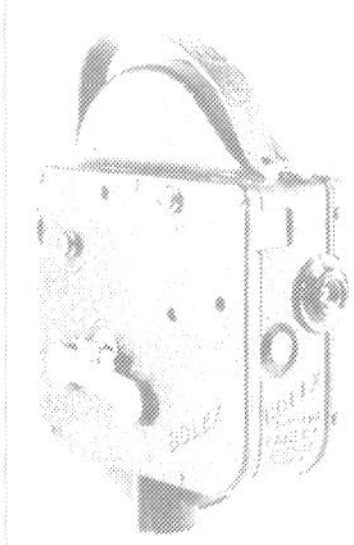

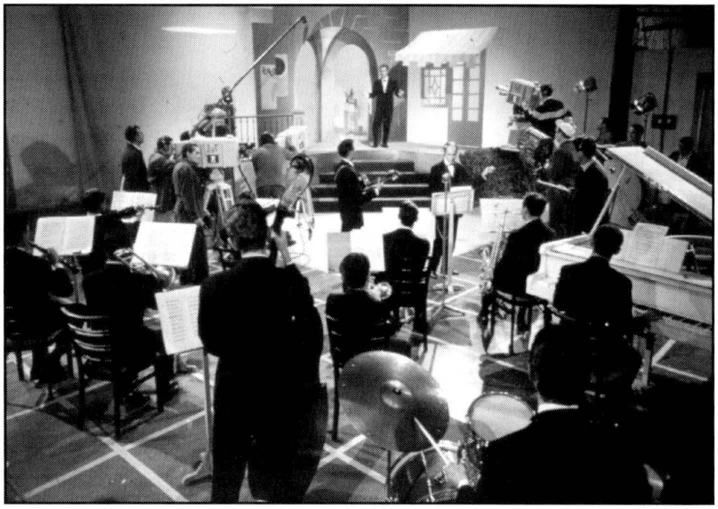

Fig. 1. La nueva Cenicienta, *George Sherman, 1964. [Courtesy of Photographic Archives, Filmoteca Española.]*

of television, was released in Madrid in October 1963. Two months later, one of the most popular and long-lasting film magazines in Spain, *Fotogramas*, announced, in capital letters, 'STARTING FROM TODAY, TELEVISION IN *FOTOGRAMAS*', and explained, 'A new artistic and informative medium, television holds today an important place in our lives … Our main purpose is to inform, comment, and steer opinion in so far as our capabilities and prestige will allow … We shall always work for the betterment of Televisión Española'.[13] From now on, the television set plays a series of changing but definite roles over the years. No longer merely an aspiration, television is a more or less desirable, but material, cinematographic reality.

In *Se necesita chico* television is devoted a whole sequence (sequence 14, from 50' to 58'). As a result of a misunderstanding, the main character, a young boy who is working as an assistant in a florist's, has been given a wreath to deliver to a television star during a live broadcast. In the preceding sequence, filmed on location, we see the boy approach the building where the first television studios were established in Madrid, in the Paseo de la Habana. He asks for the television performer. The film cuts to the interior of the building: in a television studio, the singer to whom the boy is supposed to present the wreath is being filmed in a rather sexy outfit (this reference to her feminine nature will later be used to justify her hysterical reactions). The boy places the wreath in a part of the set which is not in shot. Then the star and the compère move, and the wreath appears on screen in the background. The compère notices the wreath and tries to hide it while the singer is still unaware. Realising his mistake, the boy tries to crawl and draw the flowers off camera, but comes into shot. The film cuts to a scene of parallel action as people watch television in a coffee-shop. They see the lugubrious flowers and the boy crawling, and start to laugh. Back in the studio, the singer eventually sees the boy and the wreath, and starts to cry. The film cuts to the audience laughing in the coffee-shop. The broadcast is interrupted and a commercial appears on the television screen.

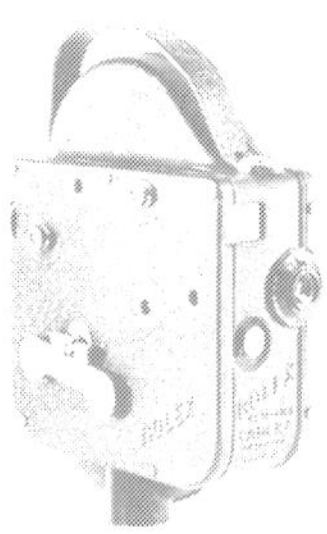

While still not associated with domestic viewing, television is, in this instance, primarily identified with the unintended entertaining consequences that arise from live broadcasts. The sharply ironic tone with which Mercero describes the adult world which surrounds the boy's careless behaviour reaches its climax in this sequence. It is quite instructive that critics in the catholic-inspired specialist journal, *Film Ideal* denounced the 'extreme vulgarity of the television episode'.[14] Now, the journal seems to imply, we know what television is: the aura which surrounded it when we could not get hold of it has vanished. In this context, cinema represents television as an apparatus of difference, the main function of which is not its potential to tell audio-visual stories but its ability to show situations, persons and sounds that occur simultaneously in distant places.

The identification of the attraction of television with live performance, and its comic, but less destructive, potential is also the way in which the new electronic medium is thematised in *La nueva Cenicienta* (New Cinderella, 1964, Fig. 1). The first film to deal with the issue of television that was shot in colour, *La nueva Cenicienta* was directed by the American director, George Sherman, and is clearly modelled on the conventions of the Hollywood musical with the added attractions of Spanish dancer, Antonio, and popular film singer, Marisol. Once again, data concerning attendance and box office receipts are not available, but a tentative deduction from its reasonably long exhibition in the cinemas in which it was released in Madrid (the Palacio de la Música and Cine Barceló for, respectively, thirty-two days and fourteen days),[15] suggests moderate receipts at the box office. Given the popularity of the two stars, we may propose that the film's income was probably lower than expected.

In *La nueva Cenicienta*, television constitutes a recurring narrative background throughout the film. The medium is first mentioned at ten minutes into the film: an American singer is asked to fly to Spain to appear in a television show with the Spanish dancer, Antonio. When he shows up, the two artists engage in a search for a girl who can accompany Antonio's performance. After a series of rather fantastic events, they eventually run into Marisol, a cheerful poor girl who lives in a big house along with her father and a bunch of would-be performers, one of whom performs with a trained chimpanzee. Thanks to Marisol's talent and beauty, they are invited to appear on television. At fifty-five minutes into the film, Bob and Antonio rehearse in a television studio while waiting for Marisol and friends who, because of a misunderstanding, eventually show up late while the programme is being broadcast live on Eurovision (61' to 63'). As Antonio explains, fifty million people are supposed to be watching the broadcast. As anticipated, Marisol and her friends' appearance with the chimpanzee degenerates into chaos, the transmission is interrupted, and eventually things are sorted out so that, as Antonio tells Marisol, 'nobody realised what was going on but for the people who were here in the studio'.

As in the previous film, television is here identified with its non-cinematographic form of production, the live broadcast, widened to include a more international audience (in this case, the broadcast is transmitted to the USA, France and Italy). Nevertheless, no use is made of parallel montage to show audience reactions. Production personnel and technical competence are good enough to overcome the potential disruption of a live programme, and prevent the audience from viewing any undesired images. Television is portrayed as a mature medium, clearly identified with modernity and Americanisation, and, being different from cinema, can be regarded as being as responsible a medium.

Involving a more precise portrayal of television, *Historias de la televisión* (Television Stories, 1964) not only includes a reference to the medium in its title, but marks a watershed in its cinematographic identity.[16] The film was a sequel to an extremely popular film about radio made in 1955 by the same director, the widely-known José Luis Sáenz de Heredia.[17] The claim is explicitly made that television is the central concern of the film. In a quasi-documentary, newsreel-like prologue an omniscient voice-over explains that the two stories which will be told come from 'that real and ... attractive world' of television which is described as 'a strange and fascinating phenomenon' that has changed the normal patterns of behaviour. Earlier in the film, two television audiences were selected as typical of the new medium: a family absorbed in viewing a local tv thriller during dinner, and a group of friends who gather to watch a football match on television. In this film, reference is explicitly made to the reality of television as a sort of counter-argument to *El pobre García*. Indeed, the 1961 film with which I opened this essay is referred to in *Historias de la televisión*, although the reference primarily arises because of a lack of new ideas than as an allusion to some critical interpretation. Actor Tony Leblanc plays the main character in the first episode. And, being as impoverished as 'el pobre García', he also tries to get money through performing a peculiar type of spontaneous advertising, hoping that those he might benefit (the directors of the firm whose product is being promoted) will pay him for his services.

In the first episode, alternation between the television studio and the audience is repeatedly deployed, and the practice of live performance is shown to be common on television. But in this case, emphasis is put on the individual story and individual viewers, not on audience reactions at large. Indeed, the threat which live performance at one time represented has clearly been exorcised. The three viewers who witness what is occurring on screen are Leblanc's father, a person, we later learn, to whom he owes money, and the person who is going to give him the money

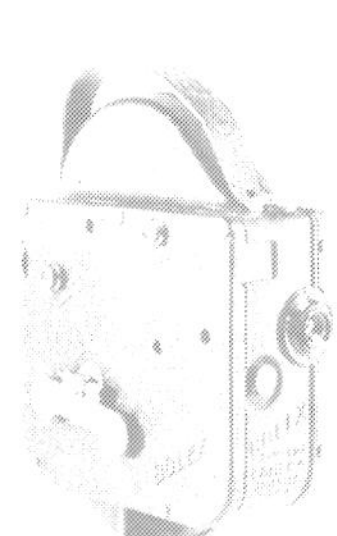

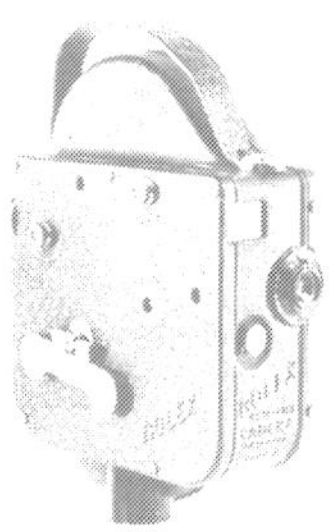

he needs. They all seem to accept that television may be used as a source of income, notwithstanding the explicit allusions to the illegality of such practice. Indeed, Leblanc's father straightforwardly interprets it as a job, albeit a rather insecure one, and this is the reason he does not completely approve of his son's performances. What the film portrays is the candour with which a society thinks that television may be the instrument with which to realise one's dreams since quiz-shows, like musical contests in the second episode, are perceived by the principal and secondary characters in the film as a volatile, but highly desirable, source of wealth and fame. Besides, participation in a television programme is a relatively easy matter: in episode 1, during the two days in which the action takes place, Leblanc appears in two shows; and towards the end of the episode (sequence 15, 43' 4"), he also appears in the news.

The second episode relates the story of a girl from a little tourist village on the Southern coast and her struggle to become a famous singer. This part of the film, a quasi-realistic fable, is introduced by a voice-over which, accompanied by images of the village and people, explains that 'the only contact people there had with the modern world was through television and a hotel' (44' 5"). A television, however, only appears towards the end of the episode, when a live programme, a Eurovision broadcast, is shown, and which the main characters purposefully interrupt in their fight against the injustice that is being perpetrated. The sequence ends very violently as the protagonists are expelled from the stage. As in the earlier episode, no visual or verbal reference is made to the audience at large; the programme just says that the broadcast must be interrupted. No visual disruption occurs on the screen nor is any allusion made to the social repercussions of the programme. The programme's producers are implicitly described as rather cool and bereft of any sense of solidarity. While television, in *Historias de la televisión*, may be represented as a fascinating and modern medium, it does not represent an adequate instrument with which to fulfil people's desires. Television can be very entertaining but it certainly is not good for the anonymous individual striving to make ends meet. No Cinderella will find her prince through television any more. While being a stable and important element for a growing number of families, television, in *Historias de la televisión*, has lost its fantastic aura.

It is quite significant that from this point on, Spanish films of the 1960s promote a plurality of discourses and, in some cases, oppositional images of television. This is all the more significant given that the development coincides with the progressive fragmentation of film audiences and the crisis in film attendance in Spain.[18] Indeed, the two following films which represented television in the second half of the 1960s bear an intense relationship with the changes that were taking place in the media world. The first was a film whose original title in Catalan, *No compteu amb els dits* (Do Not Count with Your Fingers, 1967), could not be used at the time of its release. Directed by Pere Portbella, the film was originally conceived in four episodes, and may be included amongst the first essays of the so-called Escuela de Barcelona. A complex phenomenon, explicitly influenced by the School of New York, the Escuela de Barcelona represents 'the only collective movement which pursued a search for new formal solutions and a clear break in content with the history of Spanish cinema'.[19] In order to do so, the movement promoted and accelerated the interaction of different sections of the cultural and intellectual circles active in Barcelona at the time. Reviewed as a 'parody of tv spots' and a social satire,[20] *No compteu* is both a symptom and an outcome of the enormous interest in fashion and advertising in Barcelona in the late 1960s.[21] Comprising a series of moving images which might be excerpts from some film or television commercial, the Portbella film plays with visual incongruities and with the lack of correspondence between image and sound. Conceived as a product for minorities with artistic ambition, the film was eventually released as the accompanying feature for films associated with the 'New Cinema' of the 1960s. No overall image of television is here to be decoded. What is interesting about the very idea of *No compteu* is that it is a by-product of the establishment of television as a widespread, definite, almost palpable image. In this case, commercial television is so much a part of popular culture that artists use its material to shape a contemporary iconography that is decontextualised and inserted into the logic of their work.

The second movie which clearly addresses a particular sector of the cinema-going public is *Un, dos, tres, al escondite inglés* (One, Two, Three, Stop, 1969), produced by director and script-writer José Luis Borau. This film is the directorial debut of Iván Zulueta who had previously worked in television. Indeed, the original idea for *Un, dos, tres, al escondite inglés* comes directly from Zulueta's experience as a director of an against-the-stream television programme called 'Último grito', a 'sort of weekly "magazine" which included interviews, news reports, short feature films, and material reminiscent of daring "videoclips".'[22] Completely imbued with pop aesthetics and popular culture, and deeply influenced, it has been proposed, by Richard Lester's films with The Beatles, *Un, dos, tres* tells the story of a group of young people desperately striving to boycott the organisation of a television song festival identified with the 'modern style of song'[23] which they hate and which television incessantly promotes. Frequently interrupted by a series of musical numbers representing the most popular pop songs of the moment, the film promotes an alternative popular and modern culture which is juxtaposed to the fake modernity of Franco's television. Given its low-budget production values, the film did reasonably well at the box office.[24]

From the very first scene, when the announcement of a Festival is made by an anchorman and singer, up to the third-from-last sequence (81'), television is shown as intrinsic to the lives of the main characters. In order to be able to affirm themselves and what they stand for, they have, at first, to fight against television. The obsessive presence of the electronic medium in their lives is very vividly portrayed through the repetitive insertion of fragments of live programmes as they are viewed by our heroes. Once again, the *mise-en-scène* of live programmes alternates with images of television viewers. As in *Historias de la televisión* though, no allusion is made to the existence of an impersonal collective audience whose reactions might be representative of public opinion. The television audience as a collective entity is only defined by manipulation, and can be identified with the musty, authoritarian traits of the regime's culture. From this standpoint, the last eighteen minutes represent a liberation: in sequence 44 (81'), while one girl from the group, who has managed to get through and appear in the television festival, is destroying everything on stage, her friends watch her performance live from a television set whose screen has been painted so as to make it rather difficult to see anything. In the last two sequences, action takes place in a park, where no connection with television is either possible or desired, and a large group of young people dance to celebrate the destruction of the 'Festival de Mundo Canal'.

As if in dialogue with Zulueta's movie, although from the very opposite ideological and cultural spectrum, Pedro Lazaga's *Vente a Alemania, Pepe* (Come to Germany, Pepe, 1971) does reassert, as a positive feature, the image of television as the vehicle of a conservative, pseudo-modern popular culture. A commercial and conservative product superficially dealing with the problem of migration, *Vente a Alemania, Pepe* portrays television on two occasions. In sequence 2 (0' 30" to 1' 1") we are shown a group of men watching female dancers on television in a small village bar who make resigned comments on the poor quality of the broadcast. Quite clearly, modernity (and one of its products, liberated female behaviour) is light-years away. The rest of the film is devoted to demonstrating how much better it is that the status quo will remain unchanged for as long as possible. When the restless main character eventually migrates to Germany to look for money and girls, he is faced with a tough life where neither easy money nor easy sex are to be found. The climax of the film is reached when he watches television on Christmas night in the living room of his hotel in Munich. In sequence 48 (78' to 90' 2") he is seated in front of a television when a Eurovision broadcast is announced that comprises excerpts from different countries. When 'Popular songs and

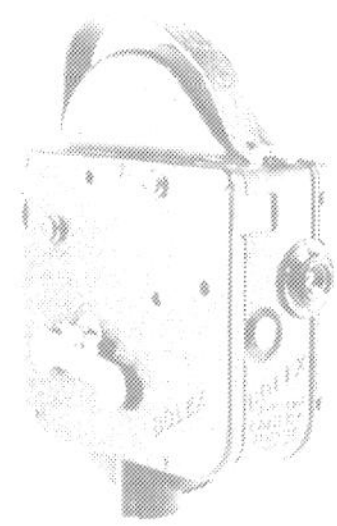

dances from Spain' appears on the screen, he resolves to return to Spain.

In the last film of this study, television has a mission: to confirm 'national' identity and cultural value, and to preserve rather than modernise, as the young characters of *Un, dos, tres* had warned. The period of its establishment as a desirable object was over by the mid-1960s. Towards the end of the decade, depending on your viewpoint, television's unyielding presence in film was either threatening or reassuringly conservative as Spanish film seemed to foreshadow a period when modernity and Francoism would become more and more incompatible.

Provisional conclusions

In the opening paragraphs, a claim was made that no comparison between social statistics and representation was to be drawn in this study. Nevertheless, there are a few elements of the 'material' world which impinge on the films I have discussed and which should be taken into account.

Regarding the market saturation of the different media, the 1960s does represent a watershed in Spain. During this decade, cinema attendance began to shrink[25] while statistics regarding television ownership marked a spectacular growth as receivers rose from 3,000 in 1956 to 2,700,000 in 1966, and 4,570,000 in 1971.[26] In percentage terms, the number of homes that owned a television grew from 1 per cent in 1960 to 32 per cent in 1966, 56 per cent in 1971, and 85 per cent by 1973.[27] Television advertising also increased dramatically as the percentage of commercials rose so that by 1974, they comprised 7.82 per cent of total output.[28] Also, the number of broadcast hours grew throughout the period covered in this study, while national coverage for television was complete by the end of 1962.

Throughout the 1960s, films seem to have provided an effective means for registering the social response to television. Moreover, the way in which television is depicted can be interpreted as reflecting attitudes typical of both minority and national audiences. At the beginning of the period, opinion-makers thought that mainstream Spanish society would surrender almost uncritically to the mesmerising power of television. As writer Vázquez Montalban states, from the late 1950s 'the subject of television had set in motion Spanish expectations, and marked a collective sentimentality'.[29] The sequence inserted in *El pobre García* by Tony Leblanc, then, effectively describes both its author's and its audience's views regarding television. It might not be too far from the truth to state that films, as hypermedia, reinforced, at some level, the huge 'process of mythification' that Vázquez Montalban describes. While *Se necesita chico* clearly reflects the era of 'flies and shawls' which characterised the production of early live broadcasts by Televisión Española,[30] the film also demonstrates the increasing popularity of television, partly due to its unpredictable institutional working practices. By the mid-1960s, the representation of television that one could expect to see in cinemas had changed. Television had lost the spontaneous charm of its early years, and had assumed a stable identity which reinforced its intrinsic frivolousness and social volatility. Trapped between the authoritarian and, with few exceptions, censorial tone of its educational and 'cultural' programmes,[31] television was soon bound to split public opinion. If in the mid-1960s films seemed to patronise television, by the end of the decade, they staged different attitudes to the modernisation of culture which the electronic medium is thought to have brought about.

As a provisional conclusion, judging from the Spanish experience of the 1960s, films did offer a model for understanding television which was consistent with the experience society had of the medium, and with the image of society which the medium constructed. Moreover, as historical and aesthetic documents, the films portrayed the interrelationship in narrative and visual terms of production processes which reveal significant, if rather unknown, fragments of a wider cultural history.

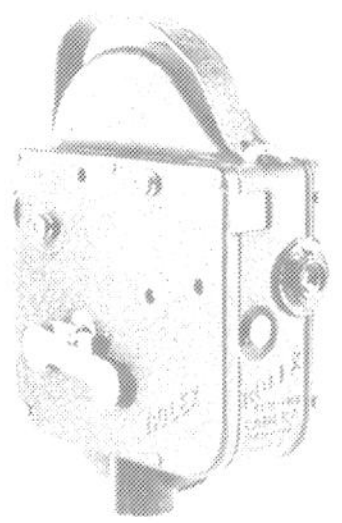

Acknowledgement: This essay is the first, provisional outcome of my participation in a group research project on 'Film consumption in Spain: an historical reconstruction' financed by the Spanish Ministry of Education (PB 96–0075). Alberto Elena, the project's director, first suggested that I could contribute with a qualitative study on the image of television in Spanish cinema. I wish to thank John Fullerton for the patience and care with which he revised my original text.

Notes

1. Pierre Sorlin, 'Stop the Rural Exodus: Images of the Country in French Films of the 1950s', *Historical Journal of Film, Radio and Television* 18 (June 1998): 195.
2. Whose suggestive hypotheses were recently referred to in a survey of film studies in Spain, see Jenaro Talens and Santos Zunzunegui, 'Introducción: por una verdadera historia del cine', in *Historia general del cine* (Madrid: Cátedra, 1998), 9–38.
3. As I have proposed in Valeria Camporesi, 'La télévision du franquisme', in *La grande aventure du petit écran* (Paris: BDIC/Musée d'Histoire Contemporaine), 100–101.
4. The term is used to describe the Italian situation in the 1950s in Maria Grazia Franchi and Roberta Lietti, 'Il ruolo del cinema nella formazione dell'identità italiana lungo gli anni '50', *Memoria e ricerca* 5 (December 1997): 11. Giovanna Grignaffini referred to the same phenomenon in an earlier analysis of contemporary cinema within which she sees 'the establishment of a "container form" (*forma-contenitore*) as a specific procedure through which, today, the relationships among different communicative systems and languages are defining themselves' and which justifies, in her view, the principle of reciprocal quotation, see Giovanna Grignaffini, 'Rosso sangue', *Cinema & Cinema* 16 (January–August 1989): 19.
5. The films were identified from an examination of contemporary and later written sources including the specialist press and popular magazines as well as critical and historical works. Of course, it should be considered a provisional set of data. The films include (in chronological order): *El pobre García* (1961), *Atraco a las tres* (1962), *Se necesita chico* (1963), *La nueva Cenicienta* (1964), *Historias de la televisión* (1965), *No compteu amb els dits* (1967), *Un, dos, tres al escondite inglés* (1969), and *Vente a Alemania, Pepe* (1971).
6. According to the standard reference work on Spanish film actors, Leblanc was then 'a real box office hit', see Carlos Aguilar, Jaume Genovés, *Las estrellas de nuestro cine. 500 biofilmografías de intérpretes españoles* (Madrid: Alianza, 1996), 321. In the early 1960s he built up his own production company whose first film was *El pobre García*. The company financed two further films.
7. In spite of its lack of minimally objective criteria, it is interesting to note that even the official 'Junta de Clasificación y Censura' (which decided the amount of public support awarded each film) classified the film in the lowest group, the so-called 2° B (bound to receive 0.5 millions of pesetas, while the best classified, 1ª A, received 4), see Antonio Vallés Copeiro del Villar, *Historia de la política de fomento del cine español* (Valencia: Filmoteca de la Generalitat, 1992), 87. Indeed, of the fifty-nine films first exhibited in 1961 (including co-productions), *El pobre García* was one of the five which received the worst verdict. As was usual in these cases, the film was first exhibited in second-run cinemas. In spite of official disdain, the producers may have nurtured the hope that it might still appeal to an acceptable segment of the cinema-going public during the early 1960s since the film was released contemporaneously in three theatres in Madrid.
8. See *Anuario español de cinematografía, 1960–61* (Madrid: Sindicato Nacional del Espectáculo, 1962), 713.
9. L. Gómez Mesa, '*El pobre García* de Tony Leblanc', *Fotogramas* (3 November 1961): 21. The reference to television is also highlighted in *Historia del cine español* in which the Francoist historian Fernando Méndez Leite observed, 'It is remarkable for the accuracy of the television session set up for the son', see Fernando Méndez Leite, *Historia del cine español* (Madrid: Rialp, 1965), II, 469.
10. As critics noted, *Atraco a las tres* was produced in response to the success of *I soliti ignoti* (Usual Unknown Persons, 1958), an Italian film released in Spain shortly before *Atraco a las tres*.
11. It was classified as 'Para todos los públicos'. Released in December 1962, the film was seen by 114,581 spectators and its revenue amounted to 2,401,626 ptas (data from Biblioteca de Filmoteca Española).
12. No official statistics on revenue or audience figures for *Se necesita chico* are available. What is

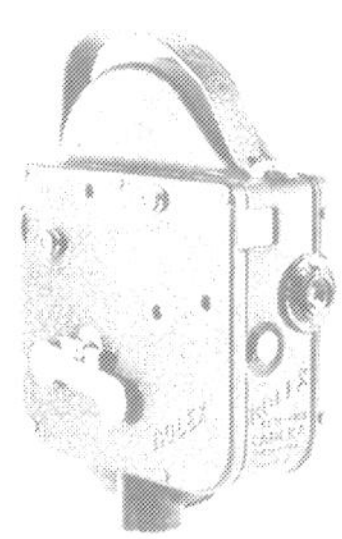

known is that the film was exhibited for only seven days in the Madrid theatre in which it was released, a fact that suggests rather poor public response.

13. *Fotogramas* 790 (6 December 1963): 201. This patronising attitude was also adopted in the article published in *Film Ideal* by the then Director General de Cinematografía, the main inspirer of the birth of the 'New Spanish Cinema' in the 1960s, José María Escudero, see, for instance, J.M. García Escudero, 'El porvenir de la tv', *Film Ideal* 72 (15 May 1961):10–11.
14. L.C. in *Film Ideal* 137 (1 February 1964): 104.
15. See *Anuario español de cinematografía, 1963–68* (Madrid: Sindicato Nacional del Espectáculo, 1969), 749. The film played for a total of 25 days in two theatres in Barcelona; in Valencia, 14; in Seville, 35; in Bilbao, 12, and in La Coruña, 8, see *Cineinforme*, IV (March 1965): 39.
16. The film also had the best box office receipts of those considered in this essay: 1,171,272 spectators which resulted in an official revenue of 26,333,172 ptas. The film was classified for 'Todos los públicos' by the official board but received a 2 by the religious authorities (i.e. unsuitable for children under 14).
17. The cousin of the Falange's founder, José Antonio Primo de Rivera, Sáenz de Heredia directed his first feature films during the Second Republic. He came later to be known for directing *Raza* (Race, 1942) for which the Caudillo, Francisco Franco wrote an original screenplay.
18. While conclusive data on this phenomenon have not been systematically analysed, this process is currently being researched by the 'Film consumption in Spain' research project.
19. Esteve Riambau and Casimiro Torreiro, *Temps era temps. El cinema de l'Escola de Barcelona i el seu entorn* (Barcelona, Departament de Cultura-Generalitat de Cataluña, 1993), 75.
20. See Jaime Picas in *Fotogramas* (1 December 1967).
21. For a discussion of this phenomenon together with information on the film, see Esteve Riambau and Casimiro Torreiro, *Temps era temps. El cinema de l'Escola de Barcelona i el seu entorn* (Barcelona, Departament de Cultura-Generalitat de Cataluña, 1993), 75, 149, 195. The authors have recently published a revised edition of this book in Spanish: Esteve Riambau and Casimiro Torreiro, *La Escuela de Barcelona. El cine de la 'gauche divine'* (Barcelona: Anagrama, 1999).
22. Carlos Heredero, 'Iván Zulueta: la mirada poliédrica', in *Tiempos del cine español* (San Sebastián: Ayuntamiento, 1990), 25.
23. Carlos Heredero, 'Iván Zulueta: la mirada poliédrica', in *Tiempos del cine español* (San Sebastián: Ayuntamiento, 1990), 75.
24. According to information given by Filmoteca Española, *Un, dos, tres al escondite inglés* earned 5,538,148 ptas (representing a sale of 262,038 tickets).
25. Information on this process can be found in Román Gubern, José Enrique Monterde, Esteve Riambau, Julio Pérez Perucha, Casimiro Torreiro, *Historia del cine español* (Madrid: Cátedra, 1995), 255–263; and in Valeria Camporesi, *Para grandes y chicos. Un cine para los españoles 1940–1990* (Madrid: Turfán, 1993), 67–94.
26. Dirección General de Radiodifusión y Televisión, *Organización. Competencia. Objetivos* (Madrid: Ministerio de Cultura, 1978), 23.
27. See Luis E. Alonso and Fernando Conde, *Historia del consumo en España* (Madrid: Editorial Debate, 1994), 200, and José Castillo Castillo, *Sociedad de consumo a la española* (Madrid: Eudema, 1987), 103, 111.
28. See José María Baget Herms, *Historia de la televisión en España* (Barcelona: Feedback, 1993), 272–286.
29. Manuel Vázquez Montalbán, *El libro gris de Televisión española* (Madrid: Ediciones 99, 1973), 42.
30. The expression refers to the primitive technology and approach to television and the oppressive censorship of the female body during the first period of Televisión Española (up to 1964).
31. For an extensive treatment of the history of Spanish television, see Jaime Barroso and Rafael Tranche, 'Televisión en España', *Archivos de la Filmoteca Valenciana 23/24* (June–October 1996): special issue, and Manolo Palacio, *Una historia de la televisión en España. Arqueología y modernidad* (Madrid: Consorcio Capital Europea de la Cultura, 1992).

Video Pleasure and Narrative Cinema: Luc Besson's *The Fifth Element* and Video Game Logic

Warren Buckland
Screen Studies, Liverpool John Moores University, UK

> [V]ideo pleasure requires the player's total immersion in the electronic text, trust in the existence of a code imposed by invisible experts, and the self-affirming and empowering experience of its incremental mastery. Predicated on the ability to rapidly discriminate between circulating signs (some hostile, some neutral, and some friendly), and to appropriately respond to them, such a pleasurable mastery involves a skilful and rapid navigation in a chaotic electronic text, a navigation propelled by strategic violent moves administered digitally.[1]

Introduction

Jerome Bruner reminds us that narrative is not a contingent, optional dimension of society, but is an essential, ecologically necessary structure with which individuals make sense of social complexity.[2] We do not need to subscribe to ontological structuralism – which argues that narrative is a timeless structure that transcends society – to accept this level-headed reminder. In the process of structuring social experience, narrative necessarily reinvents itself in each epoch, offering an historically specific experience. Fredric Jameson's analysis of the novels of Balzac and Conrad is exemplary in this respect. Jameson identifies in the narrative of *Lord Jim* the emergence of two incompatible discourses – high literary modernism and popular literature – and historises the emergence of this type of narrative by relating it to the rise of modern imperialist capitalism, dominated by reification and fragmentation.[3]

In this essay I analyse the emergence of digital narratives in contemporary society. Digital media offer potentially new technological practices for manifesting pre-existing narrative discourse, although the specific potentials and constraints of digital media – most notably, interactivity – transform narrative structure. Except for a few comments in the conclusion, I will not be interpreting digital narrative within Jameson's ultimate semantic horizon – that is, relating it to the latest stage in the development of capitalism. Instead, my aim is more modest – to develop a poetics of digital narrative, and to analyse Luc Besson's film *The Fifth Element* (1997) as an example of the way film narrative is being digitalised. In other words, I want to analyse the way film narrative has adapted to and been transformed by other types of cultural experience, such as video games (a term I am using as a synecdoche for electronic texts and digital media in general). In the second half of this essay, I focus on the way *The Fifth Element* combines traditional narrative structures (the psychologically-motivated cause-effect narrative logic) with a logic based on video game

rules (I outline these rules in the first part of this essay). In combining traditional narrative with video game rules, films such as *The Fifth Element* construct a mode of address that attempts to engage the consumer habits and forms of pleasure specific to the experience of today's young film audience.

Digital narrative

New digital technologies are currently interacting with cinema's nineteenth century analogue technologies of optics, mechanics and photochemistry. The most prominent mark of this interaction is the use of digital special effects in blockbuster movies. Although for the most part images are still produced photographically (exceptions being films such as *Toy Story*, *Antz*, and *A Bug's Life*), this photographic image is often supplemented by digital special effects. Much has been written on the digitalised filmic image, focusing on the way the new technology has transformed cinema's analogue technology by modifying its form (for example, the filmic image is no longer restricted to depicting profilmic events).[4] Digital technology alters our epistemic relation to the image for this very reason, since what we see could have been generated on the computer, yet (in the case of invisible special effects at least) is conferred a photographic credibility.

In this essay I want to go beyond an analysis of the digital image and begin discussing the way narrative has also been digitalised in a number of contemporary films. This digitisation can be detected in the way film narrative has adopted the rules and strategies of video games. (The influence works both ways, of course, since many video games adopt the visual and aural aesthetics of science fiction movies.)

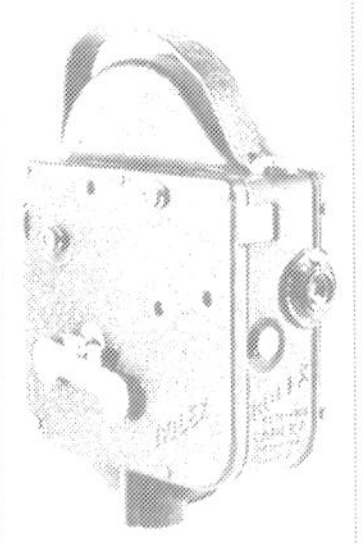

Mastering the rules of the video game

Video games possess 'an excess of visual and aural stimuli' but also 'the promise of reliable rules'.[5] These rules, which are reliable in that they are systematic and unambiguous (for they are unencumbered by morality or compassion), constitute the video game's environment, or location, which is not restrained by the laws of the physical world. The game user can experience video pleasure primarily by attempting to master these rules, that is, decipher the game's logic. Moreover, the desire to attain mastery makes video games addictive, which at times can lead to the user's total absorption into the game's rules and environment. This absorption in turn may alter the user's state of consciousness and lead to a momentary loss of self.[6]

Here I aim to outline a number of the general structures used to construct these rules. At present my list is inductive, although it covers some of the most common rules to be found in video games, including:

serialised repetition of actions (to accumulate points and master the rules);

multiple levels of adventure;

space-time warps;

magical transformations and disguises;

immediate rewards and punishment (which act as feedback loops);

pace;

interactivity.

Video games are organised around the serialised repetition of actions for several reasons, including the accumulation of points and the opportunity to master the rules of the game. 'Once players have learned a set of skills', writes Nicholas Luppa, 'they want to apply them in new situations. Elevating a game to other levels has long been a secret of good game design.'[7] In other words, users are keen to refine their newly acquired competence in game play by applying and testing it in similar but more difficult environments.

Space-time warps represent an alternative way to reach another level. They enable the player to be immediately transported to an alternative space (and time), leading to a sense of multiple fragmented spaces, with immediate transportation between them. Once the user has reached another level, his character on screen may be magically transformed into another character, or may don a disguise.

The user's accumulation of points acts as a feedback loop in the process of mastering the rules, since it represents a reward for good game play, and confers upon the user the sense that his competence is improving and the game is progressing. In similar fashion, the loss of points or a life acts as an immediate punishment for failing to master the rules. A repetition of this punishment leads to the user's premature death and an early end to the game. Serialised repetition therefore involves repeating the same stages of the game – usually at a faster pace, or moving up to another similar (but more difficult) level. According to Nicholas Luppa, pace is one of the most important features of video games: '[video games] require pacing and the beats that are being counted in that pacing are the beats between interactions.'[8] The important point here is that the player controls the beats via interaction, which confers upon the player the feeling of control – the manipulation of a character in a usually hostile digital environment. Moreover, the interactions need to be immersive – that is, must focus the user's concentration, and must be multiple and varied. Interaction is also dependent on the interface design, which must be detailed, but also easy to use.

Video pleasure, created by a user's addiction to and immersion in a game, is therefore not simply a matter of a heightened stimulation generated by high quality graphics, audio and animation, but is also – and, I would argue, primarily – a function of these rules, of the user's success at mastering these rules. Of course, in discussing the popularity and pleasure of video games, we should not leave out their content and themes, which can in fact be summarised in one word – violence: 'The central organising assumption of videology is unarguably that of violence. ... The only relevant question posed by videology is not whether a particular situation calls for negotiation or violence but how efficiently can violence be administered.'[9] Moreover, the violent enemy to be destroyed is invariably coded as 'other'. The content of video games therefore perpetuates Western ideological myths, and playing games – especially by oneself – is said to lead to anti-social behaviour. Yet beyond these social concerns there is what Fiske calls important semiotic pleasures to be gained from video games.[10] Hopefully, the above discussion goes some way to outlining how that pleasure is generated.

The Fifth Element

Just as the filmic image is a hybrid between the photographic and the digital, I argue here that the narrative in *The Fifth Element* is a hybrid between the classical (psychologically-motivated cause-effect) narrative logic and video game logic. In the following I shall focus on those moments in the film when video logic becomes apparent.

The film begins in a conventional manner with a prologue set in Egypt in 1914, which serves an expositional function (it introduces us to the five elements and the threat to earth that will arrive in 300 years). The film then jumps forward 300 years. It singles out the main protagonist, Korben Dallas (played by Bruce Willis), his helpers, including father Cornelius (Ian Holm) and his assistant David; the antagonists – 'pure evil' (called 'Mr Shadow'), represented as a ball of fire in outer space that rushes towards earth, aided by Zorg (Gary Oldman) and the mangalores. The prized object is established early on – the four stones representing the four elements (earth, wind, fire, water), and the fifth element, personified as 'the perfect woman', Leeloo (played by Milla Jovovich). The film also sets up a narrative conflict that needs to be resolved within a strict time limit: the evil ball of fire returns to earth every five thousand years to destroy everything. It can only be stopped by the four stones in combination with the fifth element. The protagonist is sent on a quest – to return the five elements to earth in order to save mankind – all within forty-eight hours. Meanwhile, the antagonists also attempt to obtain the stones. As expected, the protagonists win out. For this, Dallas receives the reward of consummating his love with the fifth element, the perfect woman. The moral of the film is transparent and banal: war and evil are bad, and can only be overcome through love.

What is significant about *The Fifth Element* is the way the banal story and its transparent moral are structured and conveyed to spectators. In his discussion of the film in his essay 'Infinite City', Nigel Floyd mentions the influence of graphic artists on the film's visual design.[11] But what is more significant than this is the influence of the rules of video games on the way the film is structured.

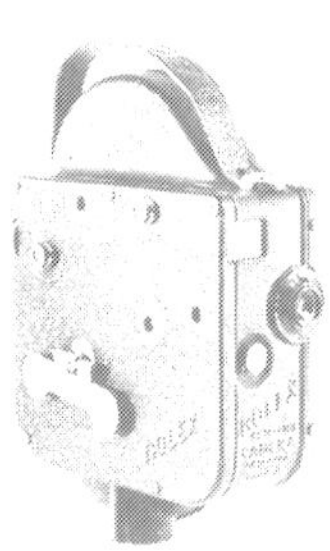

It is easy to look at the content of the film and begin identifying the rules of video game logic metaphorically: for example, we could argue that the cityscape metaphorically represents the film's different levels of play, or the scene where the mangalores shoot down the spaceship carrying the fifth element is a metaphorical representation of the game user's interactivity. However, I do not intend to read the film's content in such metaphorical terms. I am primarily interested in the way that content is literally organised according to video game logic.

The first half of the film sets up the characters, their relationship, and their aim – to obtain the four stones and save/destroy the earth. Korben is the primary protagonist, although he is aided by Cornelius, Leeloo and 'the Federation'. Zorg is the main antagonist who works directly for 'Mr Shadow', and is initially aided by the mangalores, who later set out on their own quest to obtain the stones. There are a few video game moments in the first half of the film: for example, like a video game character, Leeloo is magically reconstructed/transformed (she gains another life); the mangalores seem to have 'won the game' early on by stealing the case of stones (the four elements) for Zorg; however, when Zorg opens the case, it is empty, which then initiates a serialised repetition of the action to capture the stones; as immediate punishment, Zorg blows up the mangalore leader (or allows him to blow himself up). The mangalores and Zorg haven't mastered the rules of the game at this stage, and so they are punished (Zorg is empty handed, the mangalore leader blows himself up).

Once the characters and their aim have been established, more prominent elements of video game logic become apparent. Approximately 50 minutes into the film, we find out that the stones are with the Diva on the paradise planet. This planet represents the film's next level of adventure, and the immediate aim is for the protagonists and antagonists to get on the crowded flight that goes to the next level. A competition to win two tickets to the planet is rigged by the Federation so that Dallas wins. Upon hearing this news on the radio, Cornelius and Leeloo visit Dallas at his apartment. Leeloo removes Dallas' name from his door and puts it on the neighbour's door. Soon afterwards the police, acting under Zorg's instructions, attempt to arrest Dallas (Zorg has obviously found out as well that Dallas has two tickets to the planet). But because Leeloo put Dallas' name tag on the neighbour's door, the neighbour is arrested. The mangalores in turn kidnap 'Dallas' from police custody. Soon afterwards, Cornelius steals the tickets from the real Dallas, goes to the airport and gives the tickets to his assistant David, and to Leeloo. David pretends to be Dallas, but Dallas turns up at the airport to take his rightful place. A mangalore then turns up at the airport and pretends to be Dallas. Significantly, he has changed his appearance, and looks like Dallas' neighbour. The mangalore has disguised himself – but as the wrong person! He fails to get onto the plane and has to run from the police (who presumably think he is Dallas). Then Zorg's helper turns up at the ticket desk and also pretends to be Dallas. The flight assistant, who by this time is becoming a little weary of this serialised repetition of action and disguises, prevents Zorg's helper from getting onto the plane. Upon hearing this news Zorg immediately punishes him by killing him. Meanwhile, Cornelius becomes a stowaway on the plane. The plane takes off, transporting Dallas, Leeloo (both in disguise as tourists), and Cornelius at the speed of light to the next level of the game.

These scenes in the film are dominated by the following rules of video game logic: serialised repetition, disguises, the attempt to move up to the next level, a feedback loop (in which unsuccessful characters are immediately eliminated and successful ones rewarded), and by a space warp. However, once at this new level, we discover that the mangalores are there already (the leader, who blew himself up earlier, is now disguised as a waiter). Zorg also travels to the planet in his own ship. The same quest that took place on the previous level (locate the four stones) now takes place on the new level. How the mangalores get there, how the mangalore leader survived the explosion earlier in the film, why Zorg has Dallas arrested when Zorg can make his own way to the paradise planet, why Leeloo changes the name tag on Dallas' door, and why there was such a fuss at the airport is only of concern to those who read the film in classical narrative terms. Within video game logic, the film makes sense. The game play on the first level is going to be repeated on the next level with essentially the same characters; except that now it is going to be played at a faster pace.

Only one narrative event takes place on the paradise planet – the protagonists retrieve the stones and return to earth (the film's final level). But this event is subordinated to the action sequences and to the Diva's singing. Furthermore, those unsuccessful at retrieving the stones are killed before the next level is played out.

Zorg suddenly turns up and steals the box from the Diva's room as she is singing. He plants a bomb, departs the planet and opens the box to find it empty. This is a repetition of action performed earlier, when the mangalores delivered the (empty) box to Zorg. Zorg returns to the planet again, but this time is punished for his lack of success – although he switches off his own bomb, he is killed by a bomb left by the mangalores (which presumably kills all the mangalores as well).

The film ends in a conventional narrative manner – the formation of a heterosexual couple. But the ultimate objective of the narrative is not the transformation of characters, but survival – as is the case with video games. Character identification is achieved primarily by means of action – again, as in video games – rather than narrative development. Dallas does not so much develop as a character; he masters the rules of the game and proceeds to the next level, cashes in his points and wins the ultimate prize – the perfect woman.

Conclusion: spectacle and interactivity

> Story telling. It's not a game, it's an adventure. Every medium has its bias, the stories it cannot tell. ... We will want to tell the stories that demand interaction. But what does that mean? ... What is the story that can best be told with your meddling?[12]

One of the primary potentialities offered by digital media is that it introduces the dimension of interactivity to narrative discourse. Narrative progression and outcome can be altered by the reader's (or participant's) input. Yet what does interactivity add to narrative? And are the participant's choices merely fixed in advance by the computer software? One defining characteristic of narrative is that it has a predetermined structure based on specific outcomes. A participant's input can simply destroy that structure. The predetermined structure of narratives excludes the possibility of interactivity – that is, interactivity is incompatible with narrative structure. This is why *The Fifth Element* employs all the video game rules listed above except interactivity.

One way to understand the opposition between narrative and interactivity is to think of narrative as spectacle. This may at first sound confusing, because film theorists from the 1970s onwards have opposed spectacle to narrative (most famously articulated by Laura Mulvey in 'Visual Pleasure and Narrative Cinema' and by Stephen Heath in 'Narrative Space'). Yet I am not referring to spectacle in this narrow sense. Instead, I am using the term more globally, in Guy Debord's sense of the 'society of the spectacle', a society (modern capitalism) in which direct experiences are replaced with represented experiences. Film is therefore a spectacle in this sense because it is a representation based on spatio-temporal displacement, which creates an irreducible distance between film and spectator. And it is this irreducible distance that minimises interaction between film and spectator.

For a medium to become interactive requires it to go beyond spectacle. Daniel Dayan and Elihu Katz distinguish 'spectacle' from 'ceremony' and 'festival'.[13] In contrast to spectacle, ceremony and festival involve interaction, although to different degrees: festival involves a strong interaction between audience and the event they experience (as in carnivals), whereas ceremonies involve a measured interaction (reactions such as shouting and applause).[14] The recent emergence of interactivity as a cultural practice seems to suggest that we have gone beyond 'the society of the spectacle' – that is, beyond modern capitalism, towards what Jameson calls late or multinational capitalism. As an

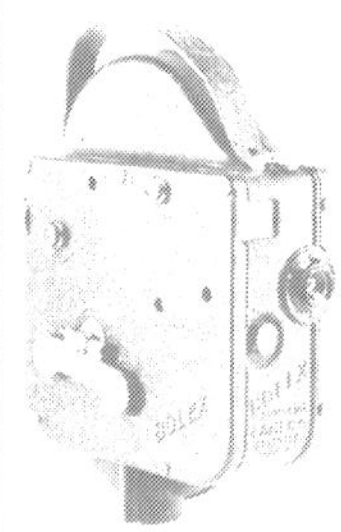

invention that emerged from modern capitalism, film is firmly rooted within the society of the spectacle. Interactive digital media, by contrast, need to be conceived along the lines of festivals and ceremonies.

Notes

1. Simon Gottschalk, 'Videology: Video-Games as Postmodern Sites/Sights of Ideological Reproduction', *Symbolic Interaction* 18, 1 (1995): 13.
2. See Jerome Bruner, *Acts of Meaning* (Cambridge, Mass.: Harvard University Press, 1992) and 'The Narrative Construction of Reality', *Critical Inquiry* 18, 1 (1991): 1–21.
3. Fredric Jameson, *The Political Unconscious: Narrative as a Socially Symbolic Act* (Ithaca, New York: Cornell University Press, 1981), Chapter 5.
4. I analyse the new digital imaging technology in 'Between Science Fact and Science Fiction: Spielberg's Digital Dinosaurs, Possible Worlds, and the New Aesthetic Realism', *Screen* 40, 2 (1999).
5. Simon Gottschalk, 'Videology', 13.
6. See John Fiske, *Reading the Popular* (London: Routledge, 1989), Chapter 2.
7. Nicholas V. Luppa, *Designing Interactive Digital Media* (Boston: Focal Press, 1998), 32.
8. Ibid., 34.
9. Gottschalk, 'Videology', 7.
10. Fiske, *Reading the Popular,* Chapter 2.
11. Nigel Floyd, 'Infinite City', *Sight and Sound* 7, 6 (June 1997): 6–8.
12. Douglas Crockford, 'Quest into the Unknown', http://www.communities.com/people/crock/quest1.txt
13. Daniel Dayan and Elihu Katz, 'Electronic Ceremonies: Television Performs a Royal Wedding', in Marshall Blonsky (ed.), *On Signs* (Baltimore: The Johns Hopkins University Press, 1985), 16–32.
14. A film can become a festival in Dayan and Katz's sense of the term (as in screenings of *The Rocky Horror Picture Show*), and on occasions films are applauded (in which they momentarily become ceremonies), but these are exceptions to the norm.

Space and Character Representation in Interactive Narratives

Björn Thuresson

Department of Cinema Studies, Stockholm University, Sweden

In a broad sense, I want to bridge the gap between traditional narratives and the new expectations of a high level of interactivity, and the call for hypermedia structured material. I'm trying to bring some of the experiences from cinema, radio and television to the production of material which is to be viewed on a computer screen. The leading questions are embarrassingly simple:

- How does an interactive narrative (I'm deliberately using the term 'narrative' instead of 'fiction') work and function?[1] What does it look like? How do you construct it?
- How does it affect the use of the material and the user?
- What examples exist to date?
- How do these examples relate to traditional narratives?

I started off by trying to pin down the essentials of what constitutes a narrative. I wanted to present a toolkit to which any producer, script writer, or graphic artist working in a digital environment could relate, and find possible applications in their own work. This is by no means revolutionary, but I hope that the combination of habits, conventions and traditions, drawn from a wide range of disciplines, can contribute in some way.

As a point of departure: I do not find it fruitful to try to determine whether this or that is a narrative. I believe that narrative is a continuum the boundaries of which are utopian. The five elements that I suggest constitute a narrative are apparent in narratives to different degrees. They are not necessarily essential. In different contexts, the narration focuses on different elements *in* the narrative. What then is narrative? A narrative takes place somewhere, denotes a time frame, includes something like characters that act in some way, and these actions (or events) are organised in some motivated order and, finally, the combination connotes and promotes personification with the events and characters by a receptee.[2] Expressed in a condensed way, the five elements that constitute narrative are: spatiality, temporality, causality, dramaturgy and personification. In this context, I will consider some of the elements of interactive narratives, and focus on different uses of specific environments and the ways in which we, as users/players, are represented in interactive narratives.

Space for events, characters and stories

A narrative takes place somewhere. There has to be an arena of some kind. This could be a thoroughly defined realistic environment or an abstract one. The events and characters help to define their surroundings. In the visual arts, the environment is often crucial, and can be used to make the presentation more efficient. What would the car chase in *Bullitt* (1968) be without the San Francisco setting?

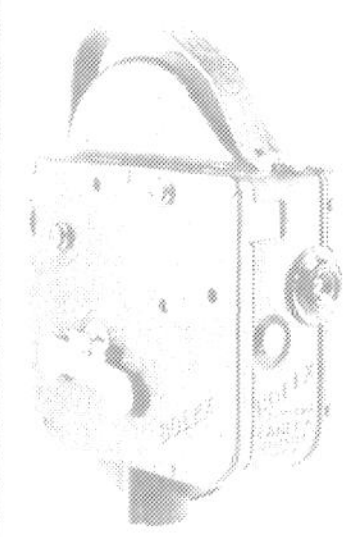

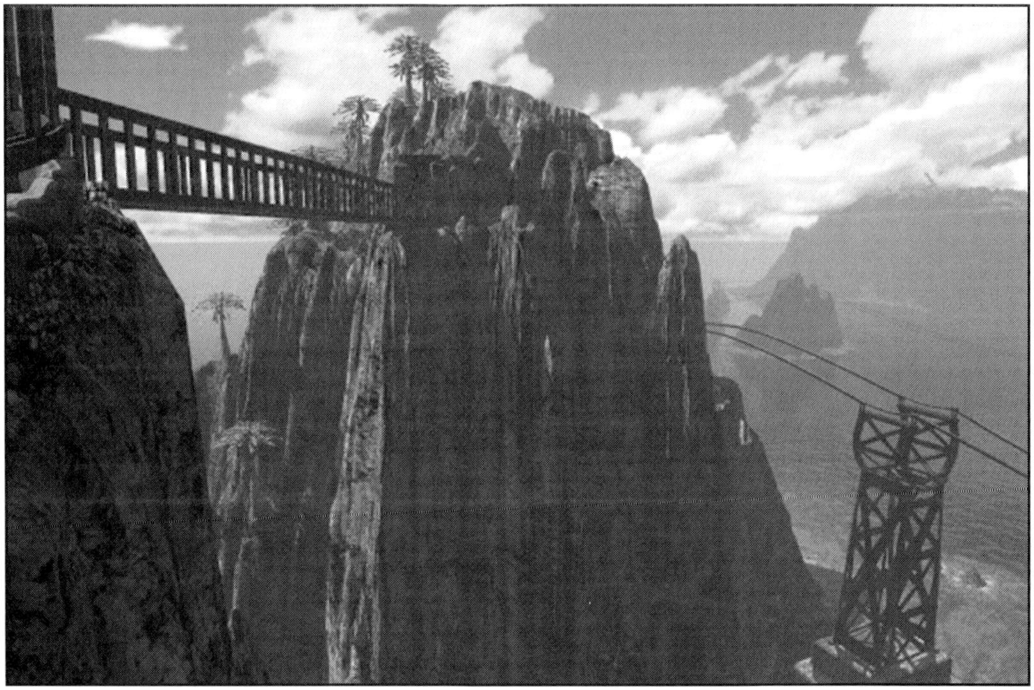

Fig. 1. Riven, *Rand and Robyn Miller, and Richard Vander Wende, 1997.*

Would any of the finales in the James Bond series be exciting if you didn't know exactly how far it is from point A to B? Would a chamber drama be comprehensible if you didn't know that, for instance, a daughter can hear her father confess an act of incest to her step mother?

In a digital environment spatiality is equally important (and to some extent even more important). In an interview with one of the two brothers who created *Myst*, Rand Miller observed that 'the interactive story design followed two paths: the linear and the non-linear':[3]

> The linear was the back story and the history, all those elements that followed a very strict time line. The non-linear was the design of the worlds and was more like architectural work. Like building a world without the time element at all – a snapshot of an age. Now the struggle was to try to merge the two by revealing some parts of the linear story during the exploration of the non-linear world, while maintaining the explorer's feeling that he/she can go anywhere and do anything they please.[4]

In *Myst*, the producers believed that the space had the possibility to 'blur' the story,[5] but also the built-in potential to scatter fragments of the story. Spatiality can be used for the *potentiality* of a story. In the case of *Myst* (and in a number of games inspired by *Myst* and its sequel, *Riven*, Figs. 1 and 2), the producers motivate the signs so that they are more than cultural inhabitants or markers of a specific time period[6] (or whatever semiotic connotations that may occur);[7] the signs are motivated by the story, they are part of its narrative content.

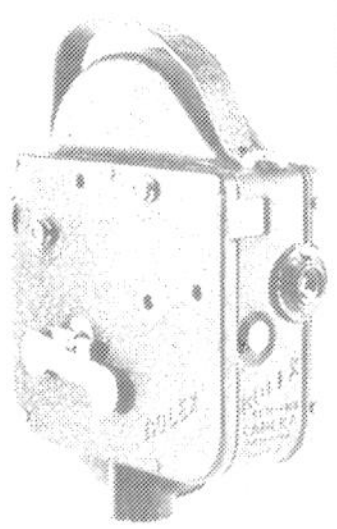

These games also situate the user/player in a special way. The window (or frame) to the world simulates your point of view. You are being handed things which leave the frame at the bottom frameline, and reappear outside the frame. You are being addressed in face-to-face conversation. The environment also reacts and remembers your actions. If you move things around and then leave, the place looks the same when you return as when you left. The navigation is based upon movements from still image to still image. The steps between them are rather small, but there is still no simulation of actual movement in space. The main reason for this is a technical one. To achieve this kind of image resolution,

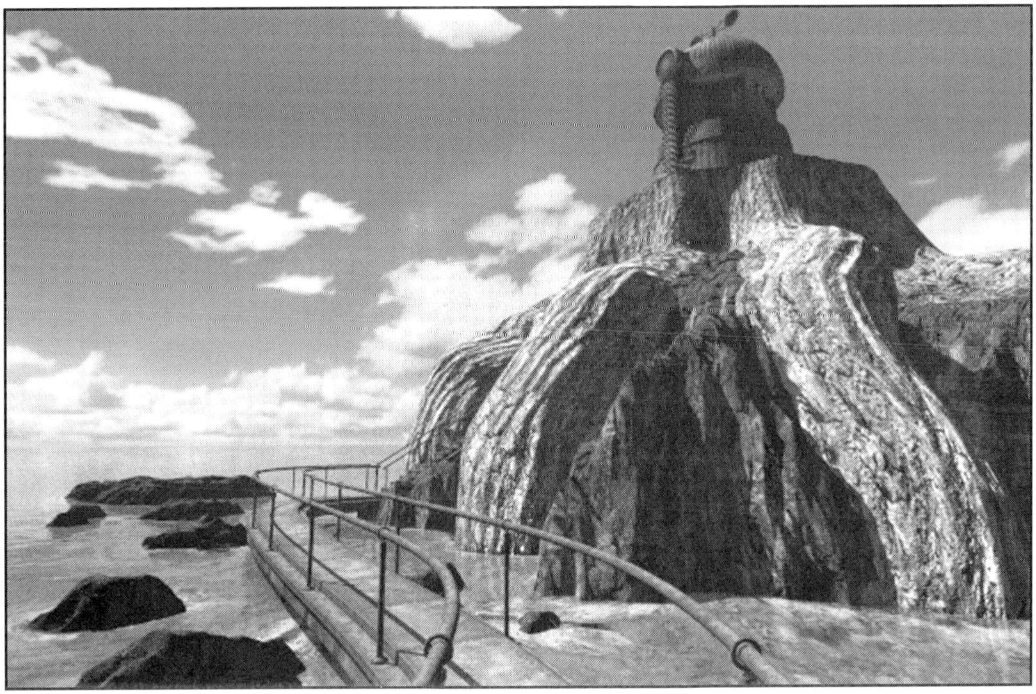

Fig. 2. Riven, *Rand and Robyn Miller, and Richard Vander Wende, 1997.*

no real-time renderer can maintain the frame-rate. But there is some 'within-the-frame animation' to simulate actions or events. That is, parts of the frame are animated to let a bell toll, doors open, or birds fly around.

In *Marathon*, *Quake* and in a whole range of similar games, there is full-screen, real-time rendering. The resolution is poorer, but the advantage (or what is sought for in the context) is fast-paced action. The producers try to situate the stories in different confined spaces (so they don't have to render faraway objects, and therefore increase speed). One runs around in attics, tunnels and strange fictional spaces in space. In the window-to-the-world you see 'your' hands together with the weapon you currently use. Since the rendering techniques are getting better all the time, new 'engines' for rendering are being introduced. One of the most successful games is *Tomb Raider I–III*. Here the heroine – Lara Croft – is seen from right behind all the time. You almost never see her face.[8] You control her, but you are *not* her.

In purer simulations (like flight simulators, or car games), you are usually able to shift the view point: inside, from the sides, from behind, or from the somewhat confusing, front view. In *Carmageddon* (as well as in *Wolfenstein* and several others), you also get a representation of how 'you' react to different events. The world in *Carmageddon* also remembers your action. If your car leaves skid marks somewhere, those marks are still there when you return.

There is definitely a difference in the ways in which characters are represented. In these games you get to 'loan' some features from a character, and then you act out the events. In games like *Riven* you invest your own knowledge and experience, and bring it to the protagonist, the 'you' in the game. It is like a constrained 'self', with a predefined action radius.

Aristotle stressed that actors (or characters) are one of the main causes of action, but also that action (to some extent) denotes characters:

> [A]n action implies personal agents, who necessarily possess certain distinctive qualities both of character and thought; for it is by these that we qualify actions themselves, and these – thought and character – are the two natural causes from which actions spring, and on actions again all success or failure depends.[9]

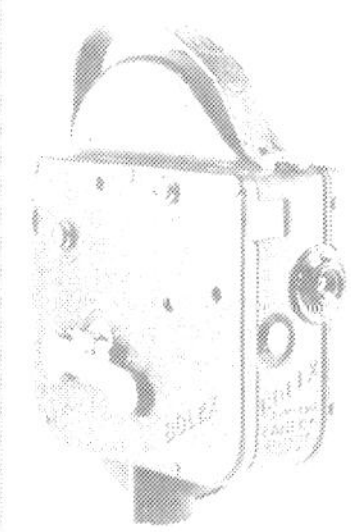

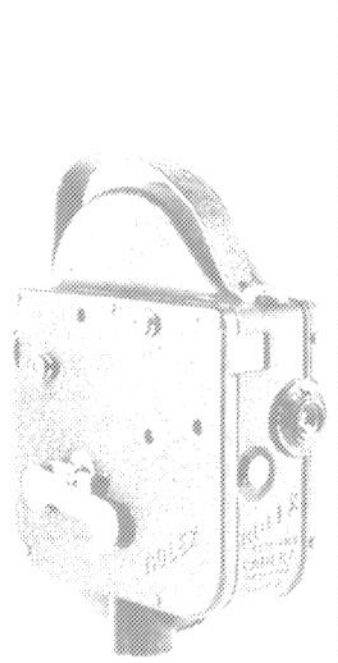

David Bordwell puts a similar notion forward when he observes that '[c]haracters are embodied; they can be assumed to perceive, think, and feel; they seem to display traits and to execute actions'.[10] But, of course, these characters do not need to be human characters. We can attribute these traits, feelings, and so forth, to basically anything. In an experiment conducted in the early 1940s,[11] a research team of psychologists made an animated movie where a couple of circles, squares and triangles of different sizes moved around in a bigger square. People watching the film were required to reconstruct what had happened. To different degrees they imposed a story structure on the presentation: 'The big triangle chases the small one; and then the small circle comes to the aid of the small triangle', for example. The other aspect of characters is the possibility (or potential) for identification or interest since, as Edward Branigan has observed, '[t]he spectator [or user/reader/perceiver] has an intrinsic interest in characters as agents since comprehending a narrative event requires at least recognising how agents interact with one another in a causal framework'.[12] Characters (and our involvement in them) are important to narratives; but how do we facilitate or integrate this element in a digital environment? Since we seem to want to impose meaning and causal relationships between events, and the settings and characters involved, the simplest anthropomorphic figure (in Barthes' terms) will be considered as a character if this figure is contextualised as a character. In John Lasseter's *Luxo Jr.* (1986) 'A larger lamp watches while a smaller, younger lamp plays exuberantly with a ball but doesn't pick up the knack of correct handling'.[13] *Luxo Jr.* is a computer animation, and the characters are completely anthropomorphic; but even at this level of recounting the plot, the age difference (together with playfulness, inability to learn) is believed to be essential to make the reconstruction comprehensible. It is then fairly easy to construct a character-like figure, but the task at hand is to contextualise the character and maintain the user's interest in the presentation.

Transfiguration of time

> We live in and with time, and it is within that temporal dimension that our ways of seeing are established.[14]

In narrative, temporality is constituted by the sequence of events, in two ways. If we are presented with a series of still photographs or sketches (as in a cartoon), a temporal relation is implied between the single images. In cinema, the temporal relationship between the single frames that constitute a shot is the actual technique that creates the effect of moving images. Temporal duality is, on the one hand, constituent of the apparatus and, on the other hand, a vehicle for narrative presentation.[15] But, as Vivian Sobchack notes, the temporal relation between successive images is less intense, 'not only because each individual image contains time only in a very indirect and coded way, but because the apparatus within which they appear is much less constraining'.[16] The sequence of events indicate that there is a temporal relationship between the events,[17] but (obviously) not a chronological succession. 'A story is *made out of* events to the extent that plot *makes* events *into* a story'.[18] The duality can also be expressed another way as 'story time' and 'actual time'.[19] The 'actual time' (the time it takes to present the story for an audience) in *North By Northwest* (Alfred Hitchcock, 1959) is 136 minutes. The 'story time' is something like two weeks, perhaps. When talking about the cinema it is fairly straightforward, but with many other forms of representation it is more difficult: how long does it take to read a book?[20] Editing (in the cinema) is 'first and foremost the ordering of units of time, units between which there are implicit temporal connections'.[21] When an intended user in a digital environment is using the material, the time of the telling can be vital. While browsing the World Wide Web, who wants to wait? Who wants to add user-time (that is, time of the telling) because of an inefficient hypermedia structure and navigation? On the other hand, if the time of the telling is efficient, the time of the story can be prolonged, and include a lot more information, 'events', if you will.

A glimpse around the virtual corner

These are some examples of what characters and environments look like now, and how our actions are defined. What can we expect in the near future? I think that full-screen rendering is sought for, and that the worlds will get more complex. I believe it is going to be important that the worlds are being flexible and have a 'memory' of action, and that predetermined temporal changes are implemented, like night or day, summer or winter.[22]

This will even extend to real-time manipulation of actions in complex environments.[23] When it comes to characters, I'm not sure. There are several options. On the one hand, rendering techniques have presented more complex movements and appearances of fictional characters. On the other hand, flesh-and-blood people are being mapped onto wire-frame models, even to the extent that an interactive digital video of real people may act out user-controlled events in virtual environments. The commonly used subjective view of the world promotes identification with a character, transforms the 'it' to a 'you'. But, is it the best way to distribute human traits? Should we rule out constructed characters as vehicles of content? I don't think so. I choose to believe that Janet Murray is correct when she notes that '[w]hen we enter a fictional world, we do not merely 'suspend' a critical faculty; we also exercise a creative faculty ... Because of our desire to experience immersion, we focus our attention on the enveloping world and we use our intelligence to reinforce rather than to question the reality of the experience.'[24]

There is a specific difference between controlling a character and being 'in' the character (or avatar). In the latter, you need to bring characteristics of 'you' (relating to your self-image in the 'real' world) to the avatar-representation in the digital environment. The subjective view then definitely promotes identification and gives a sense of presence. The case of Lara Croft in *Tomb Raider* (Fig. 3) is an example of identification with a character, enhanced with user-control. The effects are not very different from identification with characters in cinema, in literature, and so forth. In fact, these charac-

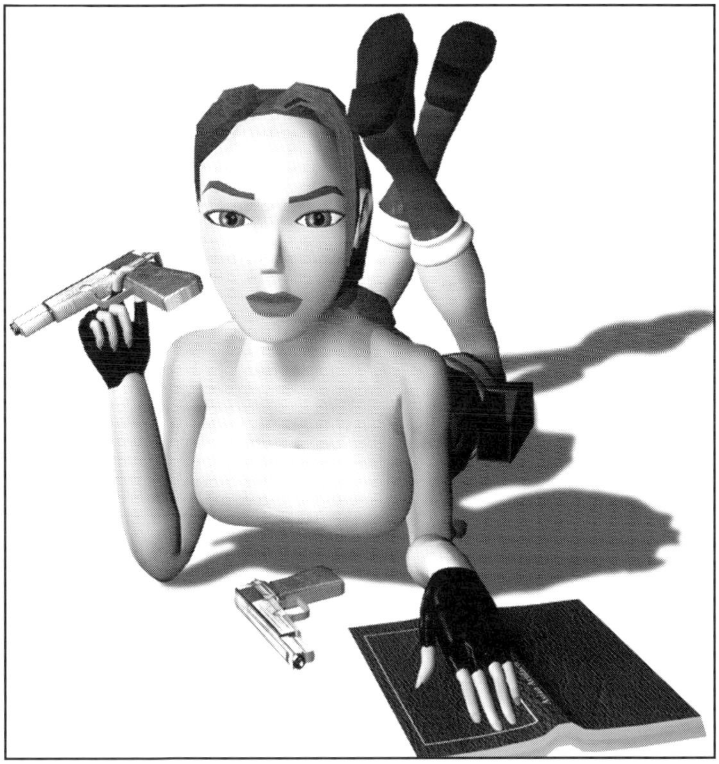

Fig. 3 (above). Lara Croft, Tomb Raider *(1996),* Tomb Raider II *(1997),* Tomb Raider III *(1998) [Core Design Limited.]*

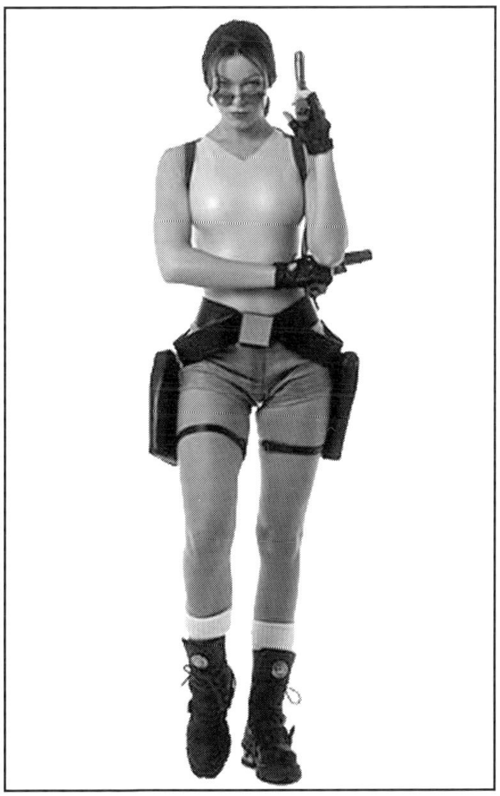

Fig. 4 (left). Nell McAndrew as Lara Croft.

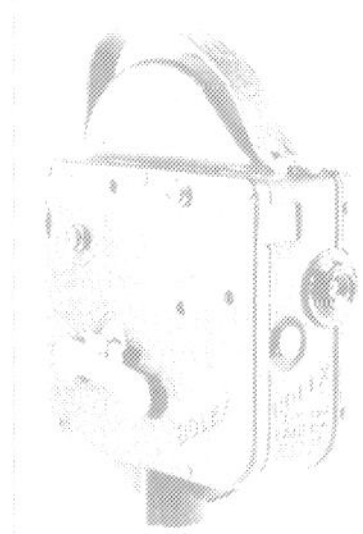

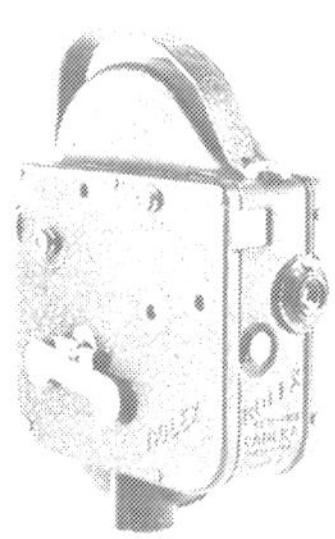

ters are treated much in the same way as stars of the silver screen with a huge amount of merchandising, fan clubs, and detailed personal information. I will end with an excerpt from an interview with Nell McAndrew (Fig. 4) who will play Lara Croft in the movie based on *Tomb Raider.*[23] Perhaps this will shed some light on how we conceptualise these 'stars of the computer screen'.

How do you feel about representing a digital character?

> Very excited! Lara Croft is such a tough, sexy character – why wouldn't I want to be her? She's larger than life and I'm proud to think that I've been chosen to play her part.

Do you think you have much in common with Lara?

> Yes. I'm very independent and I love sports – running, boxing etc. I have an athletic build like Lara and I've also got a rather large chest ... though not quite as big as Lara's!!! I also love a challenge.

Do you think Lara is a positive role model to today's young women? If so, why?

> Yes, I do think Lara is a positive model – it's great to see a woman so powerful, independent and tough, yet still sexy. I really think that Lara gives women inspiration to go out and get what they want. She's a real no-nonsense character!

What's the best thing about being Lara Croft?

> Being Lara makes me feel powerful and confident, and the reaction I've had from people so far makes me feel really popular, even loved! It's a real pleasure to meet all the fans who support Lara – they're so dedicated to her!

Acknowledgement: All Riven images and text © 1996, 1997 Cyan, Inc., and used with permission.

Notes

1. Interactive fiction refers to a wide range of writers (Michael Joyce, J. David Bolter *et al.*) whose work has artistic ambition rather than using narrative as a way of presenting and experiencing material, which is the topic of this argument.
2. This system is a combination of the structuralist approach derived from Roland Barthes, Tzvetan Todorov, Seymour Chatman, Mieke Bal, Gérard Genette, Claude Brémond, André Gaudreault *et al.*, and the phenomenological approach inspired by Martin Heidegger, Maurice Merleau-Ponty and Michel de Certeau, apparent here by Arthur Asa Berger, Don Ihde, Vivian Sobchack *et al.* The term 'personification' is loosely connected to the discussion provided by David Bordwell in *Making Meaning: Inference and Rhetoric in the Interpretation of Cinema*, (Cambridge, Mass.: Harvard University Press, 1989), 152, where he notes that 'The critic uses the schema to build up more or less 'personified' agents in, around, underneath, or behind the text. Such agents, once endowed with thoughts, feelings, actions, traits, and bodies, become capable of carrying semantic fields.'
3. Steven Jones, 'The Book of Myst in the Late Age of Print', *Postmodern Culture* 7.2, <http://www.press.jhu.edu/journals/postmodern_culture/v007>.
4. Ibid.
5. In Roland Barthes, 'Introduction to the Structural Analysis of Narratives', in *Image-Music-Text* (New York: Hill & Wang, 1977), 95, the 'catalyser' can have this function (along with acceleration, delay, anticipation). Also, in a footnote, he refers to Valéry's 'dilatory signs'.
6. This could be anything that the majority of users link to a specific time or culture. A cultural inhabitant may, for example, include a pith helmet in a rendition of the Boer War. A time marker could mean playing Sex Pistols' *God Save the Queen* in an historical piece on Britain in the late 1970s.
7. Barthes called these objects 'accepted inducers of associations of ideas', op. cit., 22.
8. In the first *Tomb Raider* the 'camera' was placed fairly close to the back of Lara's head. In *Tomb Raider II* and *III*, the 'camera' has moved back a bit. This means that you see more of Lara. In some cases you actually see quite a bit of her. One reason for moving the 'camera' back in space could be that the player then sees more of the environment. The producers had enhanced the rendering 'engine' to reveal more of the environment by simulating a wide angle view.
9. Aristotle, *Poetics,* translated by S. H. Butcher, Paragraph 6: IV.

10. David Bordwell, op. cit., 153.
11. F. Heider and M. Simmel, 'An Experimental Study of Apparent Behavior', *American Journal of Psychology* 57 (2), (April 1944): 243–259.
12. Edward Branigan, *Narrative Comprehension and Film* (London: Routledge, 1992), 101.
13. This plot summary is picked from Internet Movie Data Base <http://www.imdb.com>.
14. Jacques Aumont, *The Image* (London: BFI, 1997), 118.
15. The instrumentality of the apparatus vs. the content is expressed by many researchers in the areas of the history of technology and of the arts in different media specifications. Vivian Sobchack observes that 'while they [the film's enabling mechanisms] enable the commutation of perception and expression that is the film, neither the camera nor projector (nor lenses, editorial equipment, optical printers, sound recording and transfer equipment, screen, *et al.*) are themselves the film we experience and see, which itself visually signifies vision as visible and significant experience', *The Address of the Eye: A Phenomenology of Film Experience* (Princeton: Princeton University Press, 1992), 169.
16. Ibid., 126.
17. André Gaudreault goes even further when suggesting that '[t]here are two types of narrative in the cinema: the micro-narrative (the shot), a first level on which is generated the second narrative level; this second level more properly constitutes a filmic narrative in the generally accepted sense' in Thomas Elsaesser (ed.), *Early Cinema: Space, Frame, Narrative* (London: BFI, 1990), 71. In footnote #16, he stresses 'that every shot, taken out of its context and projected on the screen as a single object, should be regarded as a narrative in itself', 74.
18. Paul Ricoeur, 'Narrative Time', in W.J.T. Mitchell (ed.), *On Narrative* (Chicago: University of Chicago Press, 1980), 167.
19. Seymour Chatman talks about reading-time and plot-time (or, with his terminology, discourse-time and story-time) in *Story and Discourse: Narrative Structure in Fiction and Film* (Ithaca: Cornell University Press, 1978), 62. In footnote #23, he presents other terminologies describing, roughly, the same dichotomy as 'chronological' and 'pseudo-chronological' time (Mendilow), and Christan Metz's 'the time of the thing told and the time of the telling (the time of the significate and the time of the signifier)'.
20. Jacques Aumont observes in *The Image* that 'The point is to never confuse the time of the image with the time of the spectator. The spectator has the freedom to view a photograph for three seconds or three hours, but in cinema the spectator can view only as long as the projector is running', 120.
21. Ibid., 125.
22. We have already seen examples of this in the distributed 3-D environment *Active Worlds*, in strategy games like *SimCity*, and (somewhat farfetched) the Swiss clock company Swatch's Internet time: Swatch Beat <http://www.swatch.com>.
23. Like interaction in MUDs, MOOs in the 1980s; in chats, and in virtual communities like *WorldsAway* <http://www.worldsaway.com>.
24. Janet Murray, *Hamlet on the Holodeck: The Future of Narrative in Cyberspace* (New York: The Free Press, 1997), 110.
25. <http://www.tombraider.com/>.

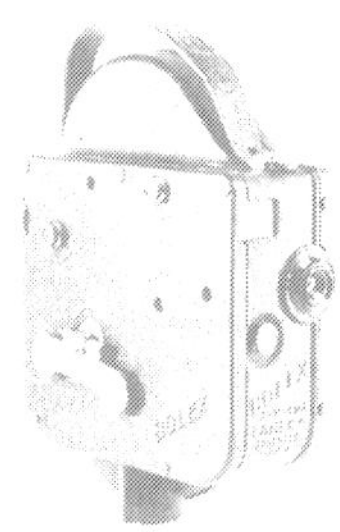

Lurking and Looking: Webcams and the Construction of Cybervisuality

Sheila C. Murphy

Program in Visual Studies, University of California, Irvine, USA

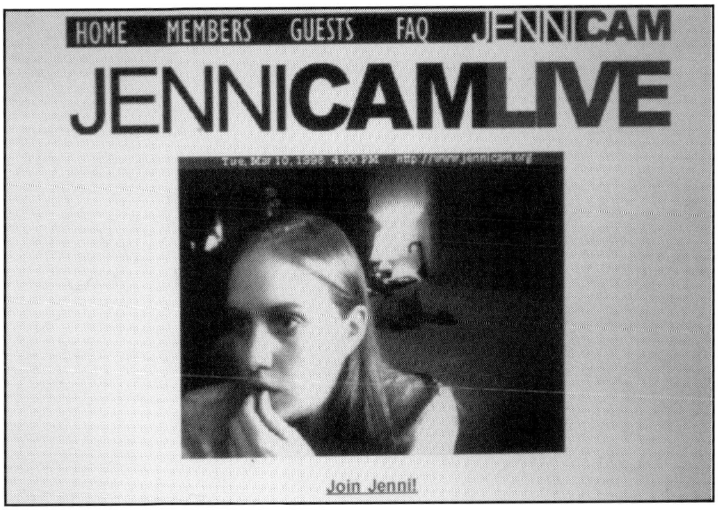

Fig. 1. Jennifer Ringley's daily life is documented on the Jennicam website. [© Jennifer Ringley, used with permission.]

The 1990s has been a decade in which communications technologies and moving image technologies have converged in multiple ways, producing new cultural 'extensions' (and fragmentations) of the human body and of subjectivity. I am particularly interested in the new mode of cultural engagement enacted by Internet users who extend their visual access to the public sphere by connecting to Internet webcam sites and becoming part of the always only partially knowable subculture of webcams (Fig. 1). Briefly, webcams, also known as SpyCams, are Internet websites that contain views of a wide range of vistas produced by connecting a video camera to a computer connected to the Internet. Webcams use either live video or still images that are automatically updated at regular intervals. But before I discuss webcams in particular, I want to note that webcams are only one example of the contemporary visual culture of surveillance. Grainy, low-resolution video images (often in black and white, and shot from the high angle 'point of view' of a surveillance video camera or a television news helicopter) have become a standard cultural marker (at least in the United States), usually signifying an unscripted and illicit performance (such as the kinds of footage used in the Fox Television series 'Caught On Tape'), acts which are not performed before a known audience but in front of a hidden camera.[1] A cultural fixation upon surveillance images and on being a subject of surveillance is evident in recent films such as *Dark City* (1998), *The Truman Show* (1998), and *Enemy of the State* (1998). This last film, directed by Tony Scott and released in the United States in November 1998, tells the story of Bobby Dean (Will Smith), a Washington labour lawyer who undergoes a transformation from naïve and trusting citizen/bourgeois individual to a scrutinised and criminalised subject (and, of course, back again). This transformation of identity is visu-

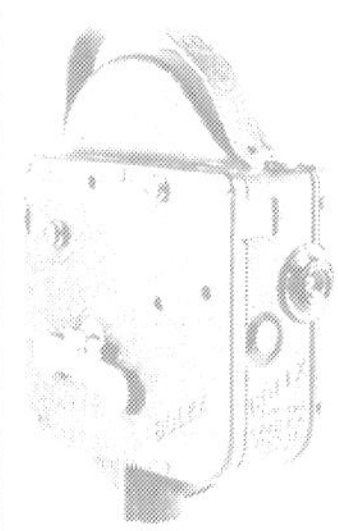

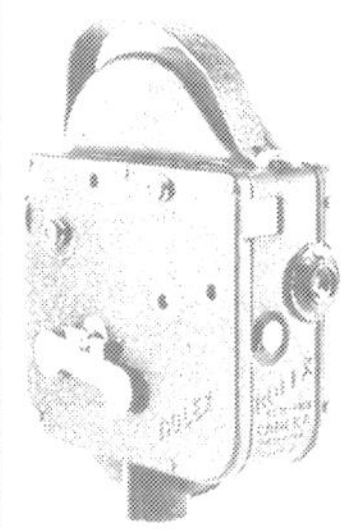

ally denoted through the use of surveillance imagery and images of the various tools of surveillance (hidden cameras, listening and tracking devices). At the end of the film, after undergoing the visual scrutiny that is focused upon an 'enemy of the state', Dean is shown watching television in his home. Clicking through the channels with his remote control, Dean comes upon a channel that displays an image (in low-res surveillance video) of Dean himself, clicking through the channels. Realising immediately that the real-time image is courtesy of his surveillance expert friend Brill (Gene Hackman), Dean waves back and speaks to Brill while watching himself on screen. He has now fully accepted that, as a US citizen in the late part of the twentieth century, he is constantly an object of surveillance. I mention this narrative episode from the film because I want to emphasise how prevalent surveillance has become as a cultural logic and, more importantly, to note how this kind of looking, and of being looked at, is increasingly accepted as a status quo, and is even welcomed in certain subcultures. This is not to say that surveillance (in general) is a celebrated or even preferred mode, but that the response to the surveillance powers of various cultural authorities (the state, the consumer marketplace that installs cameras to prevent crime, etc.) is the production of counter-surveillance. As Larry Husten, a writer for *Wired* magazine, states, 'the darker side of the information superhighway is the surveillance superhighway'.[2] The production of self-surveillance through technologies like webcams is one method that Internet enthusiasts use to counteract the surveying gaze of authorities, as well as to produce counter-narratives to normative identity categories.

This essay was originally conceived of and written as a world wide web hypertext document. I am now going to go against that grain of writing and produce for you a somewhat linear narrative about webcams. The sense of indeterminacy and multiplicity inherent in hypertext is analogous to my own hesitant sense of theoretical 'closure' regarding webcams. Because webcams are such a new cultural phenomenon, I am reluctant to provide a neat summary of their cultural impact. Instead, this is a speculative discussion that retains much of the 'hypertextuality' of my website – drawing multiple connections but leaving the final links between such connections open.

Taking my cue from the writings of Hans Magnus Enzensberger, I'd like to sketch out a 'Constituents of a Theory of Webcams'.[3] By focusing on the different historical and theoretical constituencies of the webcam apparatus and its cultural functions, I will open up a discussion of the role of surveillance in contemporary society and the new modalities of looking formed at the screen. While I do not entirely share Enzensberger's commitment to charting out the *revolutionary* potentials of new media, I do want to focus more upon *emancipatory* possibilities rather than remain mired in a paranoid, panoptic view of contemporary media culture. What follows, then, are some constituents of a theory of webcams, some speculations about webcam culture.

As I have already mentioned briefly, webcams are Internet websites that feature 'views' as their primary content. These views are produced using digital cameras, so the information is inherently digital – always composed of binary information (zeroes and ones) rather than analogue media that have been digitised, as when one scans a photograph. Webcam sites can contain static images that are manually put on-line and only updated periodically, but the majority of sites (and the ones in which I am particularly interested) utilise computer software to produce views that are automatically updated – mobile and fragmentary images of whatever is happening in front of the camera. Webcam content is diverse, and webcams can be deployed to produce personalised erotic images such as those found on the 'voyeurtv.com' site, but many webcams upload rather banal and 'empty' images to the world wide web. I find the webcam views of empty spaces especially interesting in their very lack of compelling content or activity.[4] Webcam 'views' include a vast range of subject matter. One webcam index site groups webcams by the following content categories: universities, images of water, beaches, office buildings (interior and exterior), citycams, traffic cams, nature cams, etc. Yet the common link between these various kinds of 'cams' is that they are mostly devoid of human 'presence', except for the accidental, distant, and grainy 'subject' that wanders into the camera's viewfinder. Webcam images are blurry and usually the apparatus is more readily apparent than the image; one witnesses a kind of jerky, small-screen format video image with a noticeable gap between images, despite the fact that the images are produced in 'real time'. Also, the view produced is usually roughly 3 × 4 in. in size, filling only a small part of the computer screen, leaving the rest to be filled by text written by the webcam producer or one's own documents and computer labour. Some webcams can even be 'floated' in the corner of the screen, allowing a virtual window to some other part of the world beyond the computer work space of the webcam viewer. Perhaps the most (in)famous webcam site is Jennicam,[5] a site which broadcasts a live view of the site operator's bedroom over the Internet twenty-four hours a day. In 1998, Jennicam captured the attention of the American public, and the site's operator, Jennifer Ringley, was interviewed on several television talk shows, and even appeared as a guest character on the television drama *Diagnosis Murder* in which she played a character who operated a webcam site much like Jennicam. Her virtual activities transformed her into a media celebrity.

At this point in time, it is not possible to determine whether or not webcams are a 'transitional' technology or a mode of seeing that will continue. As early as 1994, the camera technology that many webcams rely upon was used for low-quality Internet videoconferencing (CU-SeeMe). While such videoconferencing methods are still used, webcam sites are a more dominant use of the technology. At present, the technology is more suited for one-way visual communication than two-way audio-visual communication.[6]

The historical roots of webcam culture can be traced to developments in art history, cinema, television, and telecommunications. Within the tradition of twentieth-century Western art, webcams are the descendants of video art and conceptual art practices dating back to the late 1960s or early 1970s. Video and conceptual art continue to create complex media environments in which the relationship and distance between the viewer and object viewed is broken down.[7] Webcams owe their mobile views that occur within stabilised frameworks to the history of cinematic spectatorship and television, both of which establish perceptual horizons for viewing that webcam users then bring to their Internet experiences. Like television, webcams offer 24-hour programming that is continually 'live', fragmented and flowing (in Raymond Williams' sense of the word), and can be entered into on-demand by the viewer.[8] While seen on the 'viewscreen' of a computer monitor, webcam views are similar to the kind of banal cinema of duration that Andy Warhol explored in his early films or, likewise, as a kind of *cinéma vérité*.[9] This videographic 'standing reserve' is part of what Paul Virilio calls the 'false day'.[10] The views that are 'captured' by webcams are, as long as the cam is connected to the Internet, standing by; their content has been transformed into a techno-optical experience, a sight to be seen.[11]

Marshall McLuhan's axiom 'The Medium is the Message' is particularly useful for thinking about webcams. While McLuhan was particularly concerned with television and how it, as an electric technology, changed one's relation to the print technology of the book, his insistence upon considering the technological effects that are inherent in an apparatus and not only the content of the medium is important for thinking of the ways that webcams produce new temporal and spatial relations. Such relations are caught up in webcam content but are also part of the fact of the screen-user interface. It is crucial to consider these screen qualities in conjunction with what is 'on' the screen because, as McLuhan suggests, 'The effects of technology do not occur at the level of opinions or concepts, but alter sense ratios or patterns of perception steadily and without any resistance'.[12] For McLuhan, this meant that our daily lives were restructured by television (I wake up, I turn on the TV to find out what the weather will be like today, rather than looking out the (architectural) window).[13] In the case of webcams in particular and the Internet in general, the altered pattern of per-

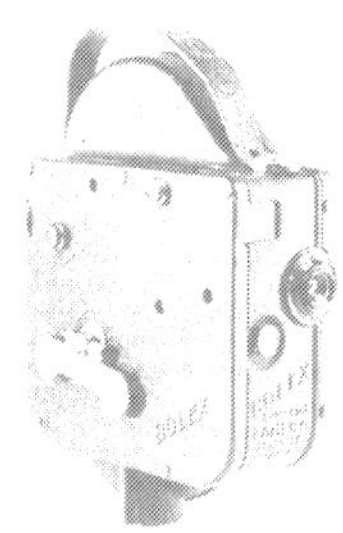

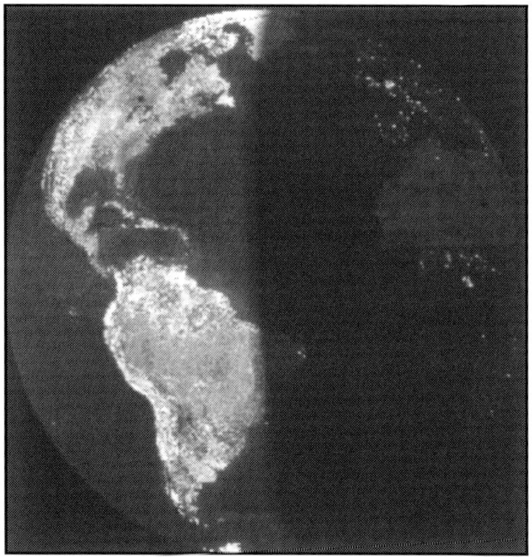

Fig. 2. The Earth and the Moon Viewer provides still images that can be virtually explored online. The Viewer was developed by John Walker, creator of AutoCAD. [© John Walker, used with permission.]

ception is enacted in the automatic reflex to login to one's favourite site/sight to check out the view.

Webcams radically alter temporal and spatial relations through their particular quality of interactive looking. Webcams allow those who are looking, engaged in what I call *lurking*, to extend their vision, their ability to see into the distance, through the screen. One connects to the Internet, selects a webcam site to view, and this site inevitably provides links or connections to other webcams which show a variety of locales even though one never leaves the physical space of the computer workstation. Webcams allow for virtual travel and spatial exploration which alter the (physical) meanings of time and distance. As one webcam index describes the contents of yet another index, 'Sick of offices in the USA? See an office in Germany!' (as if office buildings in Germany would be radically different from those in the United States). This results in what Paul Virilio describes as a condition of general arrival when he observes that:

> Currently, with the instantaneous broadcasting revolution, we are seeing the beginnings of a *'generalized arrival'* whereby everything arrives without having to leave, the nineteenth century's elimination of the journey (that is, of the space interval and of time) combining with the abolition of *departure* at the end of the twentieth, the journey thereby losing its successive components and being overtaken by *arrival* alone.[14]

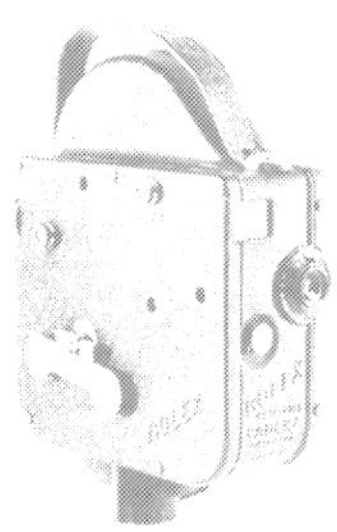

Geography and location are reconfigured at the screen through the views presented by webcams. Or as the 'T@p Voyeur Spy Cam's' website describes its T@p Daily Spy, 'Geography lesson. Who knows where Andorra is? Not me. Had to consult my atlas. It is located on the border between Spain and France. Thanks Eric (the webcam producer of this view), for showing us a part of the world I didn't even know existed.'

While McLuhan theorised that television would result in the creation of a global village replete with technological primitives connected to the network, the 'village' actually expands since the views that connect the members of webcam culture are often views at a distance. Perhaps the ultimate example of this is the Earth and Moon Viewer site (Fig. 2) which is not a live, updating cam site, but rather a site that relies upon still satellite images of the Earth from outer space. The user can choose to view the earth from the sun or from the moon, and can use their computer mouse to readjust their satellite image by entering new longitudinal or latitudinal coordinates, or clicking on the globe to shift the point of view. The arrival becomes even more general.[15]

Webcams are indicators of a shift in panoptic culture from the modern disciplinary visual power described by Foucault to a post-modern culture in which panopticism is part of the fabric of life. Foucault's panoptic model of disciplinary relations relies upon the assumption of the existence of an all-seeing (central) subject who exercises power through looking. In her essay, 'Surveying the Surveilled: Video, Space and Subjectivity', Lili Berko describes the new, dispersed panoptic relations in a post-modern culture that is increasingly occupied by surveillance technologies. Webcams are certainly an example of 'self-produced, self-surveillance' as theorised by Berko. She writes that, 'The combination of computer and communications technologies has produced the possibility for each one of us to not only be the object of surveillance, but its subject as well'.[16] Because webcam producers are taking surveillance technology into their own hands (to produce views which are often mobilised as an escape to a space outside of a *disciplinary* panoptic space) and are ironically playing with the codes of surveillance, central panoptic power through looking is diminished (Fig. 3). A (seemingly endless) chain of looking is produced in which there is no one looker, but rather many (the on-line lurker, the webcam producer who may inadvertently become a lurker when someone walks by his or her camera, the panoptic subject in the frame who may look back or undermine the pleasure of looking, etc.). Berko describes this kind of looking as a *glance* rather than an all-powerful gaze in a surveillance culture in which participants produce 'their own mini-panopticons as they go along, challenging the power of the official panoptic gaze of the invisible "Other".'[17] Perhaps it is time to consider how Foucault's panopticism functions in a culture that is inundated with technologies of seeing. I would like to suggest that the panoptic now functions in this context as an optic system, but one that is not necessarily tied to disciplinary structures. The fragmentary glance of a webcam creates (always) mobile, unpredictable, and unscripted views. These views are still a mode of surveillance but it is now surveillance as vista, as entertainment, as global spectacle rather than surveillance in a disciplinary regime. This is not to say that there is no power in looking but, rather, that such power in looking is dispersed across a more complicated network of desire-in-looking. As Jean Baudrillard writes about the televisual gaze in *Simulations*:

> This is still, if not a system of confinement, at least of scrutiny. No longer subtle, but always in a position of exteriority, playing on the opposition between seeing and being seen, even if the focal point of the panopticon may be blind.[18]

In his essay on photography, Walter Benjamin suggests that disciplinary power need not be architecturally structured.[19] The dispersion of images also functions as a dispersion of power. Benjamin's early contemplation of mechanical reproduction situates his argument about the circulation of photographic images on a slightly different register than that of his later, more widely known writings on the

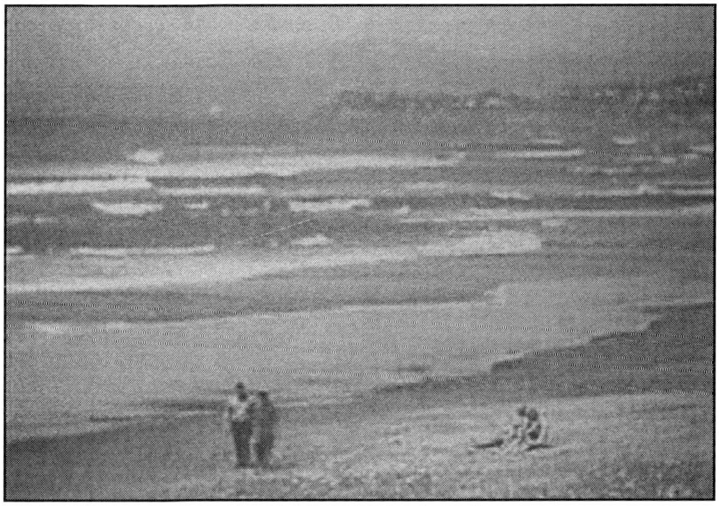

Fig. 3. The Del Mar Beach Cam webcasts a live video image of the surf in Del Mar, California. [© Connectnet.com, used with permission.]

subject. Writing in 1931, Benjamin concludes that photographs must be viewed differently from other representational forms:

> One can no longer view them as productions of individuals; they have become collective images ... the mechanical methods of reproduction are a technology of miniaturization and help man to a degree of mastery over the works without which they are no longer useful.[20]

For Benjamin, both the physical size of images and their mass circulation mark them as tokens of anonymity within a circuit of powerful exchanges. While webcams exist in a culture in which the mass circulation of images and their analogue (and digital) reproduction is nearly ubiquitous, they also function as miniaturised, collective and anonymous productions. Furthermore, the glanced-view of a webcam is not a closed and seamless narrative space, such as that found in classical Hollywood cinema, but rather, it 'insists upon the constant reopening of the gap of desire',[21] the empty, overtly technological 'space' of the webcam image that is available on-line, waiting to be seen. Enter the *lurker*.

I originally wanted to write about webcams because I think they are indicative of one of the most significant cultural reconfigurations occurring on the Internet, the activity of lurking, a form of potentially voyeuristic looking by an unseen viewer. It is commonly understood that lurking is a kind of pathological activity. This idea is reinforced by the

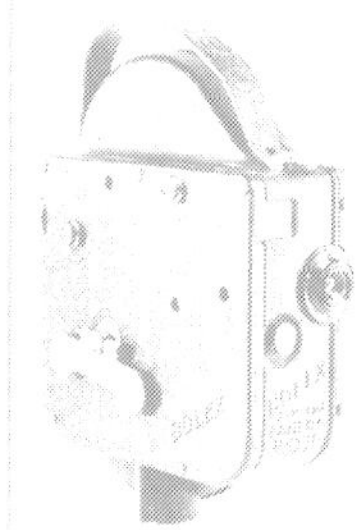

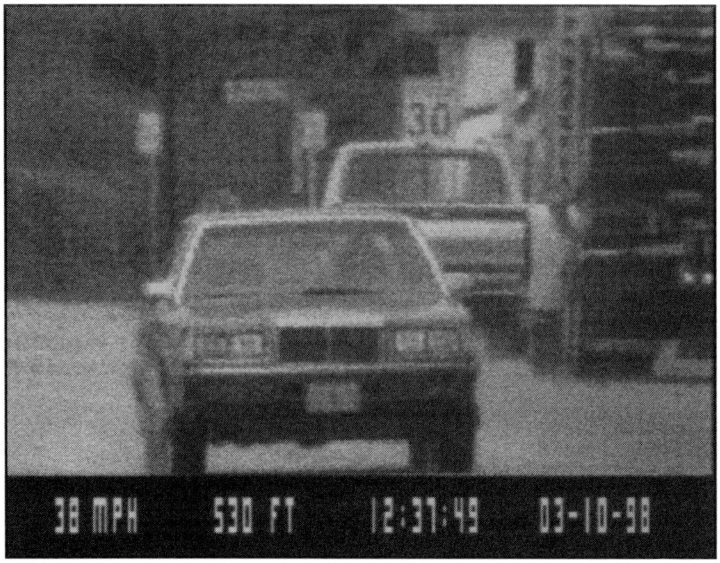

Fig. 4. Harborside Internet's webcam site shows a view of the road in Brookings, Oregon, along with the approximate speed and distance of approaching vehicles. [© Harborside Internet, used with permission.]

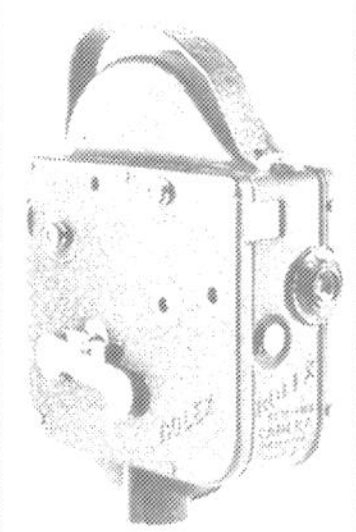

dictionary entry for 'lurk' which has a particularly negative valance: 'to skulk, sneak, prowl, suggests avoiding observation, often because of a sinister purpose'.[22] Yet on the Internet, lurking has a different resonance. Lurking is a socially acceptable, even encouraged, activity. On Usenet,[23] one of the earliest and largest sectors of the Internet, new users ('newbies') are encouraged to lurk on newsgroups (Internet forums which are hierarchically organised around different communities of interest) for several days before posting a message to the group (such initial posts/messages are often called 'delurkings'). This is so that the new user may become familiar with the etiquette and tone of the newsgroup. By lurking, the user is both caught up in the discourse of the group and, at the same time, able to remain detached from the group. As with webcams, the Usenet lurker is an always unknowable, but presumed to exist presence for the users who do post/submit the content of the newsgroup. Or to quote the famous 1993 *New Yorker* cartoon, 'On the Internet no one knows you're a dog'. The lurking public is analogous to the technologically-mediated audience in Walter Benjamin's 'The Work of Art in the Age of Mechanical Reproduction'. Writing on the transformation of performance by the cinematic apparatus, which breaks up the 'integral whole' of theatre, Benjamin insists upon a new kind of audience – actor relation:

> But now the reflected image has become separable, transportable. And where is it transported? Before the public. Never for a moment does the screen actor cease to be conscious of this fact. While facing the camera he knows that ultimately he will face the public, the consumers who constitute the market.[24]

A similar kind of unknowable public pervades much of the logic of Internet culture and is inherently connected to visuality. The growth of the World Wide Web in the recent past has changed the Internet from a textual to a largely visual Internet experience because one is able to access the web through a graphical user interface (GUI).[25] One's connection to the Internet is increasingly mediated through a visual discourse, and Internet personas are produced *visually* through websites. Webcams are part of this production of technological visual identity. After viewing many webcams and discovering that webcams are a kind of self-produced (self-)surveillance, I began to wonder what would motivate someone to construct a (banal) view of themselves or aspects of their world and put that view on-line. The answer seemed to lie in the 'and for whom?' part of the question of 'why/where/how/for whom are these views (these cyberobjects) made?' I think that they are made because the webcam producers believe that they are participating in a *global public sphere* that is composed of other webcam producers but also, significantly, of the (mostly) unknowable audiences of the views, the *lurkers* (Fig. 4). In his late interview, 'The Eye of Power', Michel Foucault states that Bentham's writing about the Panopticon was a kind of intellectual discovery – like 'Columbus's egg' – a 'theme that was in the air at the time'.[26] Webcams seem to function as a similar kind of cultural marker, a thematic that captures both the anxieties and aspirations of contemporary culture.

In attempting to draw some conclusions about how webcams perform the current cultural logic of surveillance, and what some of the consequences might be for this endless chain of lookers and lurkers, I found myself turning to the words of Jennifer Ringley, the operator of the Jennicam site. When asked about her wilful strategy of self-surveillance and the subsequent and ongoing public display of her private life, Jennifer responds in a way that perhaps best speaks to the social condition of cybervisuality:

> I don't feel like I'm giving up my privacy. Just because people can see me doesn't mean it affects me – I'm still alone in my room, no matter what. And as long as what goes on in my head is still private, I have all the space I need. On the other hand, if someone invented a TelepathyCam where you could hear everything I was thinking, I must admit I'd be a bit more squeamish.*wink*[27]

Appendix: Webcam site addresses

I list below the web addresses for the webcam sites mentioned in the essay, as well as select webcam index sites. Please note that these addresses are current as of March 1999. Web addresses, however, are not stable, fixed entities, and may have changed.

Del Mar Beach Cam
http://www.camzone.com/cams/delmar/stream.cgi

The Earth and Moon Viewer
http://www.fourmilab.ch/earthview/

Earthcam Webcam Index
http://www.earthcam.com

Harborside Traffic Cam
http://www.harborside.com:2045/cgi-bin/video

Jennicam
http://www.jennicam.org

Peek-At-You Net Cam Index
http://www.argo.net/~frank/net_cams.html

T@p Voyeur Spy Cam
http://www.spy-cams.com/default.asp

Notes

1. My connection of webcams to other forms of surveillance and technologically-motivated 'ways of seeing' has benefited from discussions with other scholars of technological and visual culture. In particular, I would like to thank Anne Friedberg, David Joselit, Gordon Kindlmann, Janice Neri, and Pascal Pinck for their input, as well as my colleagues at the Technologies of Moving Images conference, Stockholm, 1998.
2. Larry Huston, 'A Life on the Web', *Wired* 3, 10 (October 1995).
3. Han Magnus Enzensberger, 'Constituents of a Theory of the Media', in Timothy Druckery (ed.), *Electronic Culture: Technology and Visual Representation* (New York: Aperture, 1996), 62-85.
4. One could, following in the footsteps of Fredric Jameson's writings about postmodernism and multinational capital, conclude that webcams with 'empty' vistas are symptomatic of the technologically-assisted alienation and fragmentation of subjects who live within and (virtually) move among these sign-systems.
5. All webcam addresses cited here can be found in the Appendix to this essay.
6. In 1996 I participated in a pedagogical experiment using CU-SeeMe as part of D.N. Rodowick's Seminar on Digital Culture. The Internet videoconferencing capabilities proved challenging and became a topic of discussion for the class.
7. For more on performance art and subjectivity, see Amelia Jones, *Body Art/Performing the Subject* (Minneapolis, London: University of Minnesota Press, 1998).
8. Jane Feuer, 'The Concept of Live Television: Ontology as Ideology', in E. Ann Kaplan (ed.), *Regarding Television* (Fredrick, MD: University Publications of America, 1983), 12–22. Also Raymond Williams, *Television: Technology and Cultural Form* (Hanover: Wesleyan University Press), 1992.
9. Warhol's films *Sleep* (1963) and *Haircut #1* (1963) are particularly striking in this regard. In each of these films, titillating premises are 'set up' but left (visually) unfulfilled. Elsewhere I have explored how Warhol's directorial approach approximates a kind of lurking-as-production.
10. Paul Virilio, 'The Third Window', in Cynthia Schneider and Brian Wallis (eds.), *Global Television* (Cambridge, Mass.: The MIT Press, 1988), 185–197. Virilio's 'false day' is the technological day created by the video tape, which can be viewed at any time and has its own separate temporal logic from the diurnal day.

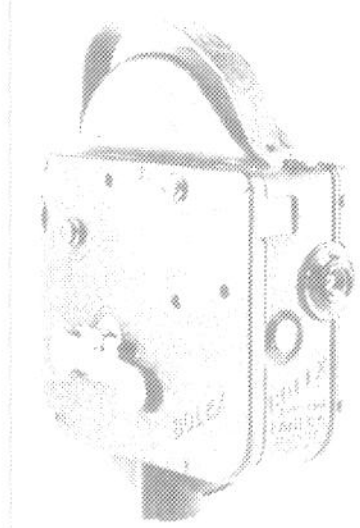

11. The technological apparatus transforms the function of the object in the way that the hydroelectric plant transforms the Rhine in Martin Heidegger's 'The Question Concerning Technology' in David Farrell Krell (ed.), *Martin Heidegger: Basic Writings* (New York: HarperCollins, 1973), 307–343. Heidegger condemns this transformation, stating that, 'In the context of the interlocking processes pertaining to the orderly disposition of electrical energy, even the Rhine itself appears to be something at our command.', 321.
12. Marshall McLuhan, *Understanding Media: The Extensions of Man* (Cambridge, Mass.: The MIT Press, 1994), 19.
13. For more on the relationship of different kinds of windows, see Anne Friedberg, *The Virtual Window* (forthcoming).
14. Paul Virilio, *Open Sky*, trans. Julie Rose, (London: Verso, 1997), 16.
15. A similar project was proposed by Vice President Al Gore in March, 1998. Gore proposed the development of a high-definition video channel and accompanying website to relay images of Earth from space. Katharine Seelye, 'Gore Proposes Video Channel to Show Earth, All the Time', *New York Times* (14 March 1998).
16. Lili Berko, 'Surveying the Surveilled: Video, Space and Subjectivity,' *Quarterly Review of Film & Video* 14, 1–2 (1992): 61–91.
17. Ibid., 63.
18. Jean Baudrillard, *Simulations*, trans. Paul Foss, Paul Patton and Philip Beitchman, (New York: Semiotext[e], 1983), 52.
19. Walter Benjamin, 'Kleine Geschichte der Photographie' (1931) reprinted as 'Walter Benjamin's Short History of Photography', *Artforum* 15 (6): 46–51.
20. Ibid., 50.
21. Berko, 74.
22. *Webster's New Universal Unabridged Dictionary,* 1989, 854.
23. Usenet is the dispersed network of Internet 'newsgroups': it consists of official forums where daily wire news reports are posted (such as Reuters), semi-professional organisations with moderators and official rules, and unmoderated communities of interest (such as alt.tv.simpsons or alt.sex.stories) where any Internet user can submit (post) items. For more on Usenet, see Judy Anderson, 'Not for the Faint of Heart: Contemplations on Usenet', in L. Cherny and E. R. Weise (eds.), *Wired Women: Gender and New Realities in Cyberspace* (Seattle: Seal Press, 1996), 126–140, and Michele Tepper, 'Usenet Communications and the Cultural Politics of Information', in D. Porter (ed.), *Internet Culture* (New York: Routledge, 1997), 39–54.
24. Walter Benjamin, 'The Work of Art in the Age of Mechanical Reproduction', in Hannah Arendt (ed.), *Illuminations* (New York: Schocken Books, 1968), 231.
25. The web's graphical user interface, most often encountered through the software programmes Netscape Navigator and Microsoft Internet Explorer, has become such a predominant mode of Internet experience that the web frequently functions metonymically for the Internet in contemporary popular culture, which omits several other aspects of Internet culture, including various forms of multiple user communities.
26. Michel Foucault, 'The Eye of Power', in Colin Gordon (ed.), *Power/Knowledge: Selected Interviews and Other Writings, 1972–77* (New York: Pantheon Books, 1980), 147–48.
27. From the Jennicam FAQ (Frequently Asked Questions): http://www.jennicam.org/faq/general.html. Jennifer recently removed this part of the Jennicam site.

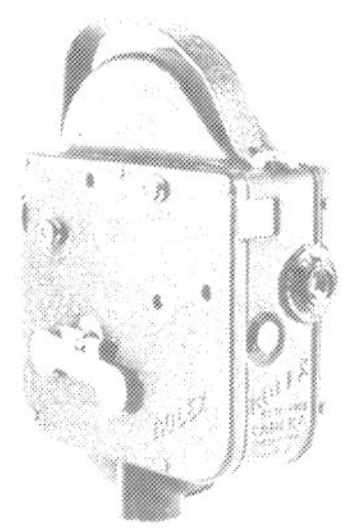

Visual Diaries: Revival of a Documentary Form in Digital Culture

Åke Walldius

Department of Cinema Studies, Stockholm University, Sweden

A broad spectrum of digital techniques is revitalising formal innovation as the technology of moving images enters its second century of development. Established narrational forms are being revised according to new expressive opportunities. The screen writers of Hollywood movies and computer games apply new technology to let us experience new exotic vistas and new zones of intimate personal virtuality.

Our daily lives are influenced by digitalisation as well. The way we write letters is gradually changing. The ways in which we take notes and pictures are changing too. It is the accommodation of one of these more prosaic forms of communication, the visual diary, that is the subject of this essay.

When examining loosely structured narrational forms such as the diary, there is an underlying question that needs to be addressed: will an emerging Cinema of Meditation evolve in parallel with the new Cinema of Digital Attractions? To focus on the personal and everyday use of digital media implies that there is indeed a continuing demand for self-reflective forms of realism, forms that may revive the surrealistic qualities of the tradition of poetic documentary.

Critics of the persuasive power of television have argued for a broader debate that makes audiences aware of the communicative strengths and weaknesses of visual media.[1] The concept of 'telepresence', of being present in the media event, is further articulating this demand for practical design knowledge that will let media users participate in active and self-reflective ways. For all disciplines engaged in the study of media, this demand for a clearer orientation *vis-à-vis* the user implies a widening of the scope of interest. There is a lot to be gained, in terms of forming common vocabularies, from stressing the inter-disciplinary quality of this new orientation. Common languages may provide a more comprehensive understanding of the relation between society and its different technologies of the image. As these common vocabularies develop, recurrent patterns across different media environments may gradually chart an evolving system of stylistic traits. Such a collaborative and piecemeal compilation of recurrent patterns may slowly make new and exciting relationships visible in concise and expressive ways.

The goal of identifying patterns of media construction is to give critical insight into the practical handling of media tools and environments. One important aspect of this search is that the patterns identified should apply to the kind of media environments that are within the reach of a broad audience. This is the context for my inquiry into the somewhat elusive genre of visual diaries. The examples will be picked from different periods and from different media to make the concepts general enough to deal with the evolving technologies of tomorrow. Each example will be treated with an emphasis on conciseness. The aim is

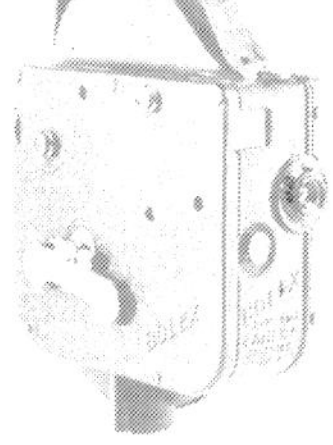

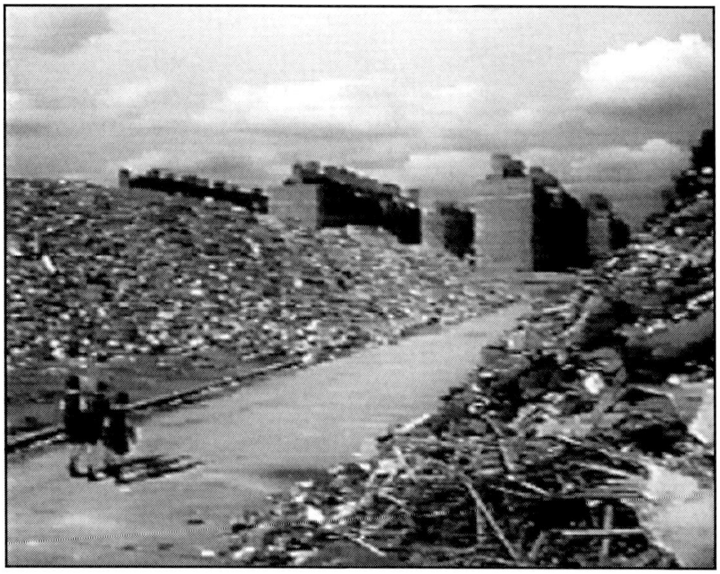

Fig. 1. A Diary for Timothy, *Humphrey Jennings, 1945.*

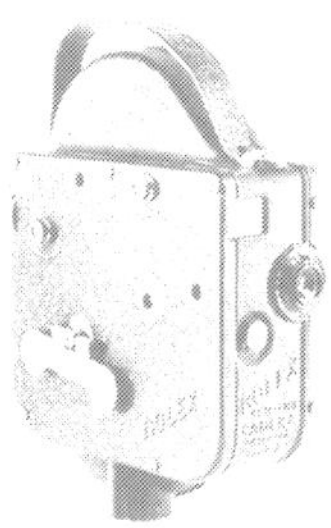

not so much to cast new light on individual examples. Rather, my primary goal is to propose an overarching perspective and a set of concepts that may make some regularities stand out in greater clarity across time and across media borders.

The three documentaries that I have chosen for my inquiry all picture everyday life in forms that exemplify the personal diary. The first example is from the era of British documentary: *A Diary for Timothy* (1945) by the surrealist poet-sociologist-filmmaker, Humphrey Jennings. The second example is picked from contemporary Swedish documentary: *Videodagboken: Erik – ett egensinnigt liv* (Videodiary: Erik, a self-willed life, 1997) by Dick Idestam-Almquist. The last of the three diaries is still 'in the making'. It is an example of 'evolving documentaries'. Its title is: *Jerome B. Wiesner: A Random Walk through the Twentieth Century*, and is directed by Glorianna Davenport and Michael Murtaugh at the MIT MediaLab.

Patterns of media construction as a tool for identifying formal innovation

On what grounds have I chosen these particular examples? In recent literature on non-fiction film, the personal diary has been analysed in terms of its performative potential. Questions about subjectivity in the private, confessional, variations of the form have led to discussions about a re-evaluation of the poetics of documentary film.[2] This focus on privacy and confession will most certainly acquire new significance when applied to the hybrids of webcam-diaries of digital culture. But an interdisciplinary approach to personal digital communication also invites us to rethink broader issues relating to contextual analysis and genre overlap. When the conventional forms of the visual diary transform and merge with other adjacent genres, the ways in which we trace underlying relationships will have to extend into the social and technological realms.

In studies of popular digital communities, Thomas Erickson and other computer scientists have shown that the concept of design patterns can be a helpful descriptive tool. The concept provides a systematic yet flexible way to identify regularities on multiple levels of abstraction when trying to grasp the evolution of specific communicative genres. In *Supporting Interdisciplinary Design: Towards Pattern Languages for Workplaces,* Erickson presents the concept of pattern languages and its application in software development and organisational design.[3] The approach stresses the cognitive and communicative power of visible patterns when trying to identify autonomous, but interlinked, entities in different design endeavours. This is not the place to summarise the specific formalism employed in the approach, but a few examplary names of patterns from the article may help to get a general understanding of their use. Erickson illustrates how the concept can be applied in an investigation of a consulting firm where different *recurrent patterns of interaction between people and the built environment* can be identified: *Maintaining Mutual Awareness* is a general pattern that exhibits other kindred patterns of smaller scale within it, such as *Blanket Email* (the custom of addressing email messages with questions or answers to large groups), *Kibitzing*, and *Doing a Walkabout*. These patterns may in turn be supported by spatial patterns such as *Open Offices, Model Shop*, and *Central Scanning Station*.[4]

In a thorough analysis, each design pattern, regardless of whether it applies to software, hardware or organisational design, typically comprises two to four pages with as many graphs depicting its context, central problematic, arguments and characteristic traits. A first step in any analysis is to identify the essential traits of the crucial entity and to find a fitting as well as a telling name for it. This inquiry, where the notion of genre is applied to an overarching perspective for the scope of the pattern language, will focus entirely on this first identifying step of the analysis. One pattern of media construction, characteristic for each film example, will be briefly discussed. To illustrate how a more thorough analysis would proceed, the presentation of each example then closes with a list of other candidates for distinctive patterns.

The three examples of visual diaries are chosen on grounds of communicative theme since they are all explicitly concerned with questions about fatherhood and inheritance. However, they differ substantially in their formal composition and in their underlying technology. By comparing examples with a marked thematic coherence, my aim is to highlight in what way the *formal traits* of each example corresponds to its characteristic *pattern of media construction*. For each example I will briefly relate the concepts of social context, authorial voice, and articulation of theme to the aspects of *formal continuity*. This will be followed by a brief discussion about the interplay between formal and technological innovation, and a tentative answer to my implied question about the prospect for self-reflective digital documentaries will be provided.

Patterns of camera work in the diaries of Timothy and Erik

In *A Diary for Timothy* (Fig. 1), the nation is at war, *mobilising* around anti-fascist cultural values. The voice of the diary is that of an elderly man relating his worries and hopes to a new-born child. The themes centre around the tensions between care versus war, memory versus oblivion. In *Erik, a self-willed life* (Fig. 2), the generation of the 1970s is *questioning* the conformity of the welfare state. The voice of the diary is a young father-to-be, who confesses his eccentricities and tells us about his commitments. The themes now centre around the tensions between playfulness versus responsibility, individuality versus the anonymous collective.

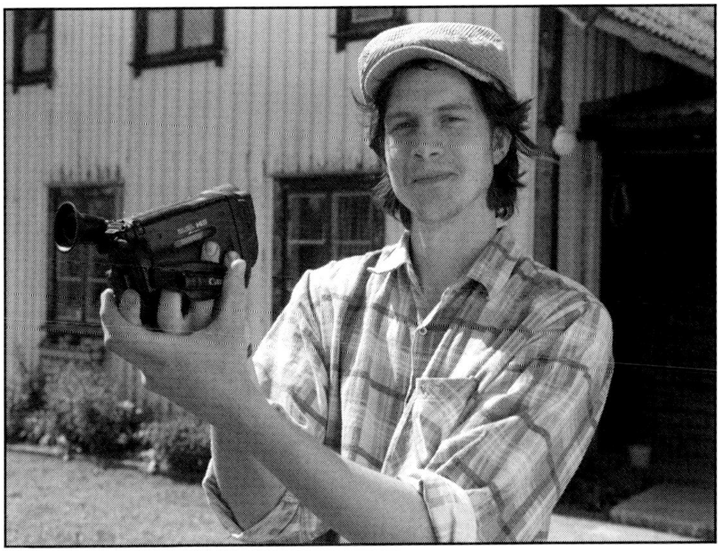

Fig. 2. Videodiary: Erik, a self-willed life, *Dick Idestam-Almquist, 1997, courtesy of Swedish Television (SVT). [Photograph by Carl-Johan Söder.]*

The scope between these two diaries has been narrowed down considerably, from the issue of national mobilisation to the issue of personal integrity. As far as *formal traits* are concerned, the comparison reveals some fundamental shifts. The *temporal continuity* of the two films remains strikingly similar. They are both told in a strictly chronological order with no browsing back in the diary. The new-born Timothy is followed during his first months which coincide with the last months of the war. In a similar manner, Erik is followed in the months before the child is born which coincide with his family moving to a new house of their own.

If the way in which the thematic scope is narrowed down is not evident in the temporal continuity of either diary, it is all the more marked as far as *spatial continuity* is concerned. In Timothy's diary, the everyday settings of home and city workshop are revealed with a calmly expository camera. The places are connected through associative montage, closely tuned to the literary qualities of the diary-notes. In Erik's diary, we experience the authentic settings of his home, the mill in which he works, his car, the barbershop, and wherever else he chooses to pick up his video camera, because Erik is indeed writing his own diary. With remarkably dynamic camera and sound, he manages to turn

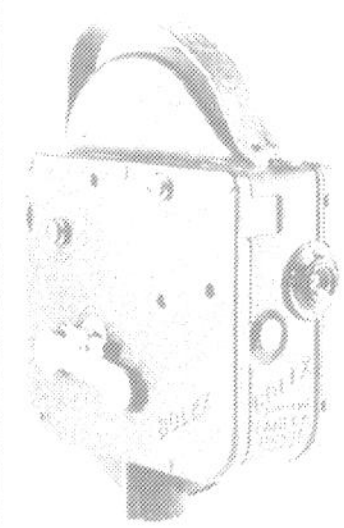

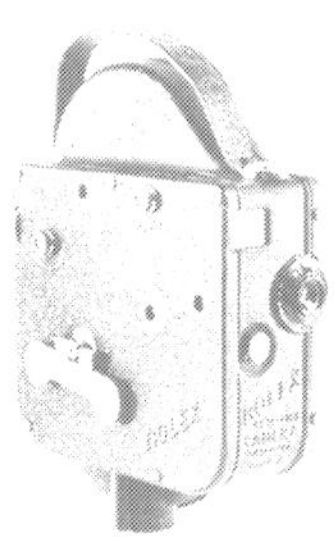

the most ordinary event into a highly personal spectacle. To me, the *Self-reflective camerawork* of Erik's diary is the most striking formal innovation of the video diary when compared with the earlier example. This formal innovation springs directly from a major shift on the practical, technological level since the principal character has taken over the role of cameraman, as well as that of director. This is indeed a new trait in the overall pattern of production. A distinctive step has been taken in the tradition of the *Documentarist as facilitator*.

The new collaborative pattern highlights the extended creative role of the documentarist as editor. Dick Idestam-Almquist, the producer at the Swedish Television Corporation, has commented that in his role as an editor, he took all the chances he could to dramatise and play the role of *auteur*. As long as Erik felt that it was still his story, Dick's role was to get it forcefully told.[5] In this film, the collaboration between the two really works. It is a wonderful example of poetic documentary in the digital era. Many of its formal qualities – its spontaneity and dynamism – are directly related to the innovative use of a digital camera and editing equipment.

Early *Documentarists as facilitators* such as George Stoney, Colin Low, Dorothy Todd Henaut, Bonnie Klein and John Gaventa are discussed by Brian Winston.[6] From his account of this tradition, institutional patterns of collaboration could be highlighted such as *Community as distributor* (the community organises internal and external distribution, and exhibition), *Media team as co-ordinator* (the team provides training and equipment), and *Director as catalyst* (the director plays a generative rather than a directional role). For further elaboration of the patterns of media construction in Jennings' film, there are many inspiring analyses. Bill Nichols writes about original devices used by Jennings for the construction of spatial and interpretative continuity such as *Public point of view* (to create social subjectivity through connecting multiple individual viewpoints, e.g. in an audience listening to a concert), and *Associative montage* (connecting on meaning rather than on causation).[7] Bjørn Sørenssen contributes with a closer analysis of compositional patterns that connect in Jennings' films, such as *Sound overlap* (connecting sound and sound), *Harmonic Mickey-mouse-editing* (connecting rhythms and chords with spatial movement), and *Poetic imagery* (connecting within the frame).[8] Other, more basic traits of construction that would become visible in a more detailed analysis of the diary of Timothy might be categorised as *Recording reconstructed events*; *Dialogue in reconstructed events* (both instanced in sequences of a mining accident and from a hospital); *Public narrator* (scenes from the BBC studio and families listening at home); *Public music* (a concert hall and a ballroom), and *Workplace music* (a factory). Similar analysis of the diary of Erik might identify patterns such as *Cameraman talking to the camera* or *Cameraman acting on camera*.

The old quest for *la camera-stylo,* heralded by Alexandre Astruc in the 1940s, seems to become less and less utopian. Technological innovations provide practical opportunities, but the creative effort lies as much in applying patterns of construction that realise these opportunities as in conceiving the technological apparatus. As far as camerawork is concerned, the spread of lightweight, digitalised video equipment has provided for a virtual socialisation of the basic skills of recording. The on-going digitalisation of the editing process calls for radically new patterns of constructive collaboration to realise the full potential of recording. In my third example of a visual diary, I will try to identify some traits that characterise such patterns of collaboration in editing.

Patterns of editing in the biographical notes of Jerome

In the portrait of Jerome Wiesner, it is the academic community of today's USA that *recollects* events from the past century. The voices are those of scientific apprentices who prepare biographical notes about their master. The themes centre around issues of decentralisation versus centralisation, dialogue versus command. If *A Diary for Timothy* is about an uncle worrying about a child, and if *Videodagboken: Erik – ett egensinnigt liv* is about a father wondering about his new role, then *Jerome B. Wiesner: A Random Walk through the Twentieth Century* is about the inheritors of a cultural paradigm who commemorate one of their antecedents.

The portrait of Jerome B. Wiesner was made to honour the memory of one of the co-founders of the MIT MediaLab, and was first presented as an interactive CD-ROM made to coincide with the tenth anniversary of the Lab. Almost everything in this biographical notebook is formally different to the earlier diaries discussed in this essay. Even the role of the Editor is taken over by us, the viewers, rather than those who are represented in the diary. *Temporally*, we are free to follow the paths taken by Jerome Wiesner throughout the century. The potential narratives are presented chronologically via a timeline, and thematically via an elastic list of keywords. But the process of narrating is user-driven and open. It is supported by a theme-corroborating montage technique inspired by the surrealist concept of *automatism*. This fascinating experiment in the integration of autonomous agents into the *telling* of stories can be enjoyed at http://ic.www.media.mit.edu/JBW/.

With regard to *spatial continuity*, we are invited to walk through most of the public scenes of the city: restaurants, concert halls, schools, university campuses, laboratories and workshops. The footage could be characterised as compiled, conventionally shot, institutional home video. But clearly the most innovative formal trait is the decentralised, user-driven and theme-oriented montage. The surrealist concept of *automatism* here takes on connotations of *decentralised* self-reflexiveness. Scenes in the evolving documentary are annotated thematically and can thereby be accessed by the continuously searching agents that take part in the process of narrating. Each editing decision, made by the viewer, sends currents in the form of decoding messages throughout the whole of the archive. This enables the viewer to see new connections in time and space, connections that up till now have been too far-reaching, or too close to visualise.

The digital archiving/accessing process is now slowly becoming as fluent as the digital recording and editing processes. A closer analysis of the diary of Jerome could focus on the innovative patterns of its underlying archiving/accessing construction. Among the characteristic traits that could be studied are: *Split screen for navigational aid* which integrates the viewer's direct encounter with the material with the experience of navigation; *Themes as links* which explicates in a textual way the wealth of narrative threads of the material; *Persons as links* which turns the characters portrayed into navigational guides; *Link layout according to chronology of material* which adheres to the encyclopaedic way of displaying navigational links; *Link layout according to chronology of exploration* which denotes the anticipatory quality of the automatist approach, extending the scope of navigation into paths expressed by the viewer in terms of order and duration of earlier paths. The list could go on; my aim here is to illustrate how a focus on practical design may provide a framework for the analysis of the formal qualities of a filmic example in a way that does not isolate the example from its thematic context.

In conclusion, the archetypal theme of 'cultural heritage' has been articulated in quite different ways in the three visual diaries examined here: 'the worries of an uncle' in the 1940s have been followed by 'the wonderings of a father' and 'the recollections of a community' in the 1990s. These shifts in perspective have been made possible by technical and formal innovation. Camerawork has become more spontaneous and fluent, allowing for a greater sense of intimacy. At the same time, the editing process has opened up new dimensions of spatial and temporal interconnectedness, allowing for the recollection of personal events on a more diversified social and historical scale. The innovative use of new media technologies has also been accompanied by new patterns of practical design. Directors have felt empowered to relinquish some of their directorial control. This process has been considerably advanced by production facilities in university settings that relinquish some of their editorial control to their designers, lecturers, and students.

In my view, the focus on patterns of design may prove to be a fruitful methodology for grasping the complex interplay between formal and technological aspects of digitalisation. Guided by this focus, continui-

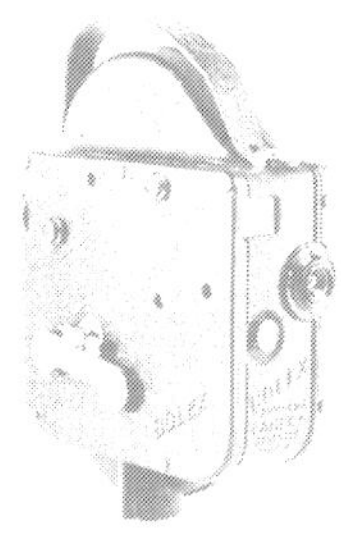

ties may be traced on different compositional levels in systematic comparisons across genres, media formats and time. In this essay I have tried to demonstrate the usefulness of this approach. I have focused on the genre of visual diaries, a genre that exemplifies the self-reflective everyday use of visual media. The rapid popular acceptance of the World Wide Web as an expressive medium has provided a global network that integrates editing facilities with archiving and accessing facilities. If Alexandre Astruc's vision of 'the camera as a pen' needs 'the screen as a paper' to become functional on a broader scale, then the World Wide Web seems to be a promising medium for the diaries of tomorrow.

Notes

1. When analysing the introduction of satellite-, cable- and video-technologies, Herbert Schiller highlights the dilemma that a critical debate about media has to evolve from within the media itself: 'The utilization of the new communications technology for human needs requires a thorough reordering of the social process which regulates the information system at all levels of personal and national existence. The dilemma is, however, that the controlling crust of the industrial state calls on the communications media to resist the social reorganization that must precede the technological reformation', Herbert Schiller, *Mass Communications and American Empire* (Boston, Mass.: Beacon Press, 1971), 150–151.
2. In an article which first appeared in 1986, Michael Renov discusses the consequences for the mediating processes of new, more intimate and more subjective relations between filmmakers and those they portray, see Michael Renov, 'Re-thinking Documentary: Towards a Taxonomy of Mediation', *Wide Angle* 8 (1986). Ten years later, Renov, focusing on *First-Person Video Confessions* and *One on One Videoletters* continues the same discussion when, with important bearings in the implications of digital culture, he observes: 'In contrast to the legacy of photographic representation as a regulatory and disciplinary apparatus, first-person video confessions of this sort afford a glimpse of a more utopian trajectory in which cultural production and consumption mingle and interact, and in which the media facilitate understanding across the gaps of human difference, rather than simply capitalize on those differences in a rush to spectacle', Michael Renov, *Resolutions – Contemporary Video Practices* (Minneapolis: University of Minnesota Press, 1996).
3. Thomas Erickson, 'Supporting Interdisciplinary Design: Towards Pattern Languages for Workplaces'. (URL http://www.pliant.org/personal/Tom_Erickson/Patterns.Chapter.html, accessed 8 March 1999). Pattern languages as a design methodology was originally proposed by the mathematician and architect Christopher Alexander in Christopher Alexander, Sara Ishikawa and Murray Silverstein, *A Pattern Language: Towns, Buildings, Construction* (New York: Oxford University Press, 1977). A background to its application in software engineering is given by Karen and Ward Cunningham (URL http://c2.com/cgi-bin/wiki?HistoryOfPatterns, accessed 8 March 1999), and an interesting example of its application in organisational design is presented by James O. Coplien (URL http://portal.research.bell-labs.com/orgs/ssr/people/cope/Patterns/Process/index.html, accessed 8 March 1999).
4. I adhere to the convention of using an initial uppercase letter for the names of design patterns throughout the essay.
5. Dick Idestam-Almquist in a presentation at the Documentary Festival, Stockholm, 9 March 1998.
6. Brian Winston, *Claiming the Real: The Documentary Film Revisited* (London: British Film Institute, 1995).
7. Bill Nichols *Representing Reality: Issues and Concepts in Documentary* (Bloomington and Indianapolis: Indiana University Press, 1991), 179–180.
8. John Corner (ed.), *Documentary and the Mass Media* (London: Edward Arnold Publishers, 1986), 57–62.

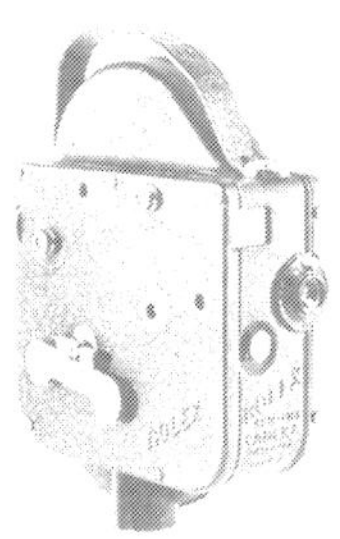

The Interactive Filmmaker's Challenge

Christopher Hales

Faculty of Art, Media and Design, University of the West of England, Bristol, UK

The earliest cinema of Méliès and Lumière has more than once been likened to the current state-of-the-art interactive 'new media', and more specifically to non-linear interactive movies. Implicit in this observation is the fact that the sophistication of interactive movie language is awaiting the passage of time and the development of technology before it matures. It has been frequently noted that computers are still waiting for the first great visionary genius to take us into a new dimension.

This essay demonstrates how computer technologies can now offer completely new interactive movie paradigms and structures; that the expectations of the audience need redefining, and that to develop these works we can inherit from the 100-year history of cinematic images. A range of the author's (and other) experimental works will be discussed to demonstrate how the interactive filmmaker's challenge is not to wait for technological change or advance, but to discover specific non-linear models and subjects that make sense for the viewer, and support their fundamental leap of faith from observer to participant. Only in this way is it possible to make movies do things they haven't done before.

Defining the ground

When cinematic images came into being towards the end of the nineteenth century, a mass medium for showing moving images became available, but at the same time there were few prior examples of how to use moving images. This meant that frames of reference had to be taken, for example, from the theatre, and there was a period of experimentation during which the earliest directors began to explore the properties of cinema that were unique to the medium. Thus cinema's development rapidly became one of searching for the most popular genres with which filmmakers could financially exploit the medium, despite the fact that cinema is, essentially, a very fixed format.

Throughout its history there has been homogeneity about the way movies are experienced by the public. Whatever the subject matter or intention, the viewer sees a relentless and successive series of images on a screen, and this paradigm has survived from its origins to the present day. Indeed, with a few notable exceptions, technological advances such as sound and colour have influenced the development of cinematic language much less than the individual ideas and free thinking of filmmakers. Those who have strayed from the mainstream to become experimental, underground, alternative, or independent film- and video-makers have opened up new possibilities. Yet these artists and their work remain almost unknown in the public eye. The mass perception of cinema (and even television) is all pervasive: films or programmes with a beginning, middle, and end, some actors, a director, film crew, and a huge budget. Filmmakers always seem to make films in the same old ways, and are taught how to do it in film schools. New technologies become a tool to make the same things in a quicker, cheaper or more convenient or visually exciting way, but never seem to be looked at as a way of making something original or different.

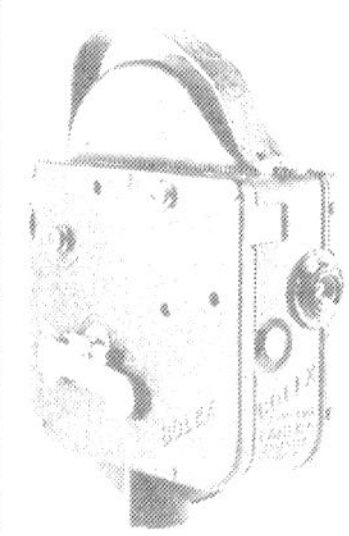

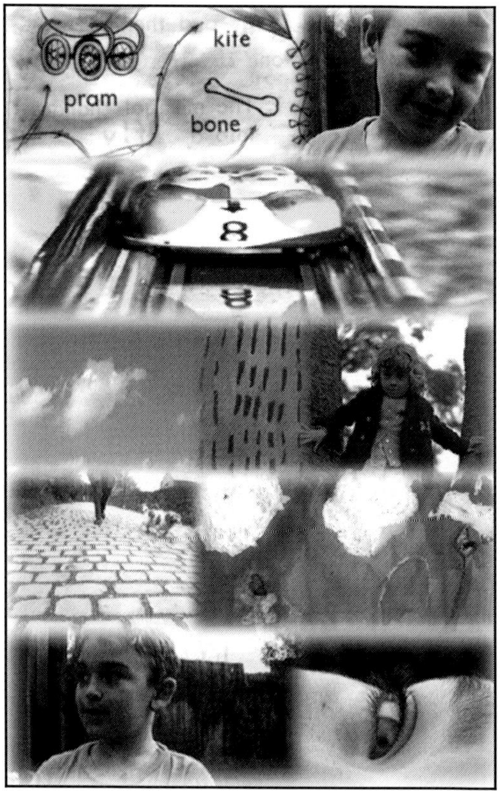

Fig. 1. Scenes from The Twelve Loveliest Things I Know.

Nowhere could a clearer parallel be drawn than with the computer multimedia and games market where a top-selling game such as *Myst* or *Tomb Raider* is followed by an 'even better' sequel (*Riven*, *Tomb Raider II*). It is vital to stress that 'even better' in this context means faster performance and more realistic characters and game play, and definitely does not imply originality. All the same, interactive computer technology does offer greater variety than cinema, with CD-ROMs, arcade games, home computers, games consoles, mouse, joystick, different types of multimedia menus, icons and buttons. But the diversification of the format is not matched by a diversification of ideas.

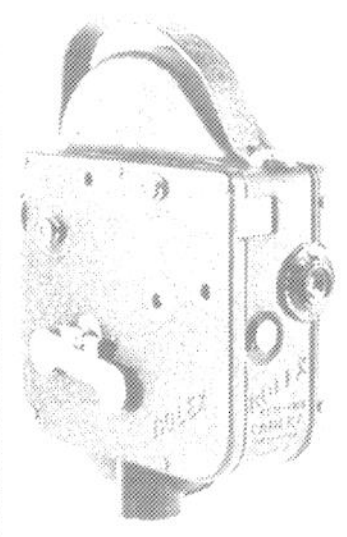

For several years, forward-looking film festivals have started to include interactive installations and CD-ROMs as part of their programme. An example is the Exploding Cinema section of the Rotterdam Film Festival. Other festivals such as Festival des Cinémas et nouveaux Médias Montréal (FCMM) include New Media explicitly as part of their title. This comes as a recognition that some artists are using interactive technologies to produce works that still bear a patronym to video and cinema, but which are new and fundamentally different because they are interactive. For example, the filmmaker Chris Marker published an interactive CD-ROM in 1997 called *Immemory*, which was a structured collection of poems, text, photographs and film. The works of the New York based filmmaker Grahame Weinbren, and Montréal's Luc Courchesne are characterised by consisting entirely of film/video segments, as is also true of my own productions. Even Bill Viola, pioneer of video art, has considered interactive structures, and has observed: 'Worlds are waiting to be explored. It is to be hoped that artists will be given their share of access to experiment with this exciting new technology.'[1] My own artworks are characterised by sequences of moving images which play continuously without pausing for the viewer to make a choice, so the visual similarity to cinema is very strong. However, these works are interactive and can be influenced in some way by the viewer (using a touchscreen or mouse) so long as they click in the right place at the right time whilst the movie is playing. A work such as this seems to occupy a grey area which is difficult to categorise because it exists somewhere between film and game, and which requires the viewer of the work to take the seemingly incompatible roles of both spectator and participant.

The fifth dimension

'In one sense the history of film is but a footnote to Lumière and Méliès', states Gene Youngblood in *Expanded Cinema*[2] as a means of expressing the fact, alluded to in the previous section, that there have been no fundamental breakthroughs in the nature of cinema since its inception. Youngblood claims that the introduction of sound and then colour could represent second and third generations in cinema's history, with the newly emerging fourth generation 'marrying basic cinematic techniques to computer and video sciences'. Any movie visit today will illustrate clearly how firmly ensconced we are in this 'fourth generation', with special effects and digital imaging being *de rigueur* for most cinema releases – often at the expense of basic screenwriting qualities. This is a manifestation of technology being used for effect and sensation, but not for a paradigm shift.

So, could the fifth dimension be interactive? Despite the fact that films have until the present consisted of a series of reels which must be changed in a cinema by a projectionist at certain predetermined moments, it does not seem to have been until 1967 that this natural choice-point was employed as a feature rather than a hindrance by the makers of the interactive movie entitled *One Man and his World*. The film was made by a group of Czechoslovakian filmmakers directed by Radusz Cincera, and was shown at the Expo 67 festival in Montréal, a milestone event for cinema and technology with demonstrations of 360-degree cinema, gigantic multiscreens and cinematic installations. As the journalist Frank Kappler wrote in *Life* magazine:

> At the fair, the majority of pavilions use movies – some of the most technically ambitious ever produced. Screens wrap around and immerse viewers. Multiple pictures make audiences understand more through feeling than through thinking. ... Film transmits facts, creates moods, tests moral judgements. It speaks at Expo 67 sometimes in poetry, sometimes in gibberish, but always in a visual blitz almost blinding in its implications for the future.[3]

The system behind *One Man and his World* was called Kino-Automat, and appears to have been more conceptual than technological since the problem of choice was overcome by having the actors appearing 'live' in front of the projected film which would stop in split-screen form at the end of the reel while the protagonist/actor (on-stage) would ask the audience to make judgments on which direction the plot should take. Lamps around the screen would light up to show each viewer's choice, and the majority vote determined which reel would be shown next.

As we may observe from Kappler's article, eulogies and predictions at the time were widespread. And yet, after the widespread world-wide press exposure during Expo 67 and hundreds of performances, the show returned to Czechoslovakia, toured the major cities ... and became nothing more than a side-show, an interesting oddity with a limited appeal. The concept of interactive audience voting was resurrected in the early 1990s when a number of small cinemas, equipped with pistol-grips by the US company Interfilm, showed movies such as *Mr Payback* and *I'm Your Man* with frequent choice points in the plot which would be determined by majority vote. Interfilm soon went out of business.

Conflict

The only logical conclusion to be drawn from the demise of these pioneering projects is that a type of interactive cinema which tries to mimic the dramatic and expositional characteristics of its well-established older sibling can only fail – or remain an outcast. Interactive movies as they have developed since 1967 seem to try either to emulate the dramatic conventions of cinema with its characters, structures and stories, or they take an existing cinema or television 'hit' (for example, James Bond or *The X-Files*) and reconform it as so-called Interactive Entertainment even though to all intents and purposes it is an action/strategy game. Even with the benefit of all our lightning-fast, interactive, computer-controlled digital technology, most interactive movies still want to do what has been done before. The protagonist still has an emotional dilemma or crisis. There is still a choice to be made by the viewer, and although this is more commonly carried out by interfacing a computer and not through a vote in a cinema theatre, the branching structure and dramatic intention of the work appears so often exactly the same as it was thirty years ago – so can it be any wonder that the interactive movie remains an interesting side-show?

The afore-mentioned works are those based on a multi-threaded narrative structure, relying on back-stories and links between characters and place. The structure can then become a non-linear matrix with a multitude of branch points and plot twists. But surely this is the essence of Hollywood's cinema, and it represents quite simply the art and craft of successful scriptwriting as it is perceived in the public eye. Robert Altman's *Short Cuts*

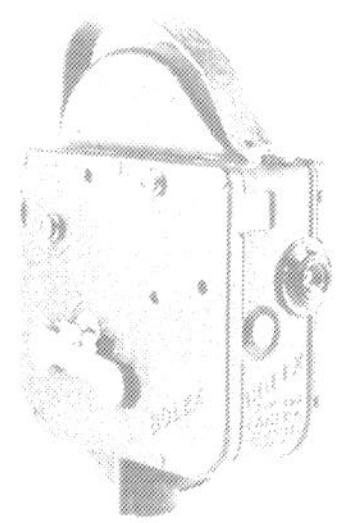

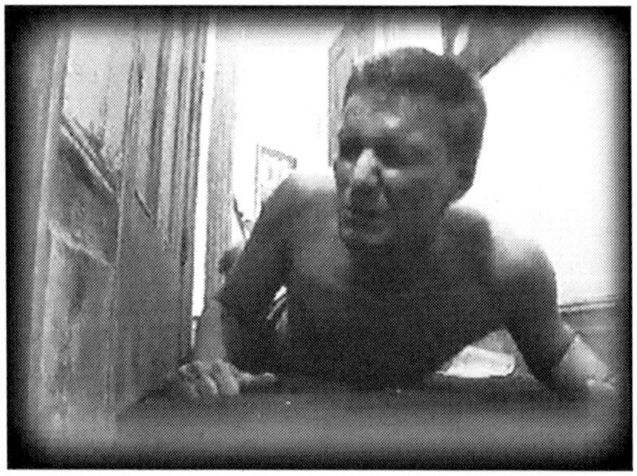

Fig. 2. Scene from Jinxed!

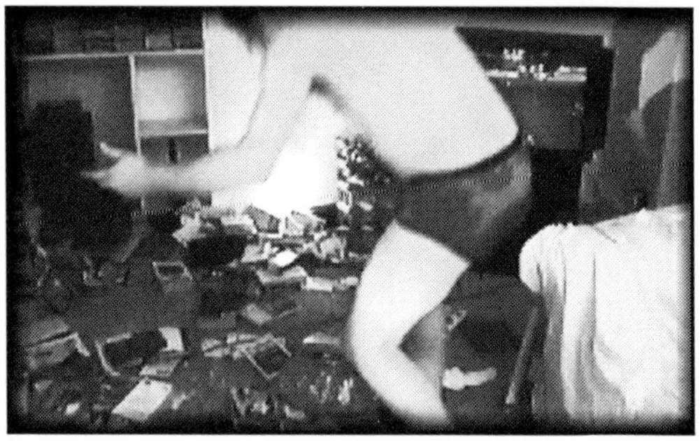

Fig. 3. Scene from Messed Up!

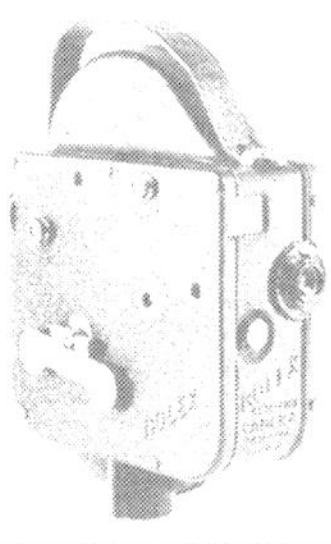

(1993) is an intricate example of this type of script. All that seems to be added by the interactive filmmakers is an element of gameplay, and the opportunity of choosing or solving a problem along the way. The interactive movie as imagined by Hollywood would be 'like its big screen counterpart [in that] it will have emotion-filled storytelling with clean production values. Hollywood developers are concentrating on the storyline, striking a middle ground between passivity and interaction'.[4] Greg Roach, founder of Hyperbole Studios (which have created several disk-based productions including the *The X-Files* interactive CD-ROM), states: 'the interactive movie ... tries to create and sustain the drama that characterises real films. The scriptwriters try to make their characters behave in a deep and realistic fashion within a properly structured plot.'[5]

During the last five years I have been making small-scale experiments using desktop Apple Macintosh technology and with limited or non-existent budgets, the production values being less important to me than the chance afforded by the technology to realise an idea quickly. Even so, the works I produce can be distributed widely as CD-ROMs, or set up as a touch-screen installation in galleries and exhibitions. Since it is very apparent that it has already been possible to make cinematic interactive movies using non-digital technology, there must be other ways to put elements of choice or change into something that on the surface resembles a coherent set of moving images. Could this become something new? Can we use digital technology to create a new paradigm for cinema, rather than use it to make the same things quicker, cheaper or more exciting?

When is a movie not a movie?

In what manner can these new paradigms be created? Experimenting with structure and interface are essential when dealing with interaction. Structure might mean the construction of branching links, or the control of forwards/backwards direction in a movie, or the juxtaposition of asynchronous multiple movie clips in a splitscreen format. Interface could be investigated by using 'buttons' or 'menus', or by specifically art-directing the movie such that objects, colour, focus or point-of-view (amongst others) become integral to the nature of the interaction. One further way is to choose those genres which seem to be most suitable to an interactive format. My slapstick comedy *Jinxed!* has no complexity of plot, and could be summarised completely in a single sentence: 'an unfortunate man trying to get ready for a job interview, but things in his apartment are jinxed and can painfully hinder him if the viewer clicks on them at the right moment with the mouse.' The effect of interacting is immediate and visual, and causes laughter and amusement, despite the apparent one-dimensionality of the action and plot. The points of interaction (objects in the man's house) are shown by a digital effect applied to the movie, and the structure is a simple branch: either he has an accident, or he carries on getting ready unobstructed. A limitation is that the film is not open to many varied readings, but my observation of its effects upon the person interacting with the film is that interest is retained through two or three viewings. After all, how many times do we enjoy a movie enough to watch it a second or third time at the cinema?

Messed Up! was a progression from *Jinxed!*, a different interactive comedy with added depth and variety. In a single sentence: 'a man relaxing in his tidy apartment, which can be progressively messed up by clicking on things, much to the annoyance of the protagonist who will try to react by tidying things up again.' This movie can be different every time, there is no fixed order for clicking on the items that can mess up the room, but the effect is the same: an amusing sense of pleasure as the hapless victim is harassed and annoyed.

Both these works do indeed have a feeling of being one-character 'dramas' although they were made with amateur actors and without any film crew or budget. Reducing the plot complexity and enhancing the visual aspects of interaction seem successful at this level. However, it seems difficult to imagine an interactive suspense film being successful in the same way since the action of clicking would already signal an anticipation in the viewer and spoil the tension. On the other hand, there are some other genres that lend themselves readily to an interactive format and one of these is the Musical. This seems suitable because verisimilitude is abstracted by the addition of song and dance routines.

My interactive film, *Tallinn People's Orchestra* was inspired both by the opening sequence of Jacques Démy's 1968 film *Les Demoiselles de Rochefort*, and Zbiegniew Rzybzchinski's Oscar-winning *Tango* (1974). It was made with a fixed camera position in the Town Hall Square of Tallinn, Estonia, and recorded the movements and activities of both residents and visitors (plus birds and aeroplanes and others) during a short period of time. Each element (person, bird, plane, car, horse) was painstakingly deconstructed and assigned a musical motif, and the computer given the ability to put them back into the Square at random, so that the people appear to 'dance' and make music together. The scene will always be composed of different combinations both of music and 'dance' (the movements of the people). Some interaction is possible because clicking on different elements removes them from the scene, empowering the user to influence the choreography of the 'actors' and their music. There is no beginning or end to this movie (as there

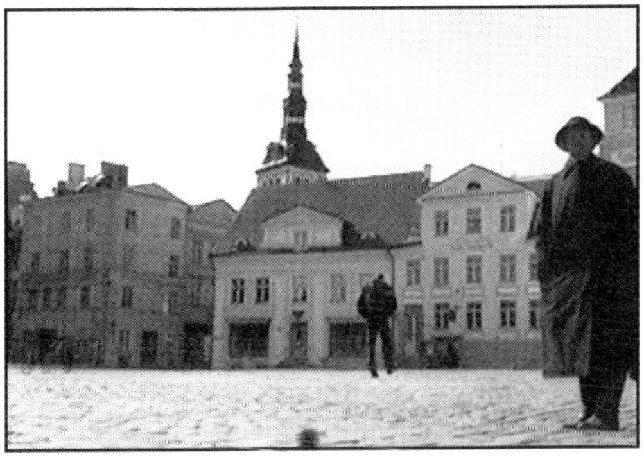

Fig. 4. Scene from Tallinn People's Orchestra.

is in *Jinxed!*); instead it is a single ever-changing and ever changeable scene, and does not suffer from the hierarchies of a branching structure.

Grahame Weinbren's early 1985 videodisk work *The Erl King* was also a musical, based upon the motif of a singer performing the traditional German folk song of the same title. Parts of the song conjure up interactive links to other video sequences which complement the song and add to its understanding. It was a type of linking which Weinbren was to pursue further with his major work from 1994, *Sonata*, and is characteristic of my documentary CD-ROM from 1995–96, *The Twelve Loveliest Things I Know*. Writing in *Millenium Film Journal*, Weinbren likened this experience to 'swimming in an ocean of streams of stories', and observed:

> Could there be a 'Story Space' … like the Ocean, in which a reader might take a dip, encountering stories and story segments as he or she flipped and dived? In these waters, turbulences created by the swimmer's own motion might cause an intermingling of the Streams of Story. The Ocean is a dynamic narrative region, a Heraclitean river into which one could never step twice, a lake of Heisenbergian uncertainty where the very attempt to examine a particular story-stream transforms it. What a goal to create such an Ocean![6]

The article goes on to demonstrate how Freud's theories of dreams (one of the central themes of *Sonata*) prove a suitable model for storytelling in which the sequence of events is not of central importance. This concept of story immersion could be likened to random

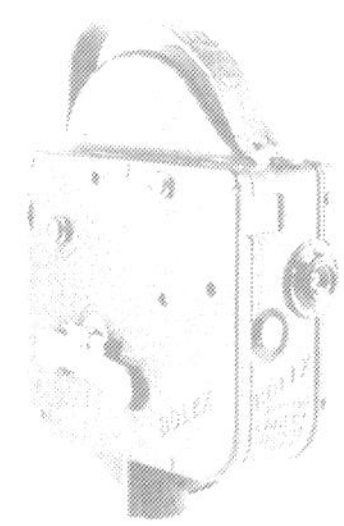

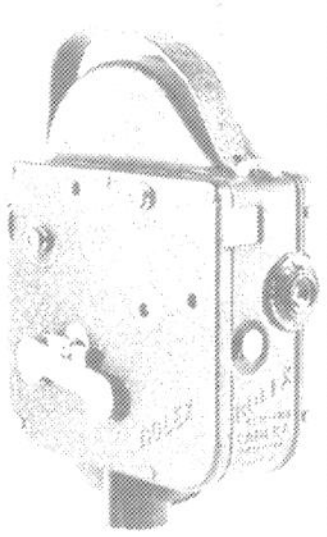

television channel surfing, or to the 'interactive television' experiments which involve the simultaneous broadcast of different aspects of a drama to different channels with the viewer free to flip through them at will.

The Twelve Loveliest Things I Know[7] was a documentary investigation in which a great deal of moving visual imagery was combined in an interactive structure. Children were interviewed, filmed, asked to make paintings and lists, in a discussion of the 'loveliest things' in their lives. Some of their responses were reconstructed as subjective filmed events, but because (as in *Sonata*) the sequence of events is not so important, the structure becomes one of serendipitous dream-like links from colour to colour and from movement to movement. Although the material was scrupulously grouped into themes such as family, achievement, nature etc., there is a cross referencing of shared sequences, and the links are triggered by visual witticisms such as those inherent in the children's own drawings. This work is characterised by not being narrative driven, simply an Ocean of sequences which make sense when read in certain ways. Immersion becomes all-important rather than the trajectory of the storylines. It seems to be a model of filmmaking far away from that of the Kino-Automat, and although by the nature of the genre both are big fish in a very small pond, they have in common an element of choice.

My experiments, and those of others, leave me in no doubt that the interactive nature of computer technology, combined with moving images, can make something new. Yet the fact that nothing substantial or revolutionary appears to be here yet is rather worrying. Observers will keep on noting that we must clearly be in that period of experimentation that characterised early cinema when it was struggling to find out what it was good at doing, before the art and science of 'filmmaking' emerged. But today we have known about the art and science of 'filmmaking' for over a century. Moving images engulf us on the television and cinema, powerful desktop computers have been around for more than a decade. We are immersed in and surrounded with such a plethora of references, histories and technologies that it is almost impossible to unburden ourselves to regain the freedom that those legendary early filmmakers enjoyed. We cannot see the wood for the trees.

Technologies of moving images have now diversified so much that nothing truly new might ever come about, only hybrids and shades of what we have already. Combining movie scenes with interactive computer structures seems to me to be a definable genre, but one which strays readily into the existing genres of film and game under the conflicting influences of interactivity and storytelling. The interactive filmmaker's challenge is to open up a middle ground which cannot be characterised as either film *or* game. Maybe this might be 'new'; at the very least it would be *original*.

Notes

1. Bill Viola, *Reasons for Knocking at An Empty House* (London: Thames and Hudson, 1995), 107–109.
2. Gene Youngblood, *Expanded Cinema* (London: Studio Vista, 1970), 156.
3. Frank Kappler, *Life* (September 1967): 39–41.
4. Hanna Liebman, 'The Big Picture', *CDROM world* (September 1994): 48.
5. Greg Roach, 'Into the Vortex', *New Scientist* (23 September 1995): 30–33.
6. Grahame Weinbren, 'In the Ocean of Streams of Story', *Millenium Film Journal,* no. 28 (Spring 1995).
7. Chris Hales, *Twelve* (London: Research Publishing, 1997), CD-ROM (Mac).

Index

Page references in **Bold type** refer to illustrations appearing in the text.

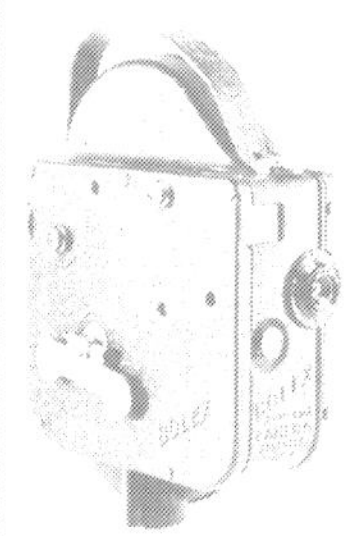

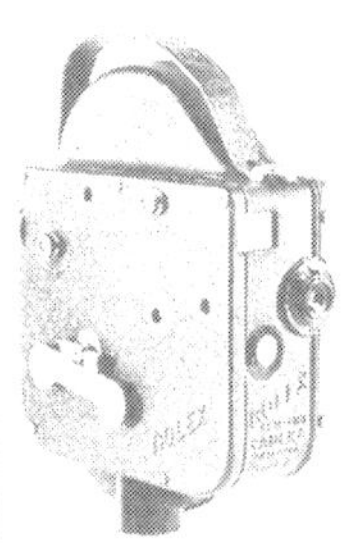

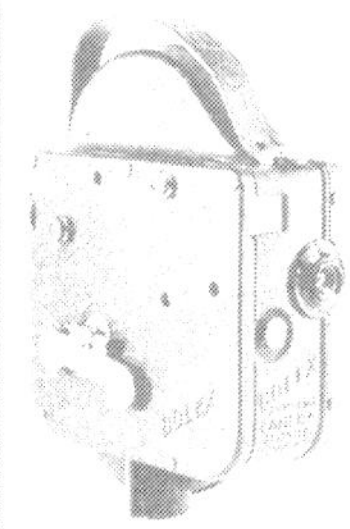

G

H

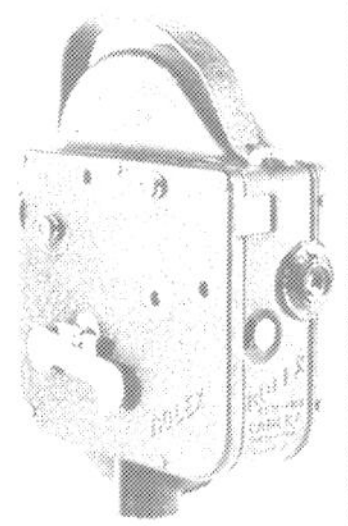

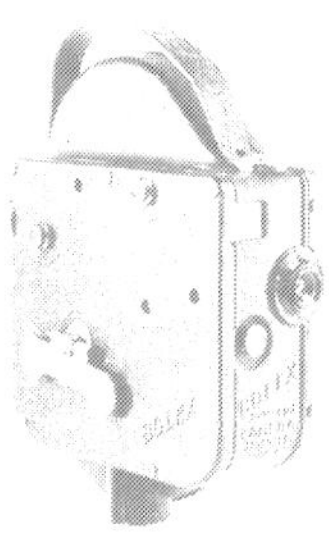

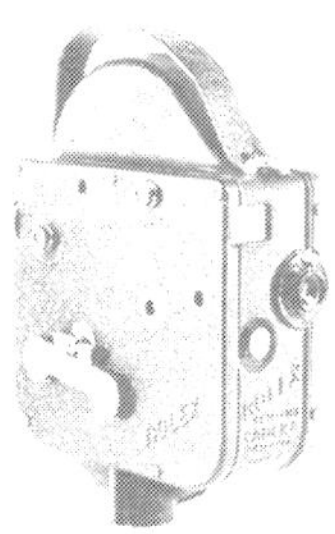

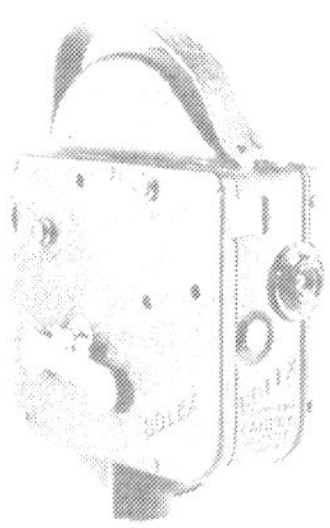

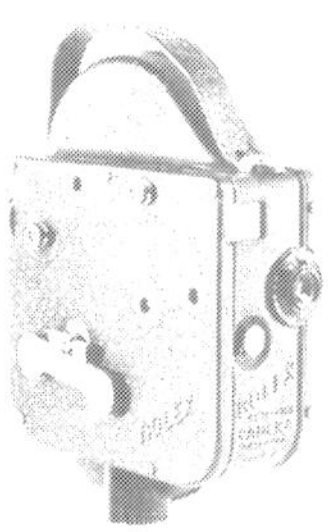

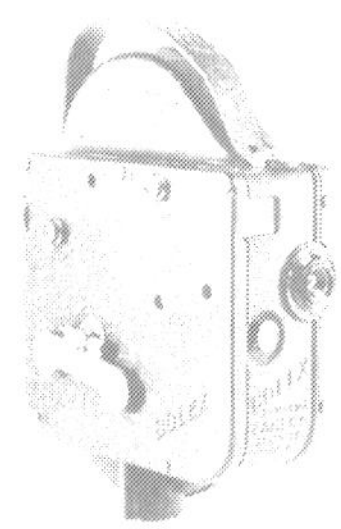